Mathematical Modeling in Systems Biology

Mathematical Modeling in Systems Biology

An Introduction

Brian P. Ingalls

The MIT Press
Cambridge, Massachusetts
London, England

This book was set in Syntas and Times New Roman by Toppan Best-set Premedia Limited, Hong Kong.

Library of Congress Cataloging-in-Publication Data

Ingalls, Brian P., 1974–
Mathematical modeling in systems biology : an introduction / Brian P. Ingalls.
  pages cm
Includes bibliographical references and index.
ISBN 978-0-262-01888-3 (hardcover : alk. paper), 978-0-262-54582-2 (pb)
1. Biological control systems. 2. Control theory. I. Title.
QH508.I46 2013
571.7—dc23
2012036809

Contents

Preface

Systems techniques are integral to current research in molecular cell biology. These systems approaches stand in contrast to the historically reductionist paradigm of molecular biology. The shift toward a systems perspective was gradual: it passed a turning point at the end of the twentieth century, when newly developed experimental techniques provided system-level observations of cellular networks. These observations revealed the full complexity of these networks and made it clear that traditional (largely qualitative) molecular biology techniques are ill-equipped for the investigation of these systems, which often exhibit nonintuitive behavior. This point was illustrated in a thought experiment proposed by Yuri Lazebnik (Lazebnik, 2002). He described a (failed) attempt to reverse engineer a transistor radio using qualitative methods analogous to those used in traditional molecular biology. Lazebnik's exercise demonstrates that without a quantitative framework to describe large networks of interacting components, the functioning of cellular networks cannot be resolved. A quantitative approach to molecular biology allows traditional interaction diagrams to be extended to mechanistic mathematical models. These models serve as working hypotheses: they help us to understand and predict the behavior of complex systems.

The application of mathematical modeling to molecular cell biology is not a new endeavor: there is a long history of mathematical descriptions of biochemical and genetic networks. Successful applications include Alan Turing's description of patterning in development (discussed by Murray, 2003), the models of neuronal signaling developed by Alan Hodgkin and Andrew Huxley (reviewed by Rinzel, 1990), and Denis Noble's mechanistic modeling of the heart (Noble, 2004). Despite these successes, this sort of mathematical work has not been considered central to (most of) molecular cell biology. That attitude is changing: system-level investigations are now frequently accompanied by mathematical models, and such models may soon become requisites for describing the behavior of cellular networks.

What This Book Aims to Achieve

Mathematical modeling is becoming an increasingly valuable tool for molecular cell biology. Consequently, it is important for life scientists to have a background in the relevant mathematical techniques so that they can participate in the construction, analysis, and critique of published models. Also, those with mathematical training—mathematicians, engineers, and physicists—now have increased opportunity to participate in molecular cell biology research. This book aims to provide both of these groups—readers with backgrounds in cell biology or mathematics—with an introduction to the key concepts that are needed for the construction and investigation of mathematical models in molecular systems biology.

I hope that after studying this book, the reader will be prepared to engage with published models of cellular networks. By "engage," I mean not only to understand these models but also to analyze them critically (both their construction and their interpretation). Readers should also be in a position to construct and analyze their own models, given appropriate experimental data.

Who This Book Was Written For

This book evolved from a course I teach to upper-level (junior/senior) under-graduate students. In my experience, the material is accessible to students in any science or engineering program, provided they have some background in calculus and are comfortable with mathematics. I also teach this material as a half-semester graduate course for students in math and engineering. The text could easily be adapted to a graduate course for life science students. Additionally, I hope that interested researchers at all levels will find the book useful for self-study.

The mathematical prerequisite for this text is a working knowledge of the derivative: this is usually reached after a first course in calculus, which should also bring a level of comfort with mathematical concepts and manipulations. A brief review of some fundamental mathematical notions is included as appendix B. The models in this text are based on differential equations, but traditional solution techniques are not covered. Models are developed directly from chemical and genetic principles, and most of the model analysis is carried out via computational software. To encourage interaction with the mathematical techniques, exercises are included throughout the text. The reader is urged to take the time to complete these exercises as they appear: the exercises will confirm that the concepts and techniques have been properly understood. (All of the in-text exercises can be completed with pen-and-paper calculations; none are especially time-consuming. Complete solutions to these exercises are posted at the book's Web site.[1]) More involved problems—mostly involving computational software—are included in the end-of-chapter problem sets.

An introduction to computational software is included as appendix C. Two packages are described: XPPAUT, a freely available program that that was written specifically for dynamic modeling; and MATLAB, which is a more comprehensive computational tool. Readers with no background in computation will find XPPAUT more accessible.

I have found that most students can grasp the necessary cell and molecular biology without a prior university-level course. The required background is briefly reviewed in appendix A: more specialized topics are introduced throughout the text. The starting point for this material is a basic knowledge of (high school) chemistry, which is needed for a discussion of molecular phenomena, such as chemical bonds.

How This Book Is Organized

The first four chapters cover the basics of mathematical modeling in molecular systems biology. These should be read sequentially. The last four chapters address specific biological domains. The material in these latter chapters is not cumulative: they can be studied in any order. After chapter 2, each chapter ends with an optional section, which is marked with an asterisk (*). These optional sections address specialized modeling topics, some of which demand additional mathematical background (reviewed in appendix B).

Chapter 1 introduces molecular systems biology and describes some basic notions of mathematical modeling, concluding with four short case-studies. Chapter 2 introduces dynamic mathematical models of chemical reaction networks. These are differential equation models based on mass-action rate laws. Some basic methods for analysis and simulation are described. Chapter 3 covers biochemical kinetics and describes rate laws for biochemical processes (i.e., enzyme-catalyzed reactions and cooperative binding). An optional section treats common approximation methods. Chapter 4 introduces techniques for analysis of differential equation models, including phase plane analysis, stability, bifurcations, and sensitivity analysis. The presentation in this chapter emphasizes the use of these techniques in model investigation; very little theory is covered. A final optional section briefly introduces the calibration of models to experimental data.

Chapter 5 covers modeling of metabolic networks. Sensitivity analysis plays a central role in the investigation of these models. The optional section introduces stoichiometric modeling, which is often applied to large-scale metabolic networks.

1. www.math.uwaterloo.ca/~bingalls/MMSB.

Chapter 6 addresses modeling of signal transduction pathways. The examples taken up in this chapter survey a range of information-processing tasks performed by these pathways. An optional section introduces the use of frequency response analysis for the study of cellular input–output systems.

Chapter 7 introduces modeling of gene regulatory networks. The chapter starts with a treatment of gene expression, then presents examples illustrating a range of gene-circuit functions. The final optional section introduces stochastic modeling in molecular systems biology.

Chapter 8 covers modeling of electrophysiology and neuronal action potentials. An optional section contains a brief introduction to spatial modeling using partial differential equations.

The book closes with three appendices. Appendix A reviews basic concepts from molecular cell biology. Appendix B reviews mathematical concepts. Appendix C contains tutorials for two computational software packages—XPPAUT and MAT-LAB—that can be used for model simulation and analysis.

The Web site www.math.uwaterloo.ca/~bingalls/MMSB contains solutions to the in-text exercises, along with XPPAUT and MATLAB code for the models presented in the text and the end-of-chapter problem sets.

Acknowledgments

The preparation of this book was a considerable effort, and I am grateful to students, colleagues, and friends who have helped me along the way. I would like to thank the students of AMATH/BIOL 382 and AMATH 882 for being the test subjects for the material and for teaching me as much as I taught them. Colleagues in math, biology, and engineering have been invaluable in helping me sort out the details of all aspects of the text. In particular, I would like to thank Peter Swain, Bernie Duncker, Sue Ann Campbell, Ted Perkins, Trevor Charles, David Siegel, David McMillen, Jordan Ang, Madalena Chaves, Rahul, Abdullah Hamadeh, and Umar Aftab. Special thanks to Bev Marshman, Mads Kaern, Herbert Sauro, and Matt Scott for reading early drafts of the manuscript and making excellent suggestions on improving the material and presentation. Thanks also to Bob Prior and Susan Buckley of the MIT Press for their support in bringing the book to completion.

Finally, I thank my wife, Angie, and our children, Logan, Alexa, Sophia, and Ruby, for their love and support. I dedicate this book to them.

1 Introduction

[T]he probability of any one of us being here is so small that you'd think the mere fact of existing would keep us all in a contented dazzlement of surprise. . . . The normal, predictable state of matter throughout the universe is randomness, a relaxed sort of equilibrium, with atoms and their particles scattered around in an amorphous muddle. We, in brilliant contrast, are completely organized structures, squirming with information at every covalent bond. . . . You'd think we'd never stop dancing.
—Lewis Thomas, *The Lives of a Cell*

1.1 Systems Biology and Synthetic Biology

Life is the most potent technology on the planet. It is also the most complex. This staggering complexity presents a fantastic puzzle to those studying its mysteries; more importantly, it offers a wealth of opportunities to those seeking to use our knowledge of biology to improve the quality of life for humanity.

Biology—the study of life—has a long and distinguished history dating back millennia, but our understanding of the *mechanisms* by which living things operate is fairly recent and is still developing. We are, of course, intimately familiar with the behavior of multicellular organisms (such as ourselves!), but the mechanisms by which living organisms function remained obscure until the 1950s. At that time, the nascent field of molecular biology began to reveal the networks of interacting molecules that drive all cellular behavior (and hence all life). These discoveries were made possible by experimental advances that allowed researchers to make observations on the tiny spatial scales of biomolecular processes. Over the past half-century, the molecular biology community has continued to uncover the details of this molecular domain. The painstaking effort involved in these nanoscale experiments necessitated a so-called reductionist approach, in which research projects often addressed individual molecules or molecular interactions.

At the turn of the twenty-first century, further breakthroughs in experimental techniques set the stage for a shift in focus. The advent of so-called high-throughput approaches allowed researchers to observe simultaneously the behaviors of large numbers of distinct molecular species. A cornerstone for these developments was the sequencing of the human genome (the first draft of which appeared in the year 2000). As a result, current molecular biology efforts have been dubbed "post-genomic." This "modern" activity is characterized by experiments that reveal the behavior of entire molecular systems and so came to be called **systems biology.**

A key feature of present-day biological studies is a reliance on computation. The Human Genome Project could not have been completed without advances in bio-informatics that allowed the processing and interpretation of vast amounts of sequencing data. In this book, we will take up a complementary use of computers in the study of molecular biology: the investigation of intracellular processes as *dynamic systems*. We will carry out these investigations by analyzing mathematical models that mimic the behavior of intracellular networks. Such modeling efforts have facilitated tremendous advances in other scientific disciplines. Use of such models in molecular biology in the past was hampered by the absence of experi-mental observations of system behavior. This is no longer the case.

In addition to their use in scientific investigation, dynamic mathematical models are used in engineering, where they play a central role in the design and analysis of engineered constructs. Biology shares several features with *engineering science*—defined as the application of the scientific method to the "made world" of engi-neered artifacts. Because engineered objects have express reasons for existing, engineering scientists are able to use *performance measures* to assess the efficiency and robustness of their function. Although biological systems are part of the natural world, they exist (that is, they have been selected) because they carry out specific functions. Consequently, performance measures can be used to assess their behavior, and biological "design principles" can be identified. There are limits to this analogy between biology and engineering: natural selection is nothing like rational engineering design. Nevertheless, there are instances in which we can be reasonably confident of a biomolecular network's primary function. In these cases, biology straddles the line between natural science and engineering science and can be described as *reverse engineering*—the unraveling (and ultimately recon-struction) of the products of an unfamiliar technology. (Historical examples of reverse engineering are primarily from wartime; e.g., the reconstruction of enemy aircraft.)

The construction, or *forward engineering*, of biomolecular networks is an aspect of **synthetic biology.** This field is focused in part on the construction of designed genetic networks. The first engineered gene circuits were announced in the year 2000. Since then, the field of synthetic biology has grown rapidly. One of its most prominent activities is the International Genetically Engineered Machine competition (iGEM), in which undergraduate student teams design, construct, and test genetic networks of their own imagining.[1]

Systems and synthetic biology represent unprecedented opportunities. In health and disease, agriculture, manufacturing, energy production, and environmental remediation, the use of biological technologies is leading to rapid progress in a wide range of human endeavors.

1.2 What Is a Dynamic Mathematical Model?

This book addresses dynamic mathematical models of biochemical and genetic networks. These models, like all models, are abstractions of reality. Models are designed to focus on certain aspects of the object of study; other aspects are abstracted away. For instance, the familiar ball-and-stick model of chemical structure focuses on a molecule's chemical bonds. It does not capture, for example, the resulting polarity in the molecule's atoms.

Biologists regularly make use of tangible "real-world" models. These can be simple, such as the molecular ball-and-stick, or complex, such as model organisms or animal disease models. Biologists also use conceptual models. These typically take the form of verbal descriptions of systems and are communicated by diagrams that illustrate a set of components and the ways in which they interact (e.g., figure 1.1). These interaction diagrams, or "cartoon" models, play a central role in representing our knowledge of cellular processes.

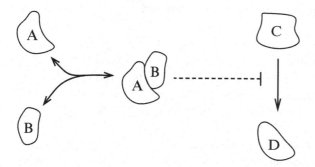

Figure 1.1
An interaction diagram, or "cartoon" model. Molecular species A and B bind reversibly to form a molecular complex. This complex inhibits the rate at which molecules of species C are converted to species D. (The blunt-ended arrow signifies inhibition. A dashed line is used to indicate that this is a regulatory interaction in which the complex is not consumed.)

1. The competition's Web site is www.igem.org/.

A drawback of these cartoon models is that they can leave significant ambiguity regarding system behavior, especially when the interaction network involves feedback. By using a mathematical description of the system, we can eliminate uncertainty in model behavior, at the cost of demanding a quantitative representation of each of the interactions in the cartoon model.

As an example, suppose that, as in figure 1.1, molecular species A and B bind to form a complex. To quantify that interaction, a numerical description of the process must be provided. In some instances, it may be sufficient to provide the equilibrium constant for the reaction. In other cases, separate rates of binding (association) and unbinding (dissociation) are needed. For a great many cellular processes, our current level of knowledge cannot support a quantitative description: we have only a qualitative understanding of the relevant molecular interactions. However, for a growing number of well-studied mechanisms, sufficient data have been collected to allow this sort of quantitative characterization.

When the relevant quantitative data are known, the interaction diagram can be used to formulate a dynamic mathematical model. The model-development process will be presented in chapter 2. The resulting model consists of a set of equations that describe how the system changes over time—the system's dynamic behavior.

Quantitative descriptions of molecular interactions typically invoke the laws of physics and chemistry. The resulting models are thus *mechanistic*—they describe the mechanisms that drive the observed behavior. Each component of a mechanistic model represents some aspect of the system being studied: modifications to model components thus mimic modifications to the real system. (Mechanistic models can be contrasted with so-called descriptive models that seek only to summarize given data sets. Descriptive models provide limited insight into system behavior.)

Investigation of mechanistic models follows two complementary paths. The more direct approach is model **simulation**, in which the model is used to predict system behavior (under given conditions). Simulations are sometimes referred to as *in silico* experiments because they use computers to mimic the behavior of biological systems. Simulations are carried out by numerical software packages and will be used heavily in this book.

Alternatively, models can be investigated directly, yielding general insight into their potential behaviors. These **model analysis** approaches sometimes involve sophisticated mathematical techniques. The payoff for mastering these techniques is an insight into system behavior that cannot be reached through simulation. Whereas simulations indicate how a system behaves, model analysis reveals *why* a system behaves as it does. This analysis can reveal nonintuitive connections between the structure of a system and its consequent behavior. Chapter 4 presents model analysis techniques that are useful in molecular systems biology.

1.3   Why Are Dynamic Mathematical Models Needed?

As mentioned earlier, interaction diagrams typically leave ambiguities with respect to system behavior, especially when feedback is involved. Moreover, as the number of components and interactions in a network grows, it becomes increasingly difficult to maintain an intuitive understanding of the overall behavior. This is the challenge of systems biology and is often summarized by saying that "cells are complex systems." We can unpack this statement by providing some definitions.

The term **system** is often used without formal definition (as in the previous section!). Its meaning is somewhat context-dependent, but it typically refers to a collection of interacting components. In his book *Out of Control* (Kelly, 1995), Kevin Kelly defines a system as "anything that talks to itself." For example, an isolated stone is not considered a system, but an avalanche of stones is: the stones in the avalanche "talk" by pushing one another around.

Besides multiple interacting components, the other defining feature of a system is a *boundary*. A system consists of a set of components: anything that is not one of these components is not part of the system and so is part of the "external environment." For example, a cell's membrane defines a boundary between the cell as a system and the extracellular environment. In certain contexts, a system is defined exclusively in terms of its interaction with this "outside world" and is then called an *input–output system*.

The term **complexity** also means different things to different people. Most would agree that a system qualifies as complex if the overall behavior of the system cannot be intuitively understood in terms of the individual components or interactions. A defining feature of complex systems is that the *qualitative* nature of their behavior can depend on *quantitative* differences in their structure. That is, behavior can be drastically altered by seemingly insignificant changes in system features. Analytical methods for the investigation of complex behavior will be presented in chapter 4.

Two essential features of complex systems are nonlinear interactions and feedback loops. Feedback can be classified as negative or positive.

Negative feedback is exhibited when system components inhibit their own activity. (A familiar example is a household thermostat that corrects for deviation of temperature from a set point.) These feedback loops generally stabilize system behavior: they are the key feature of self-regulation and homeostasis. We will see, however, that instability and oscillations can arise when there is a lag in the action of a negative feedback loop.

Positive feedback is typically associated with unstable divergent behavior. (Think of the runaway screech that occurs when a microphone and amplifier are connected in a positive feedback loop.) However, when constrained by saturation effects, positive feedback can serve as a mechanism to "lock in" a system's long-term behavior—thus providing a memory of past conditions.

1.4 How Are Dynamic Mathematical Models Used?

Dynamic mathematical models serve as aids to biological investigation in a number of ways. The act of constructing a model demands a critical consideration of the mechanisms that underlie a biological process. This rigorous, reflective process can reveal inconsistencies in a cartoon model and highlight previously unnoticed gaps in knowledge. Once a mathematical model has been constructed, it serves as a transparent description of the system and can be unequivocally communicated. Moreover, a model recapitulates system behavior: it concisely summarizes all of the data it was constructed to replicate.

Both a cartoon model and a mathematical model are manifestations of a hypothesis: they correspond to putative descriptions of a system and its behavior. The advantage of a mathematical model is that it is a "working hypothesis" in the sense that its dynamic behavior can be unambiguously investigated. Although model simulations will never replace laboratory experiments, a model can be used to probe system behavior in ways that would not be possible in the laboratory. Model simulations can be carried out quickly (often in seconds) and incur no real cost. Model behavior can be explored in conditions that could never be achieved in a laboratory. Every aspect of model behavior can be observed at all time points. Furthermore, model analysis yields insights into why a system behaves the way it does, thus providing links between network structure and behavior.

Because a model is a hypothesis, the results of model investigation are themselves hypotheses. Simulations cannot definitively predict cellular behavior, but they can serve as valuable guides to experimental design by indicating promising avenues for investigation or by revealing inconsistencies between our understanding of a system (embodied in the model) and laboratory observations. In fact, the identification of such inconsistencies is a key benefit of modeling. Because a model can be exhaustively investigated, it follows that a negative result—the inability of a model to replicate experimental observations—can be taken as a falsification of the hypotheses on which the model was built. This can lead to a refinement of the biological hypotheses and subsequently a refined model, which can then be tested against additional experiments. This iterative process leads to a continually improving understanding of the system in what has been called a "virtuous cycle."

The end goal of most modeling efforts is a fully predictive description: simulations are then guaranteed to be accurate representations of real behavior. Today's models of intracellular networks fall short of this goal, but examples abound in other sciences and in engineering. Engineers make use of accurate predictive models for *model-based design*, resulting in faster and more efficient development of engineered constructs. The Boeing 777 jet provides a compelling example: it was designed and tested extensively in computer simulations before any physical construction began.

Model-based design is also being used in synthetic biology. Although models of cellular networks have only limited predictive power, they are useful for guiding the choice of components and suggesting the most effective experiments for testing system performance. The use of model-based design in the construction of engineered genetic networks will be illustrated briefly in section 1.6.3 and is described in more detail in chapter 7.

1.5   Basic Features of Dynamic Mathematical Models

This section introduces some fundamental concepts in dynamic mathematical modeling.

1.5.1   State Variables and Model Parameters

The primary components of a dynamic mathematical model correspond to the molecular species involved in the system (which are represented in the corresponding interaction diagram). The abundance of each species is assigned to a **state variable** within the model. The collection of all of these state variables is called the *state* of the system. It provides a complete description of the system's condition at any given time. The model's dynamic behavior is the time course for the collection of state variables.

Besides variables of state, models also include **parameters**, whose values are fixed. Model parameters characterize environmental effects and interactions among system components. Examples of model parameters are association constants, maximal expression rates, degradation rates, and buffered molecular concentrations. A change in the value of a model parameter corresponds to a change in an environmental condition or in the system itself. Consequently, model parameters are typically held at constant values during simulation: these values can be varied to explore system behavior under perturbations or in altered environments (e.g., under different experimental conditions).

For any given model, the distinction between state variables and model parameters is clear-cut. However, this distinction depends on the model's context and on the timescale over which simulations run. For instance, in chapter 5, we will focus on models of metabolism, in which enzyme catalysts provide a fixed "background." In that context—and on the relevant timescale of seconds to minutes—we will treat enzyme abundances as fixed model parameters. In contrast, the models in chapter 7 describe gene regulatory networks, which are responsible for the regulation of enzyme abundance (on a slower timescale). In those models, enzyme concentrations will be time-varying state variables.

1.5.2 Steady-State Behavior and Transient Behavior

Simulations of dynamic models describe time-varying system behavior. Models of biological processes almost always arrive, in the long run, at steady behaviors. Most commonly, models exhibit a persistent operating state, called a **steady state**; in contrast, some systems display sustained oscillations. The time course that leads from the initial state to the long-time (or *asymptotic*) behavior is referred to as the **transient**. In some cases, we will focus on transient behavior, as it reflects the immediate response of a system to perturbation. In other cases, our analysis will concern only the steady-state behavior, as it reflects the prevailing condition of the system over significant stretches of time.

1.5.3 Linearity and Nonlinearity

A relationship is called **linear** if it is a direct proportionality. For example, the variables x and y are linearly related by the equation $x = ky$, where k is a fixed constant. Linearity allows for effortless extrapolation: a doubling of x leads to a doubling of y, regardless of their values. Linear relationships involving more than two variables are similarly transparent; for example, $x = k_1 y + k_2 z$. A dynamic mathematical model is called linear if all interactions among its components are linear relationships. This is a highly restrictive condition, and consequently linear models display only a limited range of behaviors.

Any relationship that is not linear is referred to (unsurprisingly) as **nonlinear**. Nonlinear relations need not follow any specific pattern and so are difficult to address with any generality. The nonlinearities that appear most often in biochemical and genetic interactions are saturations, in which one variable increases with another at a diminishing rate, so that the dependent variable tends to a limiting, or *asymptotic*, value. Two kinds of saturating relationships that we will encounter repeatedly in this text are shown in figure 1.2. Panel A shows a *hyperbolic saturation*, in which the rate of increase of y declines continuously as the value of x increases. Panel B shows a *sigmoidal saturation*, in which y initially grows very slowly with x, then passes through a phase of rapid growth before saturating as the rate of growth drops.

1.5.4 Global and Local Behavior

Nonlinear dynamic models exhibit a wide range of behaviors. In most cases, a detailed analysis of the overall, **global**, behavior of such models would be overwhelming. Instead, attention can be focused on specific aspects of system behavior. In particular, by limiting our attention to the behavior near particular operating points, we can take advantage of the fact that, over small domains, nonlinearities

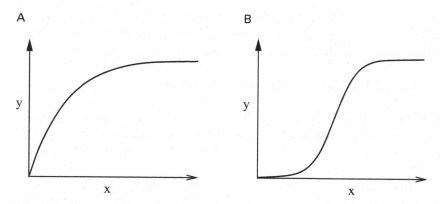

Figure 1.2
Common nonlinear relationships in cell biological processes. (A) Hyperbolic saturation. As x increases, y also increases, but at an ever-diminishing rate. The value of y approaches a limiting, or *asymptotic*, value. (B) Sigmoidal nonlinearity. The values of y show a slow rate of increase for small values of x, followed by a rapid "switch-like" rise toward the limiting value.

can always be approximated by linear relationships (e.g., a tangent line approximation to a curve). This **local** approximation allows one to apply linear analysis tools in this limited purview. Intuition might suggest that this approach is too handicapped to be of much use. However, the global behavior of systems is often tightly constrained by their behavior around a handful of nominal operating points; local analysis at these points can then provide comprehensive insight into global behavior. Local approximations are of particular use in biological modeling because self-regulating (e.g., homeostatic) systems spend much of their time operating around specific nominal conditions.

1.5.5   Deterministic Models and Stochastic Models

The notion of determinism—reproducibility of behavior—is a foundation for much of scientific investigation. A mathematical model is called **deterministic** if its behavior is exactly reproducible. Although the behavior of a deterministic model is dependent on a specified set of conditions, *no* other factors have any influence, so that repeated simulations under the same conditions are always in perfect agreement. (To make an experimental analogy, they are perfect replicates.)

   In contrast, **stochastic** models allow for randomness in their behavior. The behavior of a stochastic model is influenced both by specified conditions and by unpredictable forces. Each repetition of a stochastic simulation thus yields a distinct sample of system behavior.

   Deterministic models are far more tractable than stochastic models for both simulation and model analysis. In this text, our focus will be on deterministic models.

However, stochastic models are often called for, particularly in studies of gene regulatory networks, where thermal agitation of individual molecules is a significant source of randomness. A stochastic modeling framework is introduced in section 7.6.

1.6 Dynamic Mathematical Models in Molecular Cell Biology

A great many mathematical models of cellular phenomena have been published in the scientific literature. These are routinely archived in model repositories, such as the Biomodels database, the CellML model repository, and the JWS online repository.[2]

Before beginning our discussion of model construction in chapter 2, we briefly present four examples of modeling projects. These brief case studies illustrate the range of biological domains that will be explored in this text and demonstrate a number of uses for dynamic mathematical modeling in systems biology. Readers unfamiliar with molecular biology may find it useful to consult appendix A before continuing.

1.6.1 Drug Target Prediction in *Trypanosoma brucei* Metabolism

The single-cell parasite *Trypanosoma brucei* infects the bloodstream and causes sleeping sickness. It is a single-celled eukaryote, so is not susceptible to bacterial antibiotics. The search for efficient treatments is ongoing. In 1999, Barbara Bakker and her colleagues published a study of glycolysis in *Trypanosoma brucei* (Bakker et al., 1999). Glycolysis, an energy-producing metabolic pathway, is crucial to the metabolism of both the parasite and the host. Fortunately, the mammalian enzymes responsible for catalyzing the reactions in the host pathway are significantly different from those of *Trypanosoma*. Thus, the enzymes of the parasite can be inhibited by drugs that have little effect on the host. Bakker and her colleagues sought to identify which enzymes control the rate of glycolysis in *Trypanosoma*, with the aim of predicting ideal targets for growth-inhibiting drugs. The interaction diagram for their model, reproduced as figure 1.3, shows the metabolic reactions in the network.

Using data from previously published studies of *Trypanosoma* and from their own experiments, Bakker and her colleagues formulated a dynamic mathematical model of the pathway. Their focus was on the steady-state behavior of the system, and particularly on the rate of energy production. To predict the effects of enzyme-inhibiting drugs on the energy-production rate, they applied a *sensitivity analysis* to the model. This technique determines how sensitive the model behavior is to perturbations in the model parameters. Bakker and her colleagues identified five

---

2. These and other repositories can be accessed from www.systems-biology.org/resources/model-repositories.

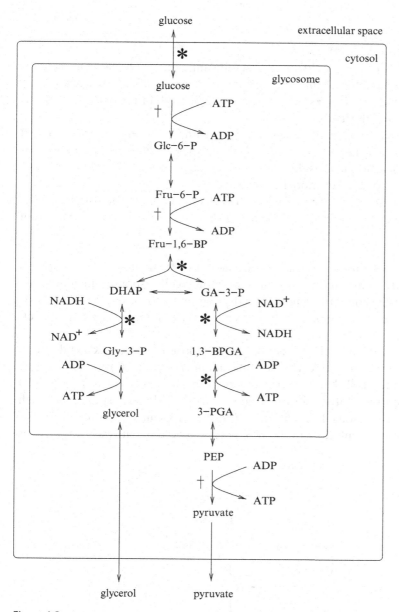

Figure 1.3
The glycolytic pathway of *Trypanosoma brucei*. The reactions occur primarily in the glycosome — a specialized organelle. The analysis performed by Bakker and her colleagues indicated that the best targets for inhibition are those steps marked by an asterisk (*). The three steps marked with a dagger (†) were commonly believed to have significant influence over the pathway. However, Bakker and colleagues found that these are poor targets — their model predicted that inhibition of these reaction steps has little effect on the overall pathway flux. Glc-6-P, glucose 6-phosphate; Fru-6-P, fructose 6-phosphate; Fru-1,6-BP, fructose 1,6-bisphosphate; DHAP, dihydroxyacetone phosphate; GA-3-P, glyceraldehyde 3-phosphate; Gly-3-P, glycerol 3-phosphate; 1,3-BPGA, 1,3-bisphosphoglycerate; 3-PGA, 3-phosphoglyceric acid; PEP, phosphoenolpyruvate; ATP, adenosine triphosphate; ADP, adenosine diphosphate; NAD+ or NADH, nicotinamide adenine dinucleotide. Adapted from figure 1 of Bakker et al. (1999).

enzymes whose inhibition would have a significant impact on pathway activity. Moreover, they demonstrated that three other enzymes that had previously been proposed as effective drug targets are in fact poor targets: inhibition of these enzymes has little impact on pathway flux.

Later, Bakker and colleagues provided experimental confirmation of several of their model predictions (Albert et al., 2005). Their experiments also provided evidence for regulatory interactions that were not included in the original model, thus opening an avenue for further model development.

We will address modeling of metabolic networks in chapter 5. Sensitivity analysis, to be introduced in section 4.5, plays a key role in the investigation of metabolism.

1.6.2 Identifying the Source of Oscillatory Behavior in NF-κB Signaling

The protein NF-κB is involved in a number of animal cell responses, including the regulation of cell division, inflammation, and programmed cell death. The NF-κB pathway plays a role in inflammatory diseases and has been implicated in the development of cancer.

In the absence of stimuli, NF-κB is inhibited by proteins called IκB (inhibitors of κB), as shown in figure 1.4A. Extracellular stimuli (such as hormones) trigger protein activity in the cell that leads to a decrease in IκB levels. This frees NF-κB from the action of its inhibitor, allowing it to stimulate a cellular response (through changes in gene expression). In addition, NF-κB activity causes IκB levels to rise, resulting in renewed inhibition of NF-κB itself. Thus, the pathway response is

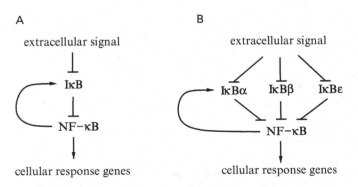

Figure 1.4
(A) The main components of the NF-κB signaling pathway. (Blunt-ended arrows signify repression.) Extracellular events trigger degradation of the inhibitor IκB. Once free of this inhibitor, NF-κB proteins trigger an appropriate cellular response. At the same time, NF-κB stimulates production of IκB proteins, leading to restored inhibition of NF-κB. (B) A more detailed diagram showing a family of IκB proteins. All three forms act as inhibitors of NF-κB. IκBβ and IκBε are unaffected by NF-κB; only the IκBα form is stimulated by NF-κB activity.

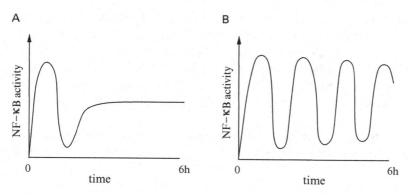

Figure 1.5
Oscillatory behavior in the NF-κB pathway. (A) The normal (wild-type) response to persistent signaling. The oscillations are quickly damped, resulting in a persistently active response. (B) The behavior of a cell in which IκBβ and IκBε are absent. The oscillations persist. Adapted from figure 1 of Cheong et al. (2008).

self-limiting. This sort of negative feedback loop typically leads to robust steady-state behavior, but it can also lead to persistent oscillations.

Indeed, NF-κB pathways exhibit a range of behaviors upon stimulation, including both damped and persistent oscillations, as sketched in figure 1.5. In 2002, Alexander Hoffmann and his colleagues presented a mathematical model of NF-κB signaling that sheds light on the system's dynamic response (Hoffmann et al., 2002). The model focuses on the roles of three distinct forms of the inhibitory IκB proteins, called IκBα, IκBβ, and IκBε. When all three of these forms are present in the cell, the pathway exhibits damped oscillations in response to stimulation (figure 1.5A). However, when cells are modified so that certain IκB proteins are absent, the response changes. When IκBα is absent, cells show pathologically high activity. Alternatively, when both IκBβ and IκBε are absent, cells respond to stimuli with sustained oscillations in NF-κB activity (figure 1.5B). This difference in behavior is a consequence of the fact that of the three IκB forms, only IκBα production is enhanced by NF-κB activity (figure 1.4B).

From a design perspective, an ideal response would be a quick rise to a steady activated level. This ideal response is closely approximated by the damped oscillations normally displayed by the cells (figure 1.5A). Hoffmann and his colleagues used their model to determine that this response is generated by the combined behavior of the IκB proteins. IκBα provides a negative feedback that quenches the quick initial rise, resulting in an appropriate steady level. However, a fast response demands a quenching signal so strong that oscillations would arise unless there were a secondary persistent quenching signal to damp the response. This secondary quenching is provided by the steady activity of IκBβ and IκBε.

Hoffmann and colleagues generated model simulations that describe the response of the pathway to stimuli of varying strengths and durations. These model predictions, verified by experiments, show that the complementary roles of the IκB proteins generate qualitatively different responses to stimuli of different durations and that these differences in signaling activity lead to distinct cellular responses. (A paper by Cheong et al., 2008, describes the results of additional NF-κB modeling efforts.)

Chapter 6 is devoted to intracellular signaling pathways. Tools to address oscillatory behavior are presented in section 4.3. A model of NF-κB activation is explored in problem 7.8.15.

1.6.3 Model-Based Design of an Engineered Genetic Toggle Switch
In the year 2000, the emerging field of synthetic biology was heralded by the simultaneous announcement of two engineered genetic networks. Both of these devices will be covered in depth in chapter 7. For now, we will briefly introduce one of these devices—a genetic toggle switch—and describe how modeling played a key role in its design.

A toggle switch is a device that transitions between two states in a user-controlled manner. Such a system is called *bistable*, meaning that the two states are persistent—the transition occurs only under intervention by the user.

The first engineered genetic toggle switch was constructed by Timothy Gardner, Charles Cantor, and Jim Collins and is commonly known as the Collins toggle switch (Gardner et al., 2000). An interaction diagram for the gene network is shown in figure 1.6. Two genes are involved. Each gene's protein product represses production of the other. This mutual antagonism results in a bistable system: in one stable

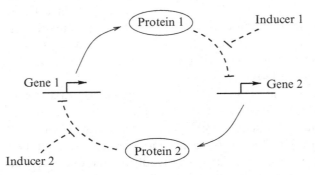

Figure 1.6
Engineered genetic toggle switch. (Dashed blunt-ended arrows indicate repression.) Each gene's protein product represses production of the other, resulting in two modes of persistent operation: one protein is abundant while the other is repressed. The proteins are chosen so that intervention by the experimenter can deactivate the abundant protein, inducing a switch in the protein levels.

condition, protein 1 is abundant and production of protein 2 is repressed; the roles are reversed in the other steady state. Gardner and his colleagues chose to use two proteins whose activity could be specifically inhibited by laboratory interventions. Inactivation of the abundant protein induces an increase in the concentration of the other, resulting in a transition between the two stable states.

If the two genes and their products have symmetric properties, intuition suggests that the switch will operate as described above. However, if it is asymmetric, the network may fail to exhibit bistability—the "stronger" protein might always dominate in their competition of mutual repression. This fact posed a challenge to Gardner and his colleagues. They could only select genetic components from the small collection of suitable genes that had been characterized in existing organisms. Whichever components they chose, there would be asymmetry in the network, and so bistability could not be ensured.

Instead of carrying out an exhaustive experimental search for combinations of components that would function properly, Gardner and his colleagues developed a dynamic mathematical model to guide their design choices. Rather than fit the model to a particular instance of the design, they constructed a simple generic model and used it to investigate the behavior of a wide variety of potential designs. Figure 1.7 shows a simulation of their model displaying the desired behavior.

Their analysis led to two useful conclusions. First, the model demonstrated that bistability cannot be achieved (even in the symmetric case) if the rates of gene

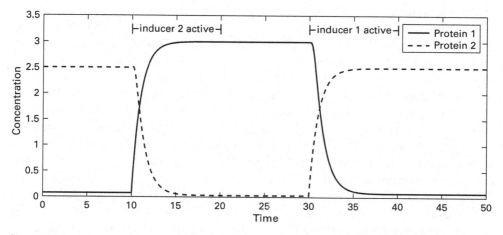

Figure 1.7
Simulation of the toggle switch model. At time zero, the system is in steady state, with protein 2 abundant and protein 1 at a low concentration. At time $t = 10$, inducer 2 is introduced, repressing protein 2 and allowing protein 1 to be expressed. Protein 1 then rises to dominance and inhibits production of protein 2. When the inducer is removed (at $t = 20$), the system remains in the high-protein-1, low-protein-2 state. At time $t = 30$, inducer 1 is introduced, causing a return to the original stable state. Units are arbitrary.

expression are too low. Second, the model indicated that nonlinearity in the protein–DNA interactions could compensate for asymmetry in the genes and proteins. Some degree of nonlinearity is critical in a functioning switch: the more significant the nonlinearity, the more forgiving the constraints on symmetry. Guided by these insights, Gardner and his colleagues were able to construct a number of successful instances of the genetic toggle switch.

Since the year 2000, a wide range of synthetic biological devices have been constructed (for a review, see Khalil and Collins, 2010). Model-based design is a common feature of these projects. This analysis often begins with a simple generic model that plays an exploratory role in addressing the possible designs. Such "toy" models are often used for the same purpose in scientific exploration—they provide proof-of-principle hypotheses that can then be improved as further experimental evidence becomes available.

The Collins toggle switch will be covered in more depth in section 7.2.3. Tools for analysis of bistable systems will be introduced in section 4.2. The nonlinearities in the toggle switch design are a result of cooperative binding effects, which are discussed in section 3.3.

1.6.4 Establishing the Mechanism for Neuronal Action Potential Generation

Neurons, the primary cells in the animal nervous system, encode information in the electrical potential (i.e., voltage difference) across the cell membrane. This information is communicated by *action potentials*—sweeping changes in membrane potential that propagate along the length of the cell.

In a series of papers published in 1952, Alan Hodgkin and Andrew Huxley, along with Bernard Katz, described a biophysical mechanism for the generation of action potentials. They confirmed the behavior of their proposed mechanism with a dynamic mathematical model (reviewed in Rinzel, 1990).

Prior to their work, it had been established that individual ions, such as Na^+ and Cl^-, are the primary carriers of electrical charge at the cell membrane and that cells maintain very different concentrations of these ions in the intracellular and extracellular spaces. Moreover, it had been hypothesized that changes in the transmembrane voltage cause changes in the permeability of the membrane to these ions. These changes in permeability can result in significant ion flux across the membrane and thus produce further changes in transmembrane potential. Hodgkin and Huxley, using newly developed laboratory techniques, carried out a series of experiments showing that membrane permeability is ion-specific. They (correctly) hypothesized that this specificity is a result of ion-specific channels that are lodged in the membrane (as illustrated in figure 1.8) and that these channels are sensitive to membrane potential.

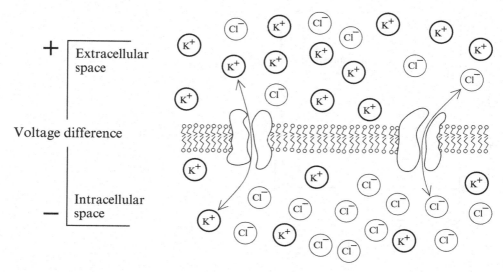

Figure 1.8
Electrical activity at the neuronal membrane. The difference in ionic charge across the membrane results in a voltage difference. The membrane contains ion-specific channels, each allowing only one type of ion to pass. These channels are voltage-sensitive—they open or close depending on the transmembrane voltage. This voltage sensitivity sets up a feedback loop: changes in voltage lead to changes in ion-transfer rates, which lead to subsequent changes in voltage, and so on. The interplay of these feedback loops can generate action potentials.

To verify that their hypothetical mechanism was capable of generating action potentials, Hodgkin and Huxley developed a dynamic model of membrane voltage and ion transport. Simulations of their model replicated neuronal behavior in a range of conditions, providing significant support for their hypothetical mechanism.

The generation of action potentials is referred to as *excitable* behavior and depends on a specific voltage threshold, as shown in figure 1.9. Small electrical perturbations cause no significant response—voltage quickly relaxes to the pre-stimulus level. However, for perturbations that exceed a certain threshold, the response is a dramatic pulse in transmembrane voltage: an action potential. Excitability is caused by coupled positive and negative feedback loops that act between the transmembrane voltage and the voltage-gated channels in the membrane. Complex dynamic behaviors such as this can only be rigorously studied through mathematical modeling.

The Hodgkin–Huxley framework for modeling of neuronal behavior is introduced in chapter 8, and the original Hodgkin–Huxley model is presented in problem 8.6.4.

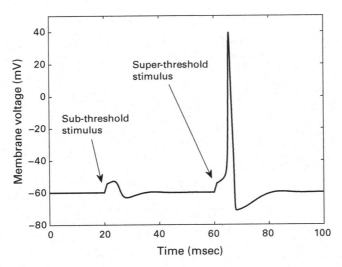

Figure 1.9
Simulation of the Hodgkin–Huxley model. The membrane voltage, originally at rest, is perturbed at time 20 milliseconds. The disturbance lasts 1 millisecond. Once removed, its effects are quickly washed away as the voltage relaxes to its predisturbance level. A second disturbance, of the same form, occurs 40 milliseconds later, and again lasts for 1 millisecond. This disturbance, which is only 3% stronger than the first, elicits a dramatically different response: this wide excursion in voltage is characteristic of an action potential.

1.7 Suggestions for Further Reading

• **Computational Systems Biology**   This book focuses on a few fundamental modeling approaches in systems biology. Wider surveys of the tools used in computational systems biology can be found in *Systems Biology: A Textbook* (Klipp et al., 2009), *System Modeling in Cellular Biology* (Szallasi et al., 2006), *An Introduction to Systems Biology: Design Principles of Biological Circuits* (Alon, 2007), and *A First Course in Systems Biology* (Voit, 2012).

• **Dynamic Modeling in Molecular Cell Biology**   Several texts focus on modeling of particular biological domains. The books *The Regulation of Cellular Systems* (Heinrich and Schuster, 1996) and *Kinetic Modeling in Systems Biology* (Demin and Goryanin, 2009) focus on modeling in metabolism. *Computational Modeling of Gene Regulatory Networks* (Bolouri, 2008) addresses modeling formalisms used to study genetic networks. The use of modeling in synthetic biology is addressed in *Engineering Genetic Circuits* (Myers, 2010). Modeling of neuronal systems is surveyed in *Mathematical Foundations of Neuroscience* (Ermentrout and Terman, 2010).

• **Mathematical Modeling**   *Modeling the Dynamics of Life* (Adler, 2004) is an introductory calculus text with an emphasis on dynamic modeling in biology. A more advanced treatment of differential equations in this context is provided in *Differential Equations and Mathematical Biology* (Jones et al., 2009). Nonlinear dynamics is introduced in the text *Nonlinear Dynamics and Chaos: With Applications to Physics, Biology, Chemistry, and Engineering* (Strogatz, 2001).

• **Mathematical Biology** Texts in mathematical biology often cover intracellular processes and typically introduce a range of modeling tools used in the field. These include *Computational Cell Biology* (Fall et al., 2002), *Mathematical Models in Biology* (Edelstein-Keshet, 2005), and *Mathematical Physiology* (Keener and Sneyd, 1998).

2 Modeling of Chemical Reaction Networks

It behooves us always to remember that in physics it has taken great [minds] to discover simple things. They are very great names indeed which we couple with the explanation of the path of a stone, the droop of a chain, the tints of a bubble, the shadows in a cup. It is but the slightest adumbration of a dynamical morphology that we can hope to have until the physicist and the mathematician shall have made these problems of ours their own. . . .
— D'Arcy Thompson, *On Growth and Form*

Models of cellular phenomena often take the form of schematic interaction diagrams, as in figure 1.1. For biochemical and genetic networks, the components in these diagrams are *molecular species*, which could be ions, small molecules, macromolecules, or molecular complexes. An interaction diagram depicts the species in a system and indicates how they interact with one another. The interactions (arrows) in the diagram can represent a range of processes, such as chemical binding or unbinding, reaction catalysis, or regulation of activity. In each case, the rate of the process depends on the abundance of certain molecular species within the network. These processes, in turn, result in the production, interconversion, transport, or consumption of the species within the network. Over time, the abundance of each species changes, leading to corresponding changes in the rates of the processes. For simple systems, we can understand the resulting behavior intuitively. However, for more complex networks—especially those involving feedback—the interaction diagram leaves ambiguity with respect to time-varying behaviors. In these cases, an accurate description of system behavior is only possible if we describe the interactions more precisely—in quantitative terms.

These quantitative descriptions can be used to construct dynamic mathematical models. In this chapter, we will address the construction of models that describe chemical reaction networks. The next chapter will introduce quantitative descriptions of biochemical processes. Together, these chapters lay a foundation for dynamic modeling of cellular behavior.

2.1   Chemical Reaction Networks

Consider a group of chemical species (i.e., chemically distinct molecules) that can undergo the following reactions:

$$A + B \rightarrow C + D \qquad\qquad D \rightarrow B \qquad\qquad C \rightarrow E + F.$$

These reactions are assumed to be *irreversible*—they only proceed in the direction indicated. (The assumption of irreversibility is necessarily an approximation. The laws of thermodynamics dictate that all chemical reactions are reversible. Nevertheless, it is often reasonable to describe a reaction as irreversible under conditions in which the reverse reaction proceeds at a negligible rate.)

A set of reactions constitutes a **chemical reaction network**. The manner in which the species interact is referred to as the network *topology* (or *architecture*). The organization of the network is apparent if we rearrange the reactions in the form of an *interaction graph*.[1] This network's interaction graph is shown in figure 2.1.

Exercise 2.1.1   Draw the interaction graph for the following reaction network:

$$A \rightarrow B + C \qquad\qquad B \rightarrow D \qquad\qquad C \rightarrow E$$

$$C \rightarrow F \qquad\qquad E + F \rightarrow G. \qquad\qquad\qquad \square$$

2.1.1   Closed and Open Networks

The reaction network considered above is **closed**, meaning that there are no reactions whose products or reactants lie outside of the network. The steady-state behavior of such networks is *thermal equilibrium*, a state in which all net reaction rates are zero.

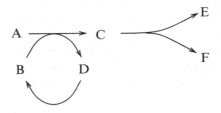

Figure 2.1
Closed reaction network.

---

1. The use of the term "graph" here is, unfortunately, different from its use in the visualization of a function or a data set. In mathematical graph theory, a graph consists of a set of objects (called *nodes*) that are connected to one another by links (called *edges*). Here, the nodes are the chemical species; the edges are the reactions.

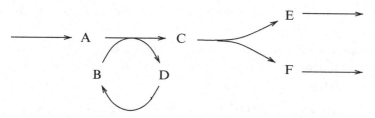

Figure 2.2
Open reaction network.

In contrast, most biochemical networks are **open** systems—they exchange material with the outside environment and reach a steady state that involves a steady flow through the network. Such a state is called a *dynamic equilibrium*. (A familiar example of a dynamic equilibrium is a steady jet of water flowing from a waterfall or faucet. Although this might appear to be an unmoving object, it is a steadily active process.) The network above could be made into an open network by adding the reactions

$$\to A \qquad\qquad E \to \qquad\qquad F \to$$

resulting in the network in figure 2.2. These additional reactions indicate that material is being exchanged with the "world outside the network" and are referred to as *exchange reactions*.

Exercise 2.1.2 Add three exchange reactions to the reaction network in exercise 2.1.1 so that the system can support steady flow through all of the species in the network. Can this be achieved by adding just two exchange reactions? □

2.1.2 Dynamic Behavior of Reaction Networks
Imagine an experiment in which the reactions in a network proceed in a fixed volume, and suppose the species are initially present at specified concentrations. To predict the time-varying changes in species concentrations, we need to know the rates at which the reactions occur. The rate of a reaction depends on the concentrations of the reactants and the physicochemical conditions (e.g., temperature, pH). We will presume that the physicochemical environment is fixed, so rate laws can be described solely in terms of reactant concentrations.

Reaction rates are usually described under two assumptions:

Assumption 1 The reaction volume is well stirred. This means that the reactants are equally distributed throughout the volume. Consequently, the rate of each reaction is independent of position in space. This allows us to refer unambiguously to *the* reaction rate in the volume (rather than having to specify different rates at different locations).

Assumption 2 There are a great many molecules of each species present, so we can describe molecular abundance by a concentration that varies continuously (as opposed to an integer-valued molecule count).

The first of these assumptions—referred to as *spatial homogeneity*—typically holds in stirred laboratory reaction vessels. It can be a good approximation in the cell, as diffusion acts quickly to mix the molecular components of this tiny "reaction vessel." However, there is a great deal of spatial structure within the cell, so that in many cases the assumption of spatial homogeneity does not hold.

The second assumption—that there are a great many reactant molecules—is referred to as the *continuum hypothesis*. It allows discrete changes in molecule number to be approximated by continuous changes in concentration: individual reaction events cause infinitesimal changes in abundance. This assumption is perfectly valid when molar quantities of reactants are involved (recall, Avogadro's number is 6.02×10^{23}) and is appropriate for cellular species with molecule counts of thousands or more. However, some cellular processes are governed by populations of molecules numbering dozens or less. In those cases, changes in molecule abundance should be treated as discrete steps in population size.

We will build our modeling framework under the assumptions of spatial homogeneity and continuously varying concentrations. The resulting models yield accurate descriptions of a wide range of biological phenomena. Modeling frameworks for addressing spatial variation and small molecule counts will be introduced in section 8.4 and section 7.6, respectively.

The Law of Mass Action In a fixed volume, under the well-stirred assumption and the continuum hypothesis, a simple description of reaction rates is provided by the **law of mass action:** *the rate of a chemical reaction is proportional to the product of the concentrations of the reactants.* Using $[\,\cdot\,]$ to denote concentration, the rate of the reaction

$$X \rightarrow P$$

is $k_1[X]$ (because there is a single reactant), whereas the rate of

$$A + B \rightarrow C$$

is $k_2[A][B]$ (two reactants), and the rate of

$$D + D \rightarrow E$$

is $k_3[D]^2$ (two identical reactants). Here, k_1, k_2, and k_3 are constants of proportionality.

Some Notes on Mass Action

1. The law of mass action has an intuitive basis: it states that the probability of a reaction occurring is proportional to the probability of the reactants colliding with one another.

2. The exponent to which each reactant appears in the rate law is called the *kinetic order* of the reactant in the reaction. For example, reactant A has kinetic order 1 in the second reaction listed above, whereas D has order 2 in the third reaction. If a reaction describes uptake from the outside environment, it can be written with no explicit reactant (e.g., $\to A$). The rate of such a reaction is constant. Because these reactions satisfy a rate law with a reactant concentration raised to the power zero ($k[S]^0 = k$), they are called *zero-order* reactions.

3. The constant of proportionality in a mass-action rate law is called the **(mass-action) rate constant** and can be indicated in the reaction formula:

$$A + B \xrightarrow{k_2} C.$$

The dimensions of the rate constant depend on the number of reactants. The rate constant for a single-reactant reaction has dimensions of time^{-1}. If a reaction has two reactants, the rate constant has dimensions of concentration$^{-1} \cdot$ time^{-1}. For a zero-order reaction, the reaction rate is equal to the rate constant, which has dimensions of concentration $\cdot$ time^{-1}.

4. In cases where the environment is not constant, the rate constant can be replaced by an *effective rate constant* that depends on factors that affect the reaction rate. In a biochemical context, effective rate constants may depend on the concentration of enzyme catalysts.

In the following sections, we will use the law of mass action to construct dynamic mathematical models of chemical reaction networks. These models will take the form of *ordinary differential equations* (ODEs).[2] We will make use of this differential equation–based framework throughout the rest of the book.

In the chapters to follow, models investigations will be carried out via computational software. However, in the remainder of this chapter, we will address elementary networks for which pen-and-paper calculations yield explicit formulas describing the time-varying species concentrations. Such formulas are called *analytic solutions* of the differential equation. The analysis of these simple cases will provide valuable insight into the more complex models to follow.

---

2. The modifier *ordinary* is used to distinguish these from *partial* differential equations (PDEs). PDE-based modeling, which addresses spatially varying behavior, is introduced briefly in section 8.4.

2.1.3 Simple Network Examples

Some readers may find it useful to review the brief summary of calculus in appendix B before proceeding.

Example I: Decay   As a first example, consider a trivial open reaction system consisting of a single species decaying at a steady rate:

$$A \xrightarrow{k}.$$

The rate of the reaction is $k[A]$. To understand how the concentration of A behaves in time, we need to consider the rate of change of the concentration. Because the reaction consumes A, we have

rate of change of $[A]$ = −(rate of reaction).   (2.1)

Let

$a(t)$ = concentration $[A]$ at time t.

Then, recalling that the derivative of a function describes its rate of change, we can rewrite statement (2.1) as

$$\underbrace{\frac{d}{dt}a(t)}_{\text{Rate of change of }[A]\text{ at time }t} = \underbrace{-ka(t)}_{\text{Rate of reaction at time }t}.$$ (2.2)

This is a **differential equation** whose solution is the function $a(t)$.

Imagine this reaction has been set up in a laboratory test-tube. As the reaction proceeds, the concentration of A will decrease over time. Experimentally, we might observe a time series of concentrations as in figure 2.3. If our mathematical model is accurate, then the solution $a(t)$ of the differential equation should describe the behavior of the system over time. That is, it should agree with the experimental measurements (figure 2.3).

To use the model to make a prediction about a particular experiment, we need to supplement the differential equation with knowledge of the concentration $[A]$ at some time. We typically know the concentration at the beginning of the experiment, at time $t = 0$. This known concentration, $a(0)$, is referred to as the **initial condition.**

There are standard solution methods for simple classes of differential equations. Because the models of biological phenomena addressed in this text are not amenable to such solution techniques, they will not be addressed here. Nevertheless, it will prove insightful to derive an explicit solution formula for this simple differential equation. To do so, we will take a direct (and rather unsatisfactory) route to the solution: we guess.

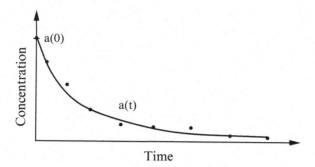

Figure 2.3
Model behavior. The points represent a hypothetical experimental time series. The curve is the corresponding model-based prediction. Discrepancies between model prediction and data may be attributed to experimental error or to inaccuracy in the model formulation.

Well, it's not quite guessing. We will begin by considering a special case of the differential equation in which $k = -1$. We do not have a chemical interpretation of the equation in this case, as rate constants are never negative. However, it will be useful momentarily to consider the resulting differential equation:

$$\frac{d}{dt}a(t) = a(t). \tag{2.3}$$

A solution $a(t)$ of this differential equation has the property that its derivative has the same value as the function itself at each time point t. That is, the function $a(t)$ is its own derivative. You may recall that the exponential function $a(t) = e^t$ has this property, as does any constant multiple of this function. Thus, any function of the form $a(t) = De^t$, for a given constant D, satisfies the differential equation (2.3).

A more relevant case occurs if we take $k = 1$, which leads to:

$$\frac{d}{dt}a(t) = -a(t).$$

Invoking the chain rule leads to the correct guess in this case: $a(t) = De^{-t}$. (By the chain rule: $d/dt(e^{-t}) = e^{-t}(d/dt(-t)) = e^{-t}(-1) = -e^{-t}$.)

Finally, consider the general case:

$$\frac{d}{dt}a(t) = -ka(t).$$

Appealing to the chain rule again, we arrive at the solution $a(t) = De^{-kt}$. How should the constant D be chosen so that this function will agree with experimental observations? Recall that we are presuming we know the initial concentration: $a(0)$. Let's call that A_0, so $a(0) = A_0$. Substituting time $t = 0$ into the solution $a(t) = De^{-kt}$, we find

$$a(0) = De^{-k \cdot 0} = De^0 = D.$$

Because we have $a(0) = A_0$, we conclude that $D = A_0$. That is, the constant D is equal to the initial concentration of A. The species concentration can then be written as a function of time:

$$a(t) = A_0 e^{-kt}. \tag{2.4}$$

This behavior is referred to as *exponential decay*.

The time-varying behavior of this family of solutions is shown in figure 2.4. The curves all approach zero as time passes. We say they *decay* to zero, or they *relax* to zero, or that their *asymptotic value* is zero. Moreover, the curves in the figure all decay at the same characteristic rate. This decay rate is characterized by the **time constant** of the process, defined, in this case, as $\tau = 1/k$. (The *half-life* $\tau_{1/2}$ is closely related to the time constant: $\tau_{1/2} = \ln 2/k = \tau \ln 2$.)

The time constant provides a useful scale for addressing the dynamics of the reaction. For example, if $\tau = 1$ second, then it is appropriate to plot system behavior on a scale of seconds. Alternatively, if $\tau = 100$ seconds, then a timescale of minutes is more appropriate. The timescale of a process determines the time interval over which

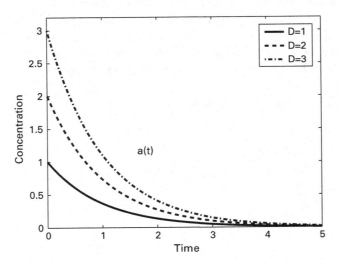

Figure 2.4
Exponentially decaying concentration profiles (equation 2.4). Parameter values are $k = 1$, $D = 1$, $D = 2$, $D = 3$.

model simulations should be run and is also used in the design of time-series experiments. For example, if the timescale of the dynamics is minutes, then useful data will be collected on that timescale. Data separated by longer periods (e.g., hours) may miss crucial aspects of the behavior; conversely, a strategy of observation at a greater frequency (say, every second) will be wasteful, as the data will be highly redundant.

Example II: Production and Decay We next consider an open network involving two reactions, production and decay:

$$\xrightarrow{k_0} A \xrightarrow{k_1} \; .$$

The first reaction is zero order; the reaction rate is equal to k_0. (Zero-order reactions are used when the concentration of the reactant is considered constant. This occurs when the reactant pool is large—so that depletion of the reactant is negligible—or when the concentration of the reactant pool is buffered by some unmodeled process. These reactions are written with the reactant absent, as above, or with a reactant whose concentration is a fixed model parameter; e.g., $X \to A$, $[X]$ fixed.)

Again letting $a(t)$ denote the concentration of A at time t, the reaction dynamics are described by

rate of change of $[A]$ = rate of production of A − rate of decay of A,

which leads to the model

$$\underbrace{\frac{d}{dt}a(t)}_{\text{Rate of change of } [A] \text{ at time } t} = \underbrace{k_0}_{\text{Rate of production}} - \underbrace{k_1 a(t)}_{\text{Rate of decay}}. \tag{2.5}$$

Before addressing the time-varying behavior of $a(t)$, we first consider the concentration that A will reach in steady state. We note that the concentration of A will remain fixed when the rate of decay is equal to the rate of production. Mathematically, this means that the steady-state concentration a^{ss} satisfies

rate of change of $[A] = k_0 - k_1 a^{ss} = 0$.

This yields

$$a^{ss} = \frac{k_0}{k_1}.$$

Exercise 2.1.3 Verify that the ratio k_0/k_1 has dimensions of concentration. ☐

We'll regularly use this procedure to find steady states without solving the corresponding differential equation. Note that in steady state, there is a nonzero flux through this network: the steady-state rate of both reactions is k_0.

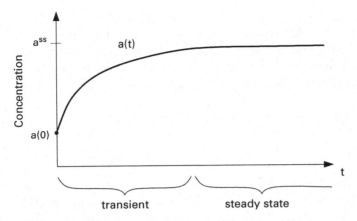

Figure 2.5
Transient and steady-state behavior. The transient occurs while the concentration relaxes to its steady-state value.

Turning now to the time-varying behavior of $[A]$, we expect to observe the concentration transitioning from its initial value to the steady-state concentration, as in figure 2.5. The time course leading to the steady state is called the **transient**.

To derive an explicit description of the time-varying concentration, we will solve the differential equation (2.5) by using another "guess": we expect the solution $a(t)$ to approach the steady-state value exponentially. If this is the case, the displacement from steady state $(a(t) - a^{ss})$ will decay exponentially to zero, and so it will satisfy a differential equation similar to our earlier model of exponential decay in equation (2.2). This insight leads to a solution of the form (details in exercise 2.1.4):

$$a(t) = De^{-k_1 t} + \frac{k_0}{k_1}. \tag{2.6}$$

As before, the constant D depends on the initial concentration.

The concentration of A relaxes exponentially to the steady-state value of $a^{ss} = k_0/k_1$. The relaxation rate k_1 is independent of the production rate k_0. This illustrates a general principle: the timescale on which the concentration of a species varies is typically determined by its decay rate rather than by its production rate.

Exercise 2.1.4 Determine the solution to equation (2.5) as follows. Let z denote the displacement of the concentration a from its steady state: $z(t) = a(t) - a^{ss} = a(t) - k_0/k_1$. Next, use equation (2.5) to verify that $z(t)$ satisfies the differential equation

$$\frac{d}{dt}z(t) = -k_1 z(t). \tag{2.7}$$

Hint: Verify that

$$\frac{d}{dt}z(t) = \frac{d}{dt}a(t) = k_0 - k_1 a(t)$$

and that $k_0 - k_1 a(t) = -k_1 z(t)$. Finally, recall that the solution of equation (2.7) is

$$z(t) = De^{-k_1 t},$$

so that from the definition of $z(t)$, we have equation (2.6). □

Exercise 2.1.5 Verify from equation (2.6) that $D = A_0 - k_0/k_1$, where $A_0 = a(0)$. Confirm that the concentration of A can be written as

$$a(t) = \left(A_0 - \frac{k_0}{k_1}\right)e^{-k_1 t} + \frac{k_0}{k_1}.$$ □

Example III: Irreversible Conversion Next we consider a closed system involving a single reaction:

$$A \xrightarrow{k} B.$$

This reaction is irreversible: molecules of B cannot be converted back to A. Because no material is exchanged with the external environment, the system is closed.

The rate of the reaction is $k[A]$. (Species B has no influence on the reaction rate.) Each reaction event consumes a molecule of A and produces a molecule of B, so we have

rate of change of $[A]$ = −(rate of reaction)

rate of change of $[B]$ = rate of reaction.

Let

$a(t)$ = concentration $[A]$ at time t

$b(t)$ = concentration $[B]$ at time t.

The reaction system can then be modeled by

$$\underbrace{\frac{d}{dt}a(t)}_{\text{Rate of change of } [A] \text{ at time } t} = \underbrace{-ka(t)}_{\text{Rate of reaction at time } t}$$

$$\underbrace{\frac{d}{dt}b(t)}_{\text{Rate of change of } [B] \text{ at time } t} = \underbrace{ka(t)}_{\text{Rate of reaction at time } t}.$$

This is a **system of differential equations**: two equations involve the two unknowns $a(t)$ and $b(t)$. Typically, it is much more difficult to solve systems of differential equations than to solve individual differential equations. However, in this case the system can be reduced to a single equation, as follows.

Because the behavior of $a(t)$ is identical to the decay reaction of example I, we know that $a(t) = A_0 e^{-kt}$, where $A_0 = a(0)$ is the initial concentration of A. To determine the concentration $b(t)$, we observe that the total concentration of A and B is conserved—every time a molecule of B is produced, a molecule of A is consumed.

Conservation is a general feature of closed systems. In this case, it says that $a(t) + b(t) = T$ (constant) for all time t. If $a(0) = A_0$ and $b(0) = B_0$, then $T = A_0 + B_0$, and we can write

$b(t) = B_0 + A_0 - a(t)$.

As time passes, $[A]$ decays to zero, and $[B]$ tends to $B_0 + A_0$; eventually, all of the molecules of species A are converted to B.

We derived this conservation from inspection of the reaction system. Conservations can also be derived from differential equation models: they appear as balances in the rates of change. In this case, the conservation $a(t) + b(t) = T$ follows from the symmetry in the rates of change for A and B. We can write

$$\frac{d}{dt}(a(t) + b(t)) = \frac{d}{dt}a(t) + \frac{d}{dt}b(t) = -ka(t) + ka(t) = 0,$$

confirming that the total concentration $a(t) + b(t)$ does not change with time.

Example IV: Reversible Conversion   Our final example is a closed system consisting of a single reversible reaction:

$$A \xrightarrow{k_+} B \qquad \text{and} \qquad B \xrightarrow{k_-} A$$

or more concisely

$$A \underset{k_-}{\overset{k_+}{\rightleftharpoons}} B.$$

Applying the law of mass action, we find that

The rate of $A \to B$ is $k_+[A]$.

The rate of $B \to A$ is $k_-[B]$.

Letting, as before, a and b denote the concentrations of A and B, we have

Rate of change of $[A]$ = (rate of production of A) − (rate of consumption of A)
$$= (\text{rate of } B \to A) - (\text{rate of } A \to B).$$

This can be written as

$$\underbrace{\frac{d}{dt}a(t)}_{\text{Rate of change of } [A] \text{ at time } t} = \underbrace{k_-b(t)}_{\text{Rate of production}} - \underbrace{k_+a(t)}_{\text{Rate of consumption}} . \tag{2.8}$$

Likewise,

$$\underbrace{\frac{d}{dt}b(t)}_{\text{Rate of change of } [B] \text{ at time } t} = \underbrace{k_+a(t)}_{\text{Rate of production}} - \underbrace{k_-b(t)}_{\text{Rate of consumption}} . \tag{2.9}$$

To begin our analysis, consider the steady-state condition, in which the rates of change of both [A] and [B] are zero. (This does not mean that both reactions have zero rates, but rather that the *net* flux between A and B is zero.) A steady-state concentration profile $[A] = a^{ss}$, $[B] = b^{ss}$ must satisfy

$$0 = k_-b^{ss} - k_+a^{ss}$$

$$0 = k_+a^{ss} - k_-b^{ss}.$$

Solving these equations (they are equivalent), we find

$$\frac{b^{ss}}{a^{ss}} = \frac{k_+}{k_-}. \tag{2.10}$$

The number $K_{eq} = k_+/k_-$ is called the *equilibrium constant* for the reaction. It is the ratio of the concentrations of the two reactants at steady state ([B]/[A]).

The concentrations can be derived by writing (from the conservation)

$$b(t) = T - a(t), \tag{2.11}$$

where $T = a(0) + b(0) = A_0 + B_0$ is the total concentration. Equation (2.8) can then be rewritten as:

$$\frac{d}{dt}a(t) = k_-b(t) - k_+a(t)$$
$$= k_-(T-a(t)) - k_+a(t) \tag{2.12}$$
$$= k_-T - (k_+ + k_-)a(t).$$

The steady-state concentration satisfies

$$0 = k_-T - (k_+ + k_-)a^{ss}.$$

Solving gives

$$a^{ss} = \frac{k_-T}{k_+ + k_-} \quad \text{and so} \quad b^{ss} = \frac{k_+}{k_-}a^{ss} = \frac{k_+T}{k_+ + k_-}. \tag{2.13}$$

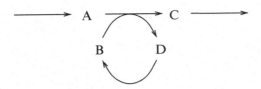

Figure 2.6
Open reaction network for exercise 2.1.7.

This system relaxes exponentially to its steady state. We can verify this by solving the differential equations (2.8)–(2.9) to yield (see exercise 2.1.6):

$$a(t) = De^{-(k_+ + k_-)t} + \frac{k_- T}{k_+ + k_-}, \tag{2.14}$$

for constant D.

The time constant for this reversible reaction, $\tau = 1/(k_+ + k_-)$, involves two rate constants, in contrast with example II (in which decay and production were uncoupled).

Exercise 2.1.6

(a) Verify that the solution (2.14) can be derived from equation (2.12) following the procedure outlined in exercise 2.1.4.

(b) Verify that the constant D in solution (2.14) is given by $D = A_0 - k_- T/(k_+ + k_-)$, where $A_0 = a(0)$ so we can write

$$a(t) = \left(A_0 - \frac{k_- T}{k_+ + k_-} \right) e^{-(k_+ + k_-)t} + \frac{k_- T}{k_+ + k_-}. \tag{2.15}$$

☐

A Remark on Conservations In examples III and IV, the total concentration of A and B was conserved throughout the reaction dynamics. This is a conservation of concentration (equivalently, of molecule count): it may or may not reflect a conservation of mass. The term **moiety** is used to describe a group of atoms that form part of a molecule, and so chemical conservation is often referred to as *moiety conservation*. More generally, chemical conservations are referred to as *structural conservations*.

Exercise 2.1.7 Identify the moiety conservation in the open system in figure 2.6. ☐

Exercise 2.1.8 The system

$$A \underset{k_-}{\overset{k_+}{\rightleftharpoons}} C + D$$

satisfies the structural conservation that the *difference* between the concentrations of C and D is constant for all time. Explain why. □

Exercise 2.1.9 The examples in this section may have given the reader the mistaken impression that all chemical systems relax exponentially to steady state. As an example of nonexponential dynamics, consider the bimolecular decay reaction

$$A + A \xrightarrow{k} .$$

The rate of the reaction is $k[A]^2$. The differential equation model is

$$\frac{d}{dt} a(t) = -2k(a(t))^2 .$$

(The stoichiometric factor 2 appears because each reaction event consumes two molecules of A.) Verify, by substituting into both sides of the differential equation, that

$$a(t) = \frac{1}{2kt + \dfrac{1}{A_0}}$$

is the solution of the differential equation that satisfies the initial condition $a(0) = A_0$. □

2.1.4 Numerical Simulation of Differential Equations

The exponential relaxation exhibited by the examples in the previous section is characteristic of *linear* systems. Nonlinear models exhibit a wide range of behaviors and do not typically admit explicit solutions such as the concentration formulas derived earlier. Differential equation models of biochemical and genetic systems are invariably nonlinear. We will resort to **numerical simulation** to investigate the behavior of these system.

Construction of Simulations We will use computational software packages to simulate differential equation models. In this section, we give a brief introduction to the algorithms used by that software.

Numerical simulations do not generate continuous curves. They produce approximate values of the solution at a specified collection of time points (analogous to an experimental time series). The first step in constructing a numerical simulation is to select this mesh of time points. The solution will be constructed by stepping from one time point to the next using an update formula. The simplest procedure for generating solutions in this manner is *Euler's method*, which is based on the following approximation. Given a differential equation of the form

$$\frac{d}{dt}a(t) = f(a(t)),$$

the derivative $d/dt(a(t))$ can be approximated by a difference quotient:

$$\frac{d}{dt}a(t) \approx \frac{a(t+h)-a(t)}{h}, \qquad \text{for } h \text{ small.}$$

(Recall, the derivative is *defined* as the limit of this quotient as h shrinks to zero.) Substituting this approximation into the differential equation gives

$$\frac{a(t+h)-a(t)}{h} \approx f(a(t)).$$

Treating this as an equality yields an update formula that can be used to determine the (approximate) value of $a(t + h)$ given the value $a(t)$:

$$a(t + h) = a(t) + hf(a(t)). \tag{2.16}$$

Euler's method consists of applying this update formula repeatedly.

To implement Euler's method, we choose a step-size h. This yields a mesh of time points $t = 0, h, 2h, 3h, \ldots, nh$, for some fixed number of steps n. Given the initial value of $a(0)$, we then use formula (2.16) to approximate the value of $a(h)$. Repeated application of (2.16) provides approximate values at the other points on the grid:

$a(0) = a(0)$ (given)

$a(h) = a(0) + hf(a(0))$

$a(2h) = a(h) + hf(a(h))$

$a(3h) = a(2h) + hf(a(2h))$

$$\vdots$$

$a(nh) = a((n-1)h) + hf(a(n-1)h).$

Because a computer can carry out these repeated calculations rapidly, the step-size h is often chosen so small that the set of points generated by this algorithm appears as a continuous curve. Figure 2.7 illustrates a case where the step-size was deliberately chosen to be large so that the discrete steps in the simulation are identifiable. The figure shows simulations generated for two different step-sizes. The simulation is more accurate—closer to the true solution—when the step-size h is chosen to be smaller (at the cost of more iterations to cover the same time interval).

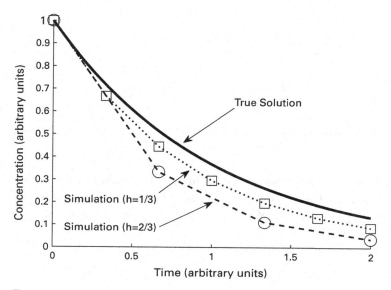

Figure 2.7
Numerical simulation of the model $d/dt\,(a(t)) = -a(t)$. The initial condition is $a(0) = 1$. For $h = 2/3$, the algorithm provides approximations of $a(2/3)$, $a(4/3)$, and $a(2)$ (open circles). The points are connected by straight lines for illustration. For $h = 1/3$, twice as many points are calculated (open squares), giving an improved approximation of the true solution.

Computational software packages that implement numerical simulation make use of sophisticated algorithms that improve on Euler's method. These details are hidden from the user, who simply passes the model to the simulation function and receives the output data. Appendix C introduces numerical simulation in the computational software packages MATLAB and XPPAUT.[3]

Numerical simulations of differential equation models are not as useful as analytic solution formulas for two reasons. First, an analytic formula is valid for all initial conditions. In contrast, each numerical simulation must be generated from a particular initial condition. Second, the dependence on the model parameters can easily be discovered from an analytic solution formula (e.g., the time constants discussed earlier). No such insights are granted by the numerical simulation, in which the parameter values must be fixed. (Computational exploration of different parameter values demands running multiple simulations.)

Nevertheless, in what follows we will rarely encounter differential equation models for which analytic solutions can be derived, and so numerical simulation will be an invaluable tool for model investigation.

3. MATLAB: www.mathworks.com/products/matlab. XPPAUT: www.math.pitt.edu/~bard/xpp/xpp.html.

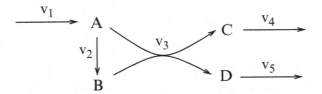

Figure 2.8
Open reaction network. The reaction rates are labeled v_i as indicated. (The v_i are not mass action rate constants.)

Network Example   To illustrate simulations of reaction network models, consider the reaction scheme in figure 2.8. The rate of each reaction is labeled v_i. (Reaction rates are commonly referred to as *velocities*.) We will follow the convention of using v_i to label reaction rates in network graphs.

Suppose the reaction rates are given by mass action, as follows:

$$v_1 = k_1 \qquad v_2 = k_2[A] \qquad v_3 = k_3[A][B] \qquad v_4 = k_4[C] \qquad v_5 = k_5[D].$$

Let a, b, c, and d denote the concentrations of the corresponding species. Taking rate constants of $k_1 = 3$ mM/s, $k_2 = 2$/s, $k_3 = 2.5$/mM/s, $k_4 = 3$/s, and $k_5 = 4$/s, the species concentrations satisfy the following set of differential equations (expressed in mM/s):

$$\frac{d}{dt}a(t) = 3 - 2a(t) - 2.5a(t)b(t)$$

$$\frac{d}{dt}b(t) = 2a(t) - 2.5a(t)b(t)$$

$$\frac{d}{dt}c(t) = 2.5a(t)b(t) - 3c(t)$$ (2.17)

$$\frac{d}{dt}d(t) = 2.5a(t)b(t) - 4d(t).$$

Note that because rate v_3 depends on the product $a(t)b(t)$, this system of equations is nonlinear. Once an initial concentration profile has been specified, numerical simulation can be used to generate the resulting concentration time courses. One such *in silico* experiment is shown in figure 2.9. In this case, the initial concentrations of all species were zero. The curves show the concentrations growing as the species pools "fill up" to their steady-state values. The concentration of A overshoots its steady state because the formation of C and D proceeds slowly until a pool of B has accumulated.

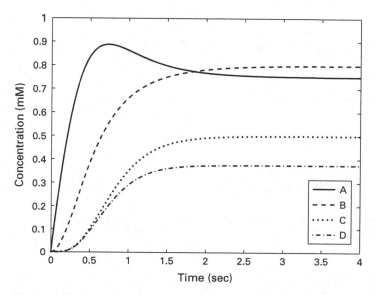

Figure 2.9
Numerical simulation of the network in figure 2.8. All species start with initial concentration of zero at time $t = 0$.

Exercise 2.1.10   Determine the steady-state concentrations for the species in model (2.17).   □

2.2   Separation of Timescales and Model Reduction

When constructing a dynamic model, one must decide which timescale to address. This choice is typically dictated by the timescale of the relevant reactions and processes. For the simple examples considered earlier, the timescales (time constants) could be deduced from the reaction system. For nonlinear processes, characteristic timescales are not so neatly defined.

Biological processes take place over a wide range of timescales. Consider, for example, a genetic network that generates a circadian rhythm. (Such networks will be taken up in section 7.3.2.) A model of this network will describe oscillatory behavior with a period of roughly 24 hours and so will incorporate processes acting on the timescale of hours. However, the network is based on gene expression, which involves the binding of proteins to DNA—these chemical processes happen on the scale of seconds. Moreover, the circadian oscillator is entrained to seasonal changes in the light–dark cycle—changes that occur on the order months. It would not be possible to resolve all of these timescales in a single model.

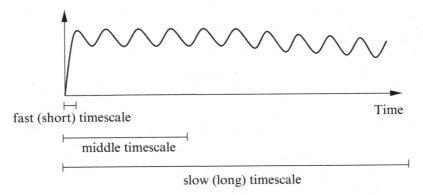

Figure 2.10
Behavior on multiple timescales. On the fast timescale, the process shows a rapid rise. This leads to an oscillatory behavior on a middle timescale. The height of the oscillations declines slowly over a longer timescale.

The same issue is faced in plotting the time-trace of a process that varies on multiple timescales. Consider figure 2.10, which shows behavior on three distinct timescales. On the short timescale, a fast rise occurs. This leads to the steady oscillatory behavior that dominates the middle timescale. On the long timescale, a slower process leads to a decreasing trend. If the differences in timescales were more extreme, a graph that focuses on the middle timescale would not reflect the other two: the fast behavior would not be properly resolved, and the slow behavior would not be captured.

To model a system that involves processes acting on different timescales, a primary timescale must be chosen. Other timescales are then treated as follows:

- processes occurring on slower timescales are approximated as frozen in time;
- processes occurring on faster timescales are presumed to occur instantaneously.

In most cases, these timescale separations are made during model construction: they often motivate the decisions as to which species and processes should be included in the model and which will be neglected. In other cases, existing models that incorporate distinct timescales can be simplified. This **model reduction** process approximates the original model by a model of reduced complexity.

Recall that for each of the closed reaction systems analyzed in the previous section, we used conservation to replace the differential equation for the concentration of B (that is, the equation for $d/dt(b(t))$) with a much simpler algebraic description, $b(t) = T - a(t)$. We thus *reduced* a model involving two differential equations to a model involving one differential equation and one algebraic equation.

Model reduction by timescale separation leads to a similar result—a differential equation describing a state variable is replaced by an algebraic equation. However, while reduction via conservation does not change the model, timescale separation leads to an approximate version of the original model. Consequently, model reduction by timescale separation should only be carried out in cases where the approximation will be accurate. A useful rule of thumb is that a difference in timescales of at least an order of magnitude (i.e., at least a factor of ten) can be used for model reduction.

We next present techniques for model reduction by separation of timescales. Elimination of slow variables is straightforward—we simply assign a constant value to each slow variable and treat it as fixed parameter, rather than as a state variable. The treatment of fast variables requires more care. We will consider two approaches that allow processes to be treated as instantaneous: the *rapid equilibrium assumption* and the *quasi-steady-state assumption*.

2.2.1 Separation of Timescales: The Rapid Equilibrium Assumption

Consider the open network

$$A \underset{k_{-1}}{\overset{k_1}{\rightleftharpoons}} B \overset{k_2}{\longrightarrow}. \tag{2.18}$$

With mass-action rate constants as indicated, the concentrations $a(t)$ and $b(t)$ satisfy

$$\frac{d}{dt}a(t) = -k_1 a(t) + k_{-1} b(t)$$

$$\frac{d}{dt}b(t) = k_1 a(t) - k_{-1} b(t) - k_2 b(t). \tag{2.19}$$

There are two processes acting here: the reversible conversion $A \leftrightarrow B$ and the decay $B \rightarrow$. As derived in section 2.1.3, the time constants of these two processes are $1/(k_1 + k_{-1})$ and $1/k_2$, respectively. If the conversion has a much smaller time constant than the decay (i.e., $k_1 + k_{-1} \gg k_2$), then the conversion reaches equilibrium quickly, compared to the timescale of the decay process. This case is illustrated by figure 2.11, in which the separation of timescales reveals itself in the concentration time courses. On a short timescale, A and B molecules interconvert until they quickly reach an equilibrium ratio: little decay occurs over this time period. On the longer timescale, the equilibrated pool of A and B molecules slowly decays.

Once the equilibrium ratio of A and B is reached, this ratio is maintained throughout the decay process. This observation suggests a strategy for model reduction: if we choose to neglect the fast timescale, we can make use of the fact that the equilibrium is maintained to relate the two concentrations. This is the **rapid equilibrium assumption**. By assuming that the conversion reaction is in equilibrium at all times,

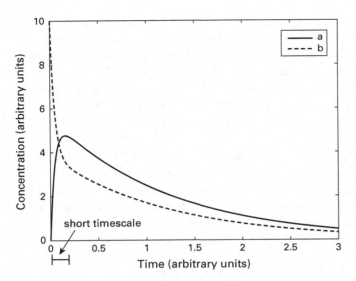

Figure 2.11
Simulation of network (2.18) with parameter values (in time^{-1}) $k_1 = 9$, $k_{-1} = 12$, and $k_2 = 2$. The time constant for the conversion is 1/21; the decay process has a time constant of 1/2. On the short timescale, the conversion comes rapidly to equilibrium (in which $[B]/[A] \approx 3/4$). On the longer timescale, the equilibrated pool of A and B molecules decays. The initial conditions are $a(0) = 0$, $b(0) = 10$.

we simplify our dynamic description of the network because one concentration is now easily described in terms of the other via the equilibrium condition. To emphasize that the model reduction leads to an approximate model that is different from the original, we introduce the notation $\tilde{a}(t)$ and $\tilde{b}(t)$ for the concentrations in the reduced model. For network (2.18), the equilibrium condition states that

$$\frac{\tilde{b}(t)}{\tilde{a}(t)} = \frac{k_1}{k_{-1}},$$

from which we have

$$\tilde{b}(t) = \tilde{a}(t)\frac{k_1}{k_{-1}}.$$

With this condition in hand, we now turn to the dynamics of the decay process, which is best described by addressing the dynamics of the equilibrated pool. The reaction network (2.18) thus reduces to:

(pool of A and B) $\rightarrow$.

Let $\tilde{c}(t)$ be the total concentration in the pool of A and B (i.e., $\tilde{c}(t) = \tilde{a}(t) + \tilde{b}(t)$). The relative fractions of A and B in the pool are fixed by the equilibrium ratio. This allows us to write

$$\tilde{c}(t) = \tilde{a}(t) + \tilde{b}(t)$$

$$= \tilde{a}(t) + \tilde{a}(t)\frac{k_1}{k_{-1}}$$

$$= \frac{k_{-1} + k_1}{k_{-1}} \tilde{a}(t).$$

Thus,

$$\tilde{a}(t) = \frac{k_{-1}}{k_{-1} + k_1} \tilde{c}(t) , \tag{2.20}$$

and

$$\tilde{b}(t) = \tilde{c}(t) - \tilde{a}(t) = \frac{k_1}{k_{-1} + k_1} \tilde{c}(t). \tag{2.21}$$

The pool decays at rate $k_2\tilde{b}(t)$. Thus, the pooled concentration satisfies

$$\frac{d}{dt}\tilde{c}(t) = -k_2\tilde{b}(t)$$

$$= -k_2 \frac{k_1}{k_{-1} + k_1} \tilde{c}(t) . \tag{2.22}$$

Schematically, we have reduced the model to a single decay reaction:

$$C \xrightarrow{\frac{k_2 k_1}{k_{-1} + k_1}} ,$$

which is the **rapid equilibrium approximation** of the original model. To predict the concentration time courses, we simulate the dynamics of the pooled concentration $\tilde{c}(t)$ in equation (2.22) and then use equations (2.20) and (2.21) to determine the corresponding concentrations $\tilde{a}(t)$ and $\tilde{b}(t)$ at each time point.

Figure 2.12 shows the behavior of the reduced model in comparison with the original model. Except for the initial fast dynamics, the reduced model provides a good approximation.

Exercise 2.2.1 Use the rapid equilibrium approximation to construct a reduced model for the network

$$\xleftarrow{k_0} A \underset{k_{-1}}{\overset{k_1}{\rightleftharpoons}} B \xrightarrow{k_2} \tag{2.23}$$

under the conditions that $k_1 + k_{-1} \gg k_2$ and $k_1 + k_{-1} \gg k_0$. □

To explore further the rapid equilibrium approximation, consider the network

$$\xrightarrow{k_0} A \underset{k_{-1}}{\overset{k_1}{\rightleftharpoons}} B \xrightarrow{k_2} . \tag{2.24}$$

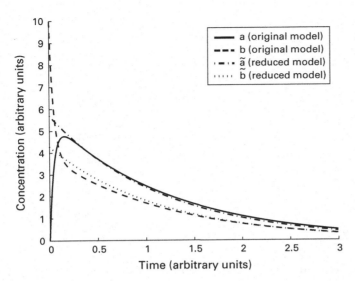

Figure 2.12
Rapid equilibrium approximation for network (2.18). The original model (2.19) was simulated with parameter values (in time^{-1}) $k_1 = 9$, $k_{-1} = 12$, and $k_2 = 2$ (as in figure 2.11). The approximate model (2.22) was simulated from initial value $\tilde{c}(0) = a(0) + b(0)$; the corresponding approximate concentrations $\tilde{a}(t)$ and $\tilde{b}(t)$ were calculated from equations (2.20) and (2.21). Initial conditions are $a(0) = 0$, $b(0) = 10$ for the original model.

This network is similar to (2.18). The zero-order reaction ($\to A$) does not affect the timescale on which the concentrations of A and B relax to their steady-state values, so, as in the previous case, a rapid equilibrium assumption is valid if $k_1 + k_{-1} \gg k_2$. In that case, the pool concentration $\tilde{c}(t) = \tilde{a}(t) + \tilde{b}(t)$ can be used to describe a reduced network

$$\xrightarrow{\ k_0\ } C \xrightarrow{\ \frac{k_2 k_1}{k_{-1} + k_1}\ }$$

with dynamics

$$\frac{d}{dt}\tilde{c}(t) = k_0 - \frac{k_2 k_1}{k_{-1} + k_1}\tilde{c}(t). \tag{2.25}$$

This approximation is illustrated in figure 2.13. The approximation is good but exhibits a persistent error in the concentration of A. This is a consequence of the fact that the original model comes to a dynamic steady state in which the conversion reaction ($A \leftrightarrow B$) is not in equilibrium.

In the next section, we will consider a model-reduction method that is guaranteed to be accurate at steady state.

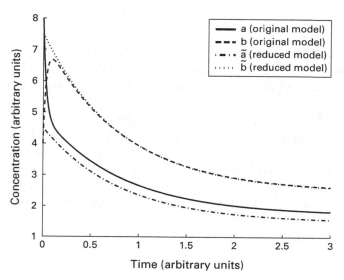

Figure 2.13
Rapid equilibrium approximation for network (2.24). Model (2.25) is used to approximate the full model for network (2.24). Parameter values (in time^{-1}) are $k_0 = 5$, $k_1 = 20$, $k_{-1} = 12$, and $k_2 = 2$. There is a persistent error in the approximation for $[A]$, caused by the fact that the conversion reaction does not settle to equilibrium in steady state. Initial conditions are $a(0) = 8$, $b(0) = 4$ (and so $\tilde{c}(0) = 12$).

Exercise 2.2.2 Develop a model for network (2.24) and determine the steady-state concentrations. Compare the steady-state ratio $[B]/[A]$ to the equilibrium constant for the conversion reaction (which was used for model reduction). Verify that the steady-state concentration ratio is not equal to the equilibrium constant k_{-1}/k_1, but the difference is small when k_{-1} is much larger than k_2. ☐

Exercise 2.2.3 The accuracy of the rapid equilibrium approximation improves as the separation of timescales becomes more significant. Derive a formula for the steady-state concentration of A in both model (2.25) and the full model for network (2.24). Consider the relative steady-state error, defined as

$$\left| \frac{[A]^{ss}_{full} - [A]^{ss}_{reduced}}{[A]^{ss}_{full}} \right|.$$

Verify that the relative steady-state error in $[A]$ is small when k_{-1} is much larger than k_2. ☐

2.2.2 Separation of Timescales: The Quasi-Steady-State Assumption
The rapid equilibrium approximation is reached by treating individual *reaction processes* as instantaneous. We now consider an alternative model-reduction method that focuses on individual *species*. Consider again the network

$$\xrightarrow{k_0} A \underset{k_{-1}}{\overset{k_1}{\rightleftharpoons}} B \xrightarrow{k_2} , \tag{2.26}$$

and again suppose $k_1 + k_{-1} \gg k_2$. Instead of focusing on the conversion reaction, we observe that all dynamic reactions involving species A occur on the fast timescale, so that, compared to the dynamics of B, species A comes rapidly to its steady-state concentration.

Following this idea, we replace our original differential equation-based description of the behavior of $[A]$ (that is, $d/dt\,(a(t)) = k_0 + k_{-1}b(t) - k_1a(t)$), with an algebraic description indicating that concentration $a(t)$ is in steady state *with respect to the other variables* in the model (in this case, $b(t)$). We introduce a *quasi–steady state* for $[A]$ in our approximate model, $\tilde{a}(t) = a^{qss}(t)$, and specify that a^{qss} "keeps up" with the transitions in any slower variables. For each time instant t, the quasi–steady state $a^{qss}(t)$ satisfies

$$0 = k_0 + k_{-1}b(t) - k_1a^{qss}(t),$$

or equivalently

$$a^{qss}(t) = \frac{k_0 + k_{-1}b(t)}{k_1}. \tag{2.27}$$

This procedure is sometimes summarized as "set $d/dt\,(a^{qss}(t))$ to zero." However, this is a problematic statement because it suggests that we are setting $a^{qss}(t)$ to a constant value, which is not the case. Instead, we are replacing the differential description of $a(t)$ with an algebraic description that says $[A]$ instantaneously reaches the steady state it would attain if all other variables were constant. Because it equilibrates rapidly, the other variables are essentially constant "from A's point of view"; that is, on its fast timescale. (A mathematically rigorous treatment of this procedure consists of a *singular perturbation* of the original model. A careful treatment of this technique provides explicit bounds on the error made in the approximation. See Segel and Slemrod (1989) for details.)

The reduced model, called the **quasi-steady-state approximation** (QSSA), follows by replacing $a(t)$ with $a^{qss}(t)$ in the original model. Again, using the alternative notation $\tilde{b}$ for the reduced model:

$$\begin{aligned}
\frac{d}{dt}\tilde{b}(t) &= k_1 a^{qss}(t) - (k_{-1} + k_2)\tilde{b}(t) \\
&= k_1 \frac{k_0 + k_{-1}\tilde{b}(t)}{k_1} - (k_{-1} + k_2)\tilde{b}(t) \\
&= k_0 + k_{-1}\tilde{b}(t) - (k_{-1} + k_2)\tilde{b}(t) \\
&= k_0 - k_2\tilde{b}(t).
\end{aligned} \tag{2.28}$$

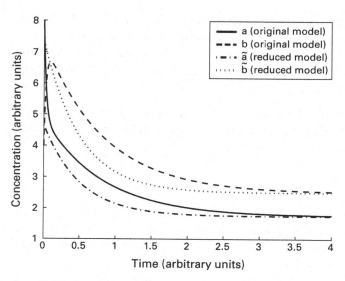

Figure 2.14
Quasi-steady-state approximation. Network (2.26) is approximated by model (2.28) and equation (2.27). Parameter values are (in time^{-1}) $k_0 = 5$, $k_1 = 20$, $k_{-1} = 12$, and $k_2 = 2$ (as in figure 2.13). The approximation exhibits an error over the transient but converges to the original model in steady state. Initial conditions are $a(0) = 8$, $b(0) = 4$, and $\tilde{b}(0) = 235/32$ (see exercise 2.2.4).

The QSSA is illustrated in figure 2.14. A significant error occurs during the transient but diminishes as the steady state is approached (in contrast with the rapid equilibrium approximation in figure 2.13). This is a general feature of the QSSA: when the system is at steady state, the quasi-steady-state description $a^{qss}(t)$ is equal to the true value of $a(t)$ (because the quasi-steady-state condition is satisfied).

In the subsequent chapters, we will use both the rapid equilibrium approximation and the QSSA for model reduction. The rapid equilibrium approximation can be easier to apply because it addresses individual reaction processes. The quasi-steady-state approximation is sometimes more difficult to justify because it typically involves multiple processes. However, the QSSA is simpler to implement mathematically and leads to better approximations over long times; for those reasons, it is often favored over the rapid equilibrium approximation.

Exercise 2.2.4 In applying the reduced model (2.28) to approximate the behavior of network (2.26), the initial condition must be chosen carefully. Suppose a simulation involves initial concentrations $a(0)$ and $b(0)$. The reduced model cannot retain the same initial concentration of B (i.e., $\tilde{b}(0) = b(0)$) because together with the corresponding quasi–steady state for A (i.e., $\tilde{a}(0) = (k_0 + k_{-1}\tilde{b}(0))/k_1$), the total initial concentration would not be in agreement with the original simulation. An improved approach maintains the total concentration $(\tilde{a}(0) + \tilde{b}(0) = a(0) + b(0))$ while respecting the concentration *ratio* dictated by the quasi-steady-state condition.

Given initial conditions $a(0)$ and $b(0)$, determine an appropriate initial condition $\tilde{b}(0)$ in the reduced model (2.28). You can check your answer by confirming the initial condition $\tilde{b}(0)$ used in figure 2.14.   □

Exercise 2.2.5 Consider a model for network (2.23). Suppose that $k_0 \gg k_2$ and $k_1 + k_{-1} \gg k_2$. Apply an appropriate quasi-steady-state approximation to reduce the model by eliminating one of the differential equations.   □

2.3 Suggestions for Further Reading

• **Calculus** There are many introductory texts covering differential calculus, some of which focus specifically on life science applications. The book *Modeling the Dynamics of Life: Calculus and Probability for Life Scientists* (Adler, 2004) is especially well-suited to the study of dynamic biological models.

• **Differential Equations** A general introduction to differential equations, including treatment of numerical simulation, can be found in *Elementary Differential Equations and Boundary Value Problems* (Boyce and DiPrima, 2008). An introduction to the theory in the context of biological applications is presented in *Differential Equations and Mathematical Biology* (Jones et al., 2009).

• **Chemical Reaction Network Theory** There is a rich literature on the dynamic behavior of chemical reaction networks. An introduction is provided by *Mathematical Models of Chemical Reactions: Theory and Applications of Deterministic and Stochastic Models* (Érdi and Tóth, 1989).

2.4 Problem Set

2.4.1 Open Reaction Network

Suppose a reaction network is composed of the following reactions

$$\xrightarrow{k_1} A \qquad A \xrightarrow{k_2} B + C \qquad B \xrightarrow{k_3}$$
$$C \xrightarrow{k_4} 2D \qquad 2D \xrightarrow{k_5} C \qquad D \xrightarrow{k_6}$$

with mass-action rate constants as indicated.

(a) Construct a differential equation model of the network.

(b) Determine the steady-state concentrations of all species as functions of the mass-action constants.

2.4.2   Open Reaction Network: Buffered Species

Consider the reaction network

$$A \xrightarrow{\ k_1\ } X \qquad\qquad X \xrightarrow{\ k_2\ } Y \qquad\qquad X + Y \xrightarrow{\ k_3\ } B,$$

where the concentrations of A and B are buffered (i.e., $[A]$ and $[B]$ are fixed model parameters).

(a) Construct a differential equation model for the dynamics of $[X]$ and $[Y]$. (The rate of the first reaction is constant: $k_1[A]$.)

(b) Determine the steady-state concentrations of X and Y as functions of $[A]$ and the rate constants. Verify that the steady-state concentration of Y is independent of $[A]$. Can you explain this independence intuitively?

2.4.3   Moiety Conservations

Consider the reaction scheme in figure 2.15.

(a) Identify two moiety conservations in the network.

(b) Consider an experiment in which the initial concentrations are (in mM) $s_1(0) =$ 3.5, $s_2(0) = 1$, $e(0) = 3$, and $c(0) = 0$. Suppose that the steady-state concentrations of S_1 and S_2 have been measured as $s_1^{ss} = 2$ mM and $s_2^{ss} = 1.5$ mM. Determine the steady-state concentrations of E and C. (Note: There is no need to consider the reaction rates or network dynamics. The conclusion follows directly from the moiety conservations.)

2.4.4   Steady-State Production Rate

Consider the reaction network

$$A + S \xrightarrow{\ k_1\ } B \qquad\qquad B \xrightarrow{\ k_2\ } A + P .$$

Suppose that the species S and P are held at fixed concentrations (i.e., $[S]$ and $[P]$ are fixed model parameters). Suppose that the reaction rates are given by mass

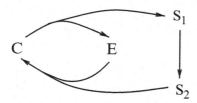

Figure 2.15
Closed reaction network for problem 2.4.3.

action, with reaction rates as indicated. If the initial concentrations of $[A]$ and $[B]$ are both 1 mM, determine the rate of production of P at steady state (as a function of k_1, k_2, and $[S]$).

2.4.5 Linear System of Differential Equations

Consider the coupled system of differential equations

$$\frac{d}{dt}x(t) = x(t) + 2y(t) \qquad \frac{d}{dt}y(t) = x(t).$$

(a) Verify that for any choice of constants c_1 and c_2, the functions

$$x(t) = c_1 e^{-t} + 2c_2 e^{2t}, \qquad y(t) = -c_1 e^{-t} + c_2 e^{2t}$$

are solutions to these equations. (This can be verified by differentiating $x(t)$ and $y(t)$ and comparing the two sides of the differential equations.)

(b) The constants c_1 and c_2 in part (a) are determined by the initial conditions for x and y. Determine the values of c_1 and c_2 that correspond to the initial conditions $x(0) = 0$, $y(0) = 1$. What is the asymptotic (long-term) behavior of the resulting solutions $x(t)$ and $y(t)$?

(c) Find a set of (nonzero) initial conditions $x(0)$, $y(0)$ for which the solution $(x(t), y(t))$ converges to $(0, 0)$.

2.4.6 Numerical Simulation

Use a software package (e.g., XPPAUT or MATLAB—introduced in appendix C) to simulate solutions to the equation

$$\frac{d}{dt}c(t) = -c(t) + 1$$

with initial conditions $c(0) = 0$, $c(0) = 1$, and $c(0) = 3$. Repeat for the system

$$\frac{d}{dt}c(t) = 5(-c(t) + 1).$$

Explain the difference in behavior between the two systems.

2.4.7 Network Modeling

(a) Consider the closed reaction network in figure 2.16 with reaction rates v_i as indicated. Suppose that the reaction rates are given by mass action as $v_1 = k_1[A][B]$, $v_2 = k_2[D]$, and $v_3 = k_3[C]$.

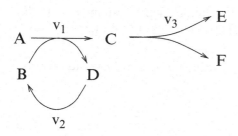

Figure 2.16
Closed reaction network for problem 2.4.7(a).

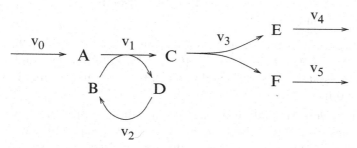

Figure 2.17
Open reaction network for problem 2.4.7(b).

(i) Construct a differential equation model for the network. Use moiety conservations to reduce your model to three differential equations and three algebraic equations.

(ii) Solve for the steady-state concentrations as functions of the rate constants and the initial concentrations. (Note that because the system is closed, some of the steady-state concentrations are zero.)

(iii) Verify your result in part (ii) by running a simulation of the system from initial conditions (in mM) of $([A], [B], [C], [D], [E], [F]) = (1, 1, \tfrac{1}{2}, 0, 0, 0)$. Take rate constants $k_1 = 3/\text{mM/s}$, $k_2 = 1/\text{s}$, $k_3 = 4/\text{s}$.

(b) Next consider the open system in figure 2.17 with reaction rates v_i as indicated. Suppose that the reaction rates are given by mass action as $v_0 = k_0$, $v_1 = k_1[A][B]$, $v_2 = k_2[D]$, $v_3 = k_3[C]$, $v_4 = k_4[E]$, and $v_5 = k_5[F]$.

(i) Construct a differential equation model for the network. Identify any moiety conservations in the network.

(ii) Solve for the steady state as a function of the rate constants and the initial concentrations.

(iii) Verify your result in (ii) by running a simulation of the system from initial conditions (in mM) of $([A], [B], [C], [D], [E], [F]) = (1, 1, \frac{1}{2}, 0, 0, 0)$. Take rate constants $k_0 = 0.5$ mM/s, $k_1 = 3$/mM/s, $k_2 = 1$/s, $k_3 = 4$/s, $k_4 = 1$/s, $k_5 = 5$/s.

(iv) Given the initial conditions and rate constants in part (iii), why would there be no steady state if we take $k_0 = 5$ mM/s?

2.4.8 Rapid Equilibrium Approximation
Consider the closed system

$$A \underset{k_{-1}}{\overset{k_1}{\rightleftharpoons}} B \underset{k_{-2}}{\overset{k_2}{\rightleftharpoons}} C,$$

with mass-action rate constants as shown. Suppose the rate constants are (in min^{-1}) $k_1 = 0.05$, $k_2 = 0.7$, $k_{-1} = 0.005$, and $k_{-2} = 0.4$.

(a) Construct a differential equation model of the system. Simulate your model with initial conditions (in mM) of $A(0) = 1.5$, $B(0) = 3$, $C(0) = 2$. Plot the transient and steady-state behavior of the system. You may need to make two plots to capture all of the dynamics (i.e., two different window sizes).

(b) It should be clear from your simulation in part (a) that the system dynamics occur on two different timescales. This is also apparent in the widely separated rate constants. Use a rapid equilibrium assumption to reduce your description of the system to two differential equations (describing one of the original species and one combined species pool) and two algebraic equations (describing the contents of the combined pool).

(c) Run a simulation of your reduced model in part (b) to compare with the simulation in part (a). Verify that the simulation of the reduced system is in good agreement with the original, except for a short initial transient. (Note that you will have to select initial conditions for the reduced system so that the initial total concentration is in agreement with part (a), and the rapid equilibrium condition is satisfied at time $t = 0$.)

2.4.9 Quasi-Steady-State Approximation
Consider the reaction network

$$\overset{k_0}{\longrightarrow} A \overset{k_2}{\longrightarrow} \qquad\qquad A \underset{k_{-1}}{\overset{k_1}{\rightleftharpoons}} B.$$

Suppose the mass-action rate constants are (in min^{-1}) $k_0 = 1$, $k_1 = 11$, $k_{-1} = 8$, and $k_2 = 0.2$.

(a) Construct a differential equation model of the system. Simulate your model with initial conditions $A(0) = 6$ mM, $B(0) = 0$ mM. Plot the transient and steady-state behavior of the system. You may need to make two plots to capture all of the dynamics (i.e., two different window sizes).

(b) It should be clear from your simulation in part (a) that the system dynamics occur on two different timescales. This is also apparent in the widely separated rate constants. Use a quasi-steady-state assumption to reduce your description of the system by replacing a differential equation with an algebraic equation.

(c) Run a simulation of your reduced model in part (b) to compare with the simulation in part (a). Verify that the simulation of the reduced system is a good approximation to the original at steady state, but not over the initial transient. (Note that you will have to select initial conditions for the reduced system so that the total concentration is in agreement with part (a) and the quasi-steady-state condition is satisfied at time $t = 0$, as in exercise 2.2.4.)

3 Biochemical Kinetics

I began to lose interest in some of the things [in mathematics] that to me seemed rather remote from reality and hankered after something more practical. I realize now that I was much better fitted to engineering than to mathematics, but physiology proved in the end to be much like engineering, being based on the same ideas of function and design.
—A. V. Hill, *The Third Bayliss–Starling Memorial Lecture*

The study of the rates of chemical reactions is called *chemical kinetics*. In the previous chapter, we used the law of mass action to establish chemical reaction rates. In this chapter, we will develop rate laws that are applicable to biochemical processes.

Individual chemical reaction events (binding, unbinding, and conversion) are called *elementary reactions*. As in the previous chapter, we will continue to use mass action to describe the rates of elementary reactions. In contrast, individual *biochemical reactions* involve small networks of elementary reactions. To develop rate laws for biochemical reactions, we will collapse these networks into single reaction events, using separation of timescale methods. The rate laws that describe these "lumped" reaction events are referred to as *biochemical kinetics*.

3.1 Enzyme Kinetics

The overwhelming majority of reactions that occur within a cell are catalyzed by **enzymes** (which are proteins). Enzymes catalyze reactions by binding the reactants (called the enzyme *substrates*) and facilitating their conversion to the reaction products. Enzyme catalysis reduces the energy barrier associated with the reaction event (figure 3.1A). Consequently, whereas enzyme catalysis increases the rate at which equilibrium is attained, it has no effect on the equilibrium itself (figure 3.1B).

The standard "lock-and-key" model of enzyme activity is illustrated in figure 3.2A. As shown for this reversible reaction, the enzyme catalyzes the reaction in both directions and is unaltered by the reaction event. The part of the enzyme that binds the substrate is called the *active* (or *catalytic*) *site*. The active site has a shape

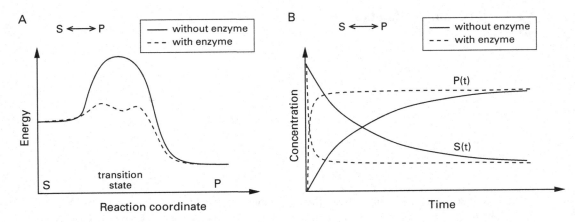

Figure 3.1
(A) Effect of catalysis on reaction energy profile. Species P occupies a lower energy state than species S, so the equilibrium favors formation of P. For the reaction to occur, the reactants must collide with sufficient force to overcome the energy barrier (corresponding to formation of a high-energy transition state). Enzyme catalysis reduces the energy barrier between reactant and product. The enzyme does not affect the energy levels of the species themselves and so has no effect on the equilibrium. (B) Effect of catalysis on reaction dynamics. The ratio of concentrations at equilibrium depends only on the reactants S and P. The rate at which this equilibrium is attained is increased in the presence of an enzyme catalyst.

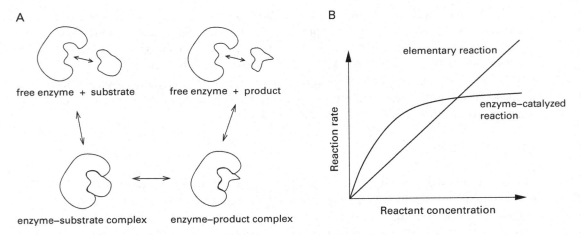

Figure 3.2
(A) Lock-and-key model of the enzyme catalysis cycle. The substrate binds a specific site on the enzyme, called the active (or catalytic) site. Once bound, the enzyme facilities formation of the product. The product dissociates from the enzyme, which is then free to catalyze another reaction event. (B) Rate laws for an enzyme-catalyzed reaction and an elementary reaction. The mass-action rate of the elementary reaction increases linearly with reactant concentration. In contrast, the rate of the enzyme-catalyzed reaction tends to a limiting value as the reactant concentration grows.

and chemical structure that is complementary to the substrate, resulting in strong binding affinity and a highly specific interaction. Most enzymes catalyze only a single reaction. This specificity of action allows each enzyme to function with remarkable efficiency—increasing reaction rates by as much as 10^7 times. The activity of enzymes within cells is highly regulated through both genetic control of enzyme abundance and biochemical modification of individual enzyme molecules.

Experimental observations of enzyme-catalyzed reactions show that they do not obey mass-action rate laws. As sketched in figure 3.2B, the rate of an enzyme-catalyzed reaction approaches a limiting value as the substrate abundance grows. This saturating behavior is caused by the fact that there is a limited amount of enzyme present: at high substrate concentrations, most of the enzyme molecules are actively catalyzing reactions, and so the addition of substrate has little effect on the reaction rate. The limiting reaction rate is reached when the entire enzyme pool is working at full capacity.

3.1.1 Michaelis–Menten Kinetics

We will use model reduction by timescale separation to formulate a rate law that describes enzyme-catalyzed reactions. The first such derivation was made in 1913 by Leonor Michaelis and Maud Menten. The resulting rate law is called **Michaelis–Menten kinetics.**

The individual chemical events involved in a single-substrate enzyme-catalyzed reaction (figure 3.2A) can be written as:

$$\underset{\text{Substrate}}{S} + \underset{\text{Free enzyme}}{E} \rightleftharpoons \underset{\text{Enzyme-substrate complex}}{C_1} \rightleftharpoons \underset{\text{Enzyme-product complex}}{C_2} \rightleftharpoons \underset{\text{Product}}{P} + \underset{\text{Free enzyme}}{E} . \qquad (3.1)$$

In our initial analysis, we will make two simplifications. First, we will lump the two complexes C_1 and C_2 together, assuming that the timescale of the conversion $C_1 \leftrightarrow C_2$ is fast compared to the timescale of the association and dissociation events. (This is a rapid equilibrium assumption.) Second, we will suppose that the product P never binds free enzyme. This makes the analysis simpler and is motivated by the fact that laboratory measurements of reaction rates are typically carried out in the absence of product. The resulting rate law describes irreversible enzyme-catalyzed reactions; the analogous rate law for reversible reactions will be presented later (equation 3.9).

These two assumptions lead to the simplified network

$$S + E \underset{k_{-1}}{\overset{k_1}{\rightleftharpoons}} C \overset{k_2}{\longrightarrow} P + E . \qquad (3.2)$$

The reaction rate k_2 is called the enzyme's *catalytic constant* (often denoted k_{cat}).

Applying the law of mass action and denoting concentrations by s (substrate), e (free enzyme), c (complex), and p (product), we have the following differential equation model:

$$\frac{d}{dt}s(t) = -k_1 s(t)e(t) + k_{-1}c(t)$$

$$\frac{d}{dt}e(t) = k_{-1}c(t) - k_1 s(t)e(t) + k_2 c(t)$$

$$\frac{d}{dt}c(t) = -k_{-1}c(t) + k_1 s(t)e(t) - k_2 c(t)$$

$$\frac{d}{dt}p(t) = k_2 c(t).$$

Let e_T denote the total enzyme concentration: $e_T = e + c$. Because the enzyme is not consumed in the reaction, e_T is constant—the enzyme is a conserved moiety. Writing $e(t) = e_T - c(t)$, we can use this conservation to eliminate the differential equation for $e(t)$, giving

$$\frac{d}{dt}s(t) = -k_1 s(t)(e_T - c(t)) + k_{-1}c(t)$$

$$\frac{d}{dt}c(t) = -k_{-1}c(t) + k_1 s(t)(e_T - c(t)) - k_2 c(t) \tag{3.3}$$

$$\frac{d}{dt}p(t) = k_2 c(t).$$

(Although we will not need to, we could also use the conservation $s + c + p$ to further simplify the model formulation.)

A simulation of this model is shown in figure 3.3A. The time courses reveal a separation of timescales. On the fast timescale, substrate S and free enzyme E associate to form the complex C. On the slower timescale, S is converted to P. We will use this separation of timescales for model reduction. Michaelis and Menten applied a rapid equilibrium approximation to the association/dissociation reaction ($S + E \leftrightarrow C$). We will present an alternative derivation developed by G. E. Briggs and J. B. S. Haldane[1] in 1925. (The rapid equilibrium derivation is treated in exercise 3.1.1.)

---

1. Haldane (1892–1964) made significant contributions to biology and was one of the founders of the field of population genetics. His published essays are full of insight and wit. An accomplished naturalist, he was once asked what could be inferred about the mind of the Creator from the works of Creation. His response: "An inordinate fondness for beetles." (Beetles represent about 25% of all known animal species.)

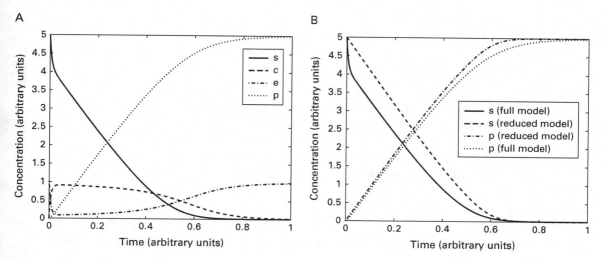

Figure 3.3
(A) Simulation of enzyme catalysis. Model (3.3) is simulated with parameter values $k_1 = 30$ (concentration^{-1} · time^{-1}), $k_{-1} = 1$ (time^{-1}), $k_2 = 10$ (time^{-1}), and $e_T = 1$ (concentration). Initial concentrations are $s(0) = 5$, $c(0) = 0$, $p(0) = 0$ (units are arbitrary). A separation of timescales is evident. On the fast timescale, the complex C reaches a quasi–steady state. The reaction $S \to P$ proceeds on the slower timescale. (B) Michaelis–Menten approximation of an enzyme-catalyzed reaction $S \to P$. Model (3.3) (full model) is simulated as in (A). The reduced model (3.4) provides a good approximation. The error in the approximation for S is caused by the sequestration of S into the complex C. In the cell, the ratio of substrate to enzyme molecules is typically much higher than in this simulation, so the sequestration effect is negligible.

The separation of timescales evident in figure 3.3A has two sources. The first is a difference in time constants for the reaction events ($1/(k_1 + k_{-1})$ for the association/ dissociation $S + E \leftrightarrow C$, and $1/k_2$ for product formation $C \to P$). The second is a distinction in concentrations. For many reactions in the cell, the substrate is far more abundant than the enzyme ($s \gg e_T$). Consequently, the enzyme complexes quickly come to equilibrium *with respect to the more abundant substrate* (see problem 3.7.2 for details). The complex C can thus be considered in quasi–steady state.

Recall that in its quasi–steady state, $c(t) = c^{qss}(t)$ is no longer an independent dynamic variable but instead "tracks" the other variables in the system according to

$$0 = -k_{-1}c^{qss}(t) + k_1 s(t)(e_T - c^{qss}(t)) - k_2 c^{qss}(t).$$

Solving for $c^{qss}(t)$, we find

$$c^{qss}(t) = \frac{k_1 e_T s(t)}{k_{-1} + k_2 + k_1 s(t)}.$$

Substituting this quasi-steady-state expression into the model (3.3), we are left with

$$\frac{d}{dt}s(t) = -\frac{k_2 k_1 e_T s(t)}{k_{-1} + k_2 + k_1 s(t)}$$

$$\frac{d}{dt}p(t) = \frac{k_2 k_1 e_T s(t)}{k_{-1} + k_2 + k_1 s(t)}.$$

$$\text{(3.4)}$$

This reduced model describes $S \to P$ as a single (nonelementary) reaction. The reaction rate is called a **Michaelis–Menten rate law**. Figure 3.3B shows the behavior of this reduced model in comparison to the original model (3.3).

Next, we define $V_{max} = k_2 e_T$ as the *limiting* (or *maximal*) *rate* and $K_M = (k_{-1} + k_2)/k_1$ as the *Michaelis* (or *half-saturating*) *constant*, and write the rate law as

$$\text{Rate of } S \to P = \frac{k_2 k_1 e_T s}{k_{-1} + k_2 + k_1 s} = \frac{V_{max} s}{K_M + s}.$$

$$\text{(3.5)}$$

This rate law, sketched in figure 3.4, is called *hyperbolic* (because the curve forms part of a hyperbola).

Exercise 3.1.1 Michaelis and Menten derived rate law (3.5) using a rapid equilibrium approximation applied to the association–dissociation of substrate and enzyme. Starting with scheme (3.2), follow this approach and re-derive the Michaelis–Menten rate law $V_{max} s/(K_M + s)$. You will end up with a different formula for the Michaelis constant. Experimental characterizations of Michaelis–Menten rate laws involve direct measurement of K_M and V_{max}, so the relation between K_M and the individual kinetic constants (k_1, k_{-1}, k_2) is not significant.   □

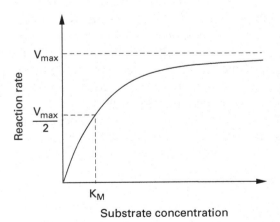

Figure 3.4
Michaelis–Menten rate law for a single-substrate enzyme-catalyzed reaction. The reaction rate approaches the limiting rate V_{max} as the substrate concentration increases. The Michaelis constant, K_M, is the substrate concentration at which the rate is equal to half of the limiting rate.

Kinetic Order   Recall that, as introduced in section 2.1.2, when a reaction rate is given by mass action, the kinetic order of a reactant is the power to which that reactant's concentration is raised in the rate law. The notion of kinetic order can be generalized as follows. If s is the concentration of a substrate for a reaction with rate $v(s)$, then the kinetic order of s is

$$\text{Kinetic order} = \left(\frac{s}{v}\right)\frac{dv}{ds}. \tag{3.6}$$

Using this definition, the kinetic order of the substrate s in the Michaelis–Menten rate law is

$$\left(\frac{s}{\frac{V_{max}s}{K_M+s}}\right)\cdot\frac{d}{ds}\left(\frac{V_{max}s}{K_M+s}\right) = \left(\frac{s(K_M+s)}{V_{max}s}\right)\left(\frac{V_{max}(K_M+s)-V_{max}s}{(K_M+s)^2}\right) = \frac{K_M}{K_M+s}. \tag{3.7}$$

In contrast with mass action, this kinetic order changes as the substrate concentration varies. In particular, when the concentration s is small compared to K_M, the kinetic order is roughly one (because $K_M + s \approx K_M$). Conversely, when s is large, the kinetic order is roughly zero. This is consistent with figure 3.4. When s is near zero, the curve grows linearly; when s is high, the curve is roughly horizontal (constant-valued). For small s, the enzyme is said to be operating in the *first-order* (or linear) regime, whereas for large s, the enzyme is saturated and is said to be acting in the *zero-order* regime.

Exercise 3.1.2

(a) Apply the definition of kinetic order in equation (3.6) to the mass-action rate law $v(s) = ks^n$ and confirm that the result is consistent with the definition of kinetic order in section 2.1.

(b) When s is small, the Michaelis–Menten rate law is approximately linear. What is the slope of the corresponding linear relationship? That is, given that

$$\frac{V_{max}s}{K_M+s} \approx ks$$

for s small, what is the corresponding constant k?   □

Reversible Reactions   When both substrate and product are present, the enzyme-catalyzed reaction scheme is

$$S + E \underset{k_{-1}}{\overset{k_1}{\rightleftharpoons}} C \underset{k_{-2}}{\overset{k_2}{\rightleftharpoons}} P + E. \tag{3.8}$$

Assuming the complex C is in quasi–steady state with respect to S and P, the reaction rate (which can now take negative values) is given by the reversible Michaelis–Menten rate law:

$$\text{Net rate of } S \rightarrow P = \frac{V_f \dfrac{s}{K_S} - V_r \dfrac{p}{K_P}}{1 + \dfrac{s}{K_S} + \dfrac{p}{K_P}}, \tag{3.9}$$

where V_f and V_r are the maximal rates of the forward and reverse reactions, and K_S and K_P are the Michaelis constants for S and P, respectively (details in problem 3.7.3). In some cases, the product rebinds to the free enzyme, but the rate of the reverse reaction is negligible. This is referred to as product inhibition (see problem 3.7.4).

Exercise 3.1.3 Verify the *Haldane relation*, which states that when the enzyme-catalyzed reaction $S \leftrightarrow P$ is in equilibrium,

$$K_{eq} = \frac{p}{s} = \frac{k_1 k_2}{k_{-1} k_{-2}}.$$

(Hint: When $S \leftrightarrow P$ is in equilibrium, both of the reversible reactions in scheme (3.8) must be in equilibrium.)   □

3.1.2 Two-Substrate Reactions

Most enzyme-catalyzed reactions involve more than one substrate. To describe the enzyme catalysis of these reactions, we must expand our description of Michaelis–Menten kinetics.

Catalysis of the irreversible two-substrate reaction

$$A + B \rightarrow P + Q$$

involves two distinct association events: each substrate must bind the catalyzing enzyme. (Trimolecular collisions in which both substrates bind the enzyme simultaneously are exceedingly rare events and so can be neglected.) The catalytic process can follow a number of different routes, including the following[2]:

• A *compulsory order* mechanism, in which substrate A binds to the free enzyme, thus forming a complex EA. Substrate B then binds, forming a ternary (i.e., three-

---

2. Here we use concatenation of species names to denote molecular complexes. This is standard practice in biochemistry and is a useful notation but unfortunately can be confused with the multiplicative product when these symbols are also used to represent concentrations. We will avoid such confusion by using distinct notation for chemical species and concentrations (e.g., $a = [A]$, $b = [B]$), so that AB is a molecular complex, whereas ab is the product of two concentrations.

molecule) complex *EAB*. This complex is then converted to *EPQ*, from which the products are released.

• A *random order* mechanism, in which either substrate can bind first. The products are released from the ternary complex *EPQ*.

• A *double-displacement* (or *ping-pong*) mechanism, in which substrate *A* first forms a complex *EA* with the enzyme. The enzyme is then modified in some manner (e.g., by gaining a functional group from *A*), and product *P* is released. The modified enzyme *E** then binds *B*, forming complex *E*B*. The enzyme recovers its original state by converting *B* to *Q*. Finally, product *Q* is released.

To develop a two-substrate Michaelis–Menten rate law, consider the compulsory order reaction scheme

$$E + A \underset{k_{-1}}{\overset{k_1}{\rightleftharpoons}} EA$$

$$EA + B \underset{k_{-2}}{\overset{k_2}{\rightleftharpoons}} EAB \tag{3.10}$$

$$EAB \overset{k_3}{\longrightarrow} E + P + Q.$$

As in the single-substrate case, a rate law can be derived by assuming that the complexes come to quasi–steady state. (This is justified if the substrates are abundant compared to the enzyme.)

Letting $C_1 = EA$, with concentration c_1, and $C_2 = EAB$, with concentration c_2, we can model the complexes by

$$\frac{d}{dt}c_1(t) = k_1 a(t)e(t) - k_{-1}c_1(t) - k_2 c_1(t)b(t) + k_{-2}c_2(t)$$

$$\frac{d}{dt}c_2(t) = k_2 c_1(t)b(t) - k_{-2}c_2(t) - k_3 c_2(t).$$

Using conservation of enzyme to substitute for the free enzyme concentration ($e(t) = e_T - c_1(t) - c_2(t)$) and applying a quasi-steady-state approximation to both complexes gives a reaction rate of

$$\text{Rate of } A + B \to P + Q = k_3 c_2^{\text{qss}}(t) = \frac{k_3 e_T a(t)b(t)}{k_{-1}\dfrac{k_{-2}+k_3}{k_1 k_2} + \dfrac{k_{-2}+k_3}{k_2}a(t) + \dfrac{k_3}{k_1}b(t) + a(t)b(t)}. \tag{3.11}$$

Exercise 3.1.4 Verify equation (3.11). ☐

This rate law for the compulsory-order enzyme-catalyzed reaction $A + B \to P + Q$ can be written more concisely as

$$v = \frac{V_{\max}ab}{K_{AB} + K_B a + K_A b + ab}.$$
(3.12)

When the same QSSA approach is applied to the other reaction mechanisms described earlier (random order and double-displacement), similar reaction rates can be derived (see problem 3.7.6).

Note that if the concentration of either A or B is held constant, then this rate law reduces to a single-substrate Michaelis–Menten expression. This simplification is commonly used when one substrate is present at a near-fixed concentration; for example, when it is a common coreactant (called a *cofactor*), such as water or ATP. In those cases, a single-substrate Michaelis–Menten rate law is used, because the reaction is effectively dependent on a single reactant concentration.

Exercise 3.1.5 Consider rate law (3.12).

(a) Verify that when the concentration of A is very high, the rate law reduces to a single-substrate Michaelis–Menten rate law for B, with maximal rate $V_{\max}$ and Michaelis constant $K_M = K_B$. Verify that this is consistent with the limiting behavior of the reaction network (3.10) (which in this case takes the form $EA + B \leftrightarrow EAB \to EA + P + Q$ because all enzyme will be continually bound to A).

(b) Verify that when the concentration of B is very high, the rate law reduces to a single-substrate Michaelis–Menten rate law for A, with maximal rate $V_{\max}$ and Michaelis constant $K_M = K_A$. Verify that this is consistent with the limiting behavior of the reaction network (3.10) (which in this case takes the form $E + A \to EAB \to E + P + Q$ because complex EA will bind B immediately after forming). $\square$

3.2   Regulation of Enzyme Activity

Enzyme activity can be regulated through a number of mechanisms. Genetic effects can cause changes in the abundance of enzymes (e_T in our models). These changes occur on the slow timescale of genetic processes—minutes to hours. A much faster means of control is provided by biochemical modification of individual enzyme molecules. We will consider two distinct mechanisms by which enzyme activity can be altered biochemically: competitive inhibition and allosteric regulation.

3.2.1   Competitive Inhibition

A competitive inhibitor is a molecule that mimics an enzyme's substrate but does not undergo a reaction, as shown in figure 3.5. This impostor molecule binds to the active site and clogs the enzyme so it is not available for catalysis. (A familiar example is the drug aspirin: it binds the active site of the enzyme cyclooxygenase

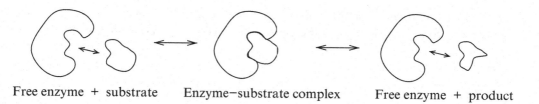

Free enzyme + substrate   Enzyme–substrate complex   Free enzyme + product

Free enzyme + inhibitor   Enzyme–inhibitor complex (inactive)

Figure 3.5
Competitive inhibition of enzyme catalysis. The inhibitor — a chemical mimic of the substrate — binds the enzyme's active site. Inhibitor-bound enzymes are not available to catalyze reactions.

and thus inhibits the production of prostaglandins, which are involved in pain pathways, blood clotting, and production of stomach mucus.)

To derive a rate law for a competitively inhibited enzyme, we construct a reaction scheme consisting of both the catalytic process and the formation of an enzyme–inhibitor complex

$$S + E \underset{k_{-1}}{\overset{k_1}{\rightleftharpoons}} C \xrightarrow{k_2} E + P$$

$$I + E \underset{k_{-3}}{\overset{k_3}{\rightleftharpoons}} C_I,$$

where I is the inhibitor and C_I is the inactive enzyme–inhibitor complex. We will apply a quasi-steady-state approximation to the two complexes. With c and c_I denoting concentrations, we have

$$\frac{d}{dt}c(t) = k_1 s(t) e(t) - k_{-1} c(t) - k_2 c(t)$$

$$\frac{d}{dt}c_I(t) = k_3 e(t) i - k_{-3} c_I(t).$$

We treat the inhibitor concentration i as a fixed quantity. (This is justified by presuming that the inhibitor is far more abundant than the enzyme, so that formation of C_I does not change i significantly.) Applying the quasi-steady-state assumption to the two complexes and using the conservation $e(t) = e_T - c(t) - c_I(t)$ yields

$$c = \frac{e_T s}{\dfrac{iK_M}{K_i} + s + K_M},\tag{3.13}$$

where $K_M = (k_{-1} + k_2)/k_1$, and K_i is the *dissociation constant* for inhibitor binding: $K_i = k_{-3}/k_3$. The rate law can then be written as

$$\text{Rate of } S \rightarrow P = k_2 c = \frac{V_{max} s}{K_M(1 + i/K_i) + s},$$

with $V_{max} = k_2 e_T$.

This rate law is sketched in figure 3.6 for various levels of inhibitor. Competitive inhibition does not influence the limiting reaction rate V_{max}. When the substrate is much more abundant than the inhibitor, the inhibition has only a negligible effect. However, the competition for enzymes affects the concentration of substrate needed to achieve a given reaction rate: the effective Michaelis constant of the inhibited reaction, $K_M(1 + i/K_i)$, increases with inhibitor abundance.

Exercise 3.2.1 Verify equation (3.13).   □

Competitive inhibition depends on the conformation of the enzyme's active site: inhibitors must be chemically similar to reactants. We next consider a more versatile

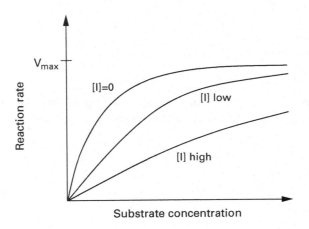

Figure 3.6
Reaction rates for competitively inhibited enzyme catalysis. The limiting reaction rate, V_{max}, is unaffected by the presence of the inhibitor. Instead, the inhibitor increases the substrate concentration required to reach the half-maximal rate.

form of regulation that does not suffer from this restriction and instead makes use of components of the regulated enzyme besides the active site.

3.2.2 Allosteric Regulation

The catalytic efficiency of an enzyme depends on the conformation of its active site. This conformation depends, in turn, on the overall configuration of the protein (its tertiary structure). This configuration, and hence the nature of the active site, can be altered by modifications to the chemical energy landscape of the protein (e.g., the position and strength of charges). Such modifications can be made by molecules that bind the protein. This mode of enzyme regulation, called **allosteric control**, was proposed by François Jacob and Jacques Monod in 1961. The term *allostery* (from Greek: *allo*, "other," and *steros*, "solid" or "shape") emphasizes the distinction from competitive inhibition—the regulating molecule need not bear any chemical resemblance to the substrate. Likewise, the site on the enzyme where the regulator binds (called the *allosteric site*) can be distinct from the active site in both position and chemical structure.

Typically, the binding of an allosteric regulator to a protein invokes a transition between a functional state and a nonfunctional state. For enzymes, this is typically a transition between a catalytically active form and an inactive form.

Allostery offers a range of strategies for the regulation of enzyme activity. For instance, conformational changes could affect substrate binding or reaction catalysis. Moreover, an enzyme molecule might simultaneously bind multiple allosteric regulators whose effects can be integrated in a variety of ways. In this section, we will consider a simple case that highlights the functional differences between allosteric inhibition and competitive inhibition.

Consider an enzyme that binds a single allosteric regulator. Suppose the regulator inhibits enzyme catalysis by blocking product formation, as shown in figure 3.7. For simplicity, we will assume that the inhibitor has no effect on substrate binding. The reaction scheme is

$$S + E \underset{k_{-1}}{\overset{k_1}{\rightleftharpoons}} ES \overset{k_2}{\longrightarrow} E + P$$

$$I + E \underset{k_{-3}}{\overset{k_3}{\rightleftharpoons}} EI$$

$$I + ES \underset{k_{-3}}{\overset{k_3}{\rightleftharpoons}} ESI \tag{3.14}$$

$$S + EI \underset{k_{-1}}{\overset{k_1}{\rightleftharpoons}} ESI \,.$$

We have assumed that the binding of substrate and inhibitor are independent. This scheme is called *noncompetitive inhibition*.

Allosteric regulator

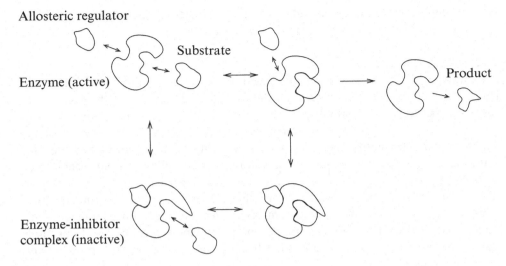

Enzyme (active)

Substrate

Product

Enzyme-inhibitor
complex (inactive)

Figure 3.7
Allosteric inhibition of enzyme activity. The enzyme has two binding sites: the active site where the
substrate binds, and the allosteric site where the allosteric regulator binds. In this example, the inhibited
enzyme can bind substrate but cannot catalyze formation of the reaction product.

To derive a rate law, the quasi-steady-state approximation can be applied to the
complexes ES, EI, and ESI. Together with the conservation $e_T = [E] + [ES] + [EI]
+ [ESI]$, this gives the reaction rate as

$$\text{Rate of } S \rightarrow P = k_2[ES] = \frac{V_{max}}{1 + i/K_i} \frac{s}{K_M + s}, \tag{3.15}$$

where $V_{max} = k_2 e_T$, $K_M = (k_{-1} + k_2)/k_1$, and $K_i = k_{-3}/k_3$. This rate law is sketched in
figure 3.8 for various levels of inhibitor. In contrast with competitive inhibition,
this allosteric inhibitor reduces the limiting rate V_{max} but does not affect the half-
saturating concentration K_M. More generally, other allosteric inhibition schemes
impact both V_{max} and K_M.

Exercise 3.2.2 An alternative to the noncompetitive inhibition scheme of figure 3.7
is *uncompetitive inhibition*, in which the inhibitor only binds the enzyme–substrate
complex (so complex EI does not occur). Apply a quasi-steady-state analysis to
verify that uncompetitive inhibition reduces K_M and V_{max} by the same factor. □

3.3 Cooperativity

The term **cooperativity** is used to describe potentially independent binding events
that have a significant influence on one another, leading to nonlinear behaviors. The

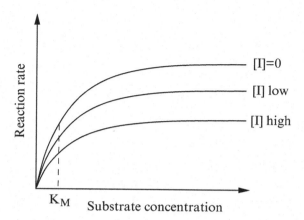

Figure 3.8
Rate laws for allosterically inhibited enzyme catalysis (noncompetitive inhibition). The limiting reaction rate is reduced by the inhibitor. The binding affinity, reflected in the half-maximal substrate concentration K_M, is unaffected.

most commonly cited example (and the first to be recognized, at the turn of the 20th century) is the binding of oxygen to the protein *hemoglobin*. Hemoglobin is the main transporter of oxygen in blood (in vertebrates). Hemoglobin is a tetrameric protein (i.e., composed of four polypeptide chains): each monomer binds one oxygen molecule.

Hemoglobin's efficiency as an oxygen carrier can be assessed by a curve showing the fraction of protein in the oxygen-bound form as a function of the abundance of oxygen. When such plots were first generated from experimental data, the binding curve for hemoglobin was found to have an S-shaped, or *sigmoidal*, character, as sketched in figure 3.9. This came as somewhat of a surprise, as most binding curves are hyperbolic, not sigmoidal.

Further insight into the oxygen-binding behavior of hemoglobin came from studies of the closely related protein myoglobin, which is used to store oxygen in vertebrate muscle cells. Structurally, myoglobin is very similar to the individual monomers that form a hemoglobin tetramer. As shown in figure 3.9, the binding curve for myoglobin is hyperbolic. This suggests that hemoglobin's sigmoidal binding curve does not result from the nature of the individual binding sites. Rather, this nonlinearity results from cooperative interactions among the four monomers.

To address cooperativity, consider the binding of a molecule X to a protein P. The generic name for a binding molecule is **ligand** (from Latin: *ligare*, "to bind"). The reaction scheme is

$$P + X \underset{k_{-1}}{\overset{k_1}{\rightleftharpoons}} PX .$$

(3.16)

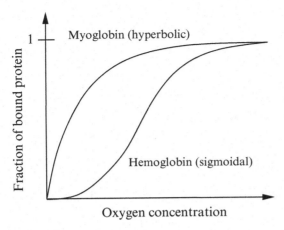

Figure 3.9
Binding curves for hemoglobin and myoglobin. Myoglobin shows a commonly observed hyperbolic
behavior. The binding curve for hemoglobin is *sigmoidal* (S-shaped). This nonlinear behavior is characteristic of cooperative binding mechanisms.

The *fractional saturation* of the pool of protein, denoted Y, is defined as the fraction
of binding sites that are occupied by ligand:

$$Y = \frac{\text{Number of occupied binding sites}}{\text{Total number of binding sites}} = \frac{[PX]}{[P]+[PX]}.$$

Letting $K = k_{-1}/k_1$ (the dissociation constant for the binding event), we find, at
steady state, $[PX] = [P][X]/K$. Then we have

$$Y = \frac{[P][X]/K}{[P]+[P][X]/K} = \frac{[X]/K}{1+[X]/K} = \frac{[X]}{K+[X]}. \tag{3.17}$$

(This functional form was seen earlier in the Michaelis–Menten rate law: the rate
of an enzyme-catalyzed reaction is proportional to the fractional occupancy of the
enzyme.)

The binding curve associated with equation (3.17) is hyperbolic—consistent with
the oxygen-binding behavior of myoglobin. To describe the binding of oxygen to
hemoglobin, we will next consider a protein P with four ligand-binding sites.

If the binding sites are identical and the binding events are independent of
one another, then the binding behavior is no different from equation (3.17). (In
this case, the tetrameric structure simply increases the number of independent
binding sites per protein.) Likewise, if the binding events are independent and the
binding sites have different affinities, then the binding curve is again hyperbolic (see
problem 3.7.9).

To address the case in which the binding sites are identical but cooperative effects occur between them, consider the following scheme:

$$P + X \underset{k_{-1}}{\overset{4k_1}{\rightleftharpoons}} PX_1$$

$$PX_1 + X \underset{2k_{-2}}{\overset{3k_2}{\rightleftharpoons}} PX_2$$

$$PX_2 + X \underset{3k_{-3}}{\overset{2k_3}{\rightleftharpoons}} PX_3$$

$$PX_3 + X \underset{4k_{-4}}{\overset{k_4}{\rightleftharpoons}} PX_4,$$

where complex PX_i has i ligand molecules bound. The rate constants depend on the number of bound ligand molecules, as follows. The first association reaction has rate k_1, the second has rate k_2, and so on. These rate constants are scaled by stoichiometric prefactors that account for the number of binding sites involved in each reaction. (For instance, there are four sites available for the first ligand to bind, so the overall reaction rate is $4k_1$. There are only three sites available for the second ligand to bind, so the rate is $3k_2$.)

Because there are four binding sites on each protein molecule, the fractional saturation is given by

$$Y = \frac{\text{Number of occupied binding sites}}{\text{Total number of binding sites}} = \frac{[PX_1] + 2[PX_2] + 3[PX_3] + 4[PX_4]}{4([P] + [PX_1] + [PX_2] + [PX_3] + [PX_4])}.$$

When the binding events are in equilibrium, the fractional saturation can be written as

$$Y = \frac{[X]/K_1 + 3[X]^2/(K_1 K_2) + 3[X]^3/(K_1 K_2 K_3) + [X]^4/(K_1 K_2 K_3 K_4)}{1 + 4[X]/K_1 + 6[X]^2/(K_1 K_2) + 4[X]^3/(K_1 K_2 K_3) + [X]^4/(K_1 K_2 K_3 K_4)}, \quad (3.18)$$

where the parameters K_i are the dissociation constants ($K_i = k_{-i}/k_i$, for $i = 1, 2, 3, 4$). This is known as the *Adair equation* for four sites. It exhibits a sigmoidal character when the affinities of the later binding events are significantly greater than those of the earlier events. This is positive cooperativity: the binding of the first ligand molecules enhance the binding of the remaining ligands.

If the final binding event has a much higher affinity than the earlier binding events (i.e., $K_4 \ll K_1, K_2, K_3$), then the fractional saturation can be approximated by

$$Y \approx \frac{[X]^4/(K_1 K_2 K_3 K_4)}{1 + [X]^4/(K_1 K_2 K_3 K_4)}.$$

This approximation is formalized in the **Hill function**

$$Y = \frac{([X]/K)^n}{1+([X]/K)^n} = \frac{[X]^n}{K^n+[X]^n}, \tag{3.19}$$

which is used to describe processes that involve cooperative binding events. The constant K is the half-saturating concentration of ligand and so can be interpreted as an averaged dissociation constant. (Note that when $n = 1$, the Hill function reduces to a hyperbolic function.)

In 1910, the English physiologist A. V. Hill proposed the function (3.19) as a convenient curve for fitting the sigmoidal binding behavior observed for hemoglobin. Hill did not attach any significance to the particular form of this function—he used it as an empirical fit to the data.

As shown in figure 3.10, the exponent n, called the *Hill coefficient*, reflects the steepness of the sigmoid and is commonly used as a measure of the switch-like character of a process. The Hill coefficient is often chosen to coincide with the number of events in a multiple-binding-event process, as in our derivation above. However, Hill functions constructed in this manner must be interpreted carefully, as empirical observations often correspond to Hill coefficients well below the number of binding events. For example, Hill himself found that the hemoglobin binding data was best fit with values of n ranging from 1 to 3.2. (When fit to empirical data, non-integer Hill coefficients are commonly used.)

Having justified the difference in binding behavior between myoglobin and hemoglobin, we can now interpret their respective biological functions. Myoglobin

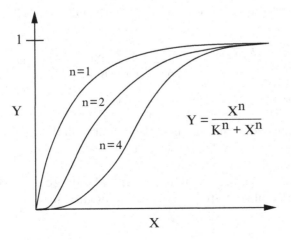

Figure 3.10
Hill functions. As the Hill coefficient n increases, the sigmoidal curve becomes more switch-like. When the Hill coefficient is 1, the curve is hyperbolic.

serves to *store* oxygen in muscle tissue and so is saturated with oxygen at all but the lowest oxygen concentrations. Hemoglobin *shuttles* oxygen and so must bind and release oxygen in the appropriate conditions. A sigmoidal binding curve has the property that a relatively small change in ligand concentration can lead to a large change in binding saturation. This allows hemoglobin to fulfill its transport task without demanding a wide difference in oxygen concentrations at its "pick up" and "drop off" points. For example, the difference in oxygen concentration between the alveoli of the lungs and the capillaries of active muscle is only fivefold. A hyperbolic binding curve cannot exhibit a wide change in ligand binding over such a narrow range in ligand availability. (Indeed, an 81-fold change in ligand concentration is needed to take a hyperbolic binding curve from 10% to 90% saturation.) Cooperative binding provides hemoglobin with a narrow, *switch-like* response to oxygen availability, resulting in efficient shuttling. In subsequent chapters, we will see a range of biological uses for such switch-like behavior.

Exercise 3.3.1 Consider a protein that binds ligand at two identical sites and derive the corresponding Adair equation:

$$Y = \frac{[X]/K_1 + [X]^2/(K_1 K_2)}{1 + 2[X]/K_1 + [X]^2/(K_1 K_2)},$$

where K_1 and K_2 are the dissociation constants for the first and second binding events. □

Exercise 3.3.2 Verify that the Hill function (3.19) has slope $n/4K$ at its half-saturation point ($[X] = K$). □

Exercise 3.3.3 The Hill function description of cooperative binding can be equivalently derived in the limiting case where n ligands bind protein P simultaneously. Confirm this fact by deriving the fractional saturation when the binding event is

$$P + nX \underset{k_{-1}}{\overset{k_1}{\rightleftharpoons}} PX_n.$$

The Hill function (3.19) can be recovered by setting $K = \sqrt[n]{k_{-1}/k_1}$. □

Exercise 3.3.4 Consider an enzyme that has two identical catalytic sites, and suppose that the substrates exhibit cooperative binding. To simplify your analysis, presume that the cooperativity is strong, so that the substrates can be assumed to bind simultaneously (as in exercise 3.3.3). Furthermore, presume that catalysis only occurs when two substrate molecules are bound. The reaction scheme is then

$$2S + E \underset{k_{-1}}{\overset{k_1}{\rightleftharpoons}} C \overset{k_2}{\longrightarrow} 2P + E.$$

Verify that in the absence of product, the reaction rate takes the form

$$v = \frac{V_{\max} s^2}{K_M + s^2}.$$ □

3.4 Compartmental Modeling and Transport

As described in chapter 2, ordinary differential equation models rely on the assumption that the reaction network occurs in a well-stirred volume. The cell is, of course, not a homogeneous mixture, but this well-stirred assumption can often be justified when applied to individual cellular *compartments*. Prokaryotic cells typically consist of a single compartment, but eukaryotic cells are composed of a collection of membrane-bound compartments (e.g., mitochondria, the nucleus, the endoplasmic reticulum, and the cytosol).

Multicompartment models can be used to describe systems that involve activity in more than one cellular compartment (figure 3.11). In such models, species concentrations in each well-mixed compartment are described separately. Transport of molecules between connected compartments is described explicitly. In this section, we will consider some common models of cross-membrane transport.

3.4.1 Diffusion
Some molecules, such as oxygen, readily diffuse across bilipid membranes. The transport of such species can be described by *Fick's law*, which states that the rate of diffusion is proportional to the difference in concentration.

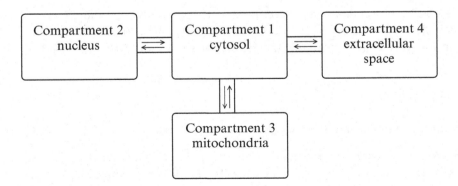

Figure 3.11
A multicompartment model. Each compartment is presumed well-mixed. Transport between compartments is restricted to the connections shown. When multiple copies of a compartment are present (e.g., mitochondria), they can be treated as a single averaged compartment.

To illustrate, suppose species S is present in two neighboring compartments and diffuses freely across the membrane that separates them. If $[S]_1$ and $[S]_2$ are the concentrations of S in compartments 1 and 2, then the rate of flow of molecules from compartment 1 to compartment 2 is given by

$$\text{Rate of flow} = D([S]_1 - [S]_2), \tag{3.20}$$

where the constant D quantifies how readily S diffuses across the membrane. To describe the resulting changes in concentration, compartment volumes must be taken into account. If V_1 and V_2 are the volumes of compartments 1 and 2, respectively, the concentrations in the two compartments can be described by

$$\frac{d}{dt}[S]_1(t) = -\frac{D([S]_1(t)-[S]_2(t))}{V_1} \qquad \frac{d}{dt}[S]_2(t) = \frac{D([S]_1(t)-[S]_2(t))}{V_2}.$$

The constant D thus has units of volume $\cdot$ time^{-1}.

In the long term, diffusion leads to equal concentrations in both compartments. The concentration in the smaller volume will approach steady state more quickly. In particular, if one compartment has a much larger volume than the other, the change in concentration in the larger compartment may be negligible, so it could be treated as a fixed pool (see problem 3.7.12).

Many molecular species, such as charged ions and macromolecules, cannot diffuse through bilipid membranes. Transport of these molecules is facilitated by specialized transporter proteins. In some cases, these proteins simply provide small holes in the membrane that allow particular molecular species to pass: these are called *channels*, or *pores*. For example, nuclear pores allow free diffusion of some species between the nucleus and cytosol of eukaryotic cells. Another example, ion-specific channels, will be taken up in chapter 8. Transport through a channel or pore is driven by diffusion: the rate of transport is described by equation (3.20). (In this case, the coefficient D is proportional to the number of channels or pores that appear in the membrane.)

We next consider more specialized transporter proteins.

3.4.2 Facilitated Transport

Many molecular species are transported across membranes in a species-specific fashion. Transmembrane carrier proteins bind specific molecules and facilitate their transport across the membrane, as in figure 3.12.

Passive Transport In some cases, the transport mechanism is reversible: the direction of transport is then determined by the concentration of ligand on either side of the membrane. This is called *facilitated diffusion* or *passive transport*, and results, in the long term, in equal concentrations in the two compartments.

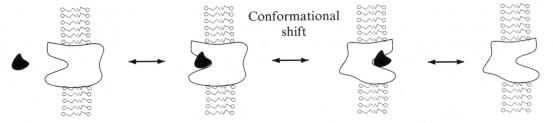

Figure 3.12
Transmembrane carrier protein. The ligand binds the protein on one side of the membrane. Through a conformational shift, the protein passes the ligand across the membrane and then releases it.

Passive transport is similar to enzyme catalysis—instead of changing a molecule's chemical identity, a transporter "catalyzes" a change in its location. A reaction scheme showing the steps in passive transport is identical to the scheme for a reversible enzyme-catalyzed reaction:

$$S_1 + T \rightleftharpoons TS_1 \rightleftharpoons TS_2 \rightleftharpoons T + S_2 ,$$

where T is the transporter and S_i is the transported species in compartment i.

If we make the simplifying assumption that the transport step ($TS_1 \leftrightarrow TS_2$) is in rapid equilibrium and then apply a quasi-steady-state approximation to the transporter–ligand complex, the rate of transport is given as

$$\text{Transport rate} = \frac{\alpha_1 [S_1] / K_1 - \alpha_2 [S_2] / K_2}{1 + [S_1] / K_1 + [S_2] / K_2},$$

where the parameters α_i are the maximal transport rates in each direction, and the K_i reflect the binding affinities of the ligand on either side of the membrane (compare with equation 3.9). This transport rate reduces to a linear expression $((\alpha_1 / K_1)[S_1] - (\alpha_2 / K_2)[S_2])$ when the transporter pool is far from saturation (i.e., $[S_1] \ll K_1$ and $[S_2] \ll K_2$).

This model describes a *uniporter*—a transporter that carries a single ligand molecule across a membrane. Other proteins transport multiple ligands simultaneously. Proteins that facilitate the transport of multiple ligands in the same direction are called *symporters*; simultaneous transport of ligands in opposite directions is facilitated by *antiporters*.

Exercise 3.4.1 Prepare a reaction scheme that corresponds to a two-ligand symporter. How does this mechanism compare with our treatment of two-substrate enzyme-catalyzed reactions in section 3.1.2? Does the analysis have to be modified to address a two-ligand antiporter?   □

Active Transport Passive diffusion acts to eliminate differences in concentration. In some cases, cellular processes maintain concentration gradients across membranes. To achieve these persistent gradients, cells expend energy to transport molecules against diffusion—from low concentration to high concentration—by *active transport*. Many active transporters, or *pumps*, consume ATP; others are co-transporters that couple the diffusion-driven transport of one ligand to the transport of another ligand against its concentration gradient.

Because intracellular ATP levels are tightly regulated, models often treat the concentration of ATP as fixed. In these cases, the dependence of a transporter on ATP can be incorporated into an effective kinetic constant. Some pumps couple the consumption of a single ATP molecule to the transport of multiple ligand molecules. For example, the sodium–potassium pump, found in animal cells, couples the consumption of a single ATP molecule to the transport of three Na^+ ions into the cell and two K^+ ions out of the cell, both against their respective concentration gradients. Another example is provided by the Ca^{2+} pumps in the membrane of the endoplasmic reticulum of eukaryotic cells. Some of these pumps transport two Ca^{2+} ions out of the cytosol each time they consume an ATP molecule. Assuming that the two calcium ions bind with strong cooperativity and that the rate of transport is proportional to the fractional occupancy of the pumps, the transport rate is then

$$\text{Transport rate} = \frac{\alpha[Ca^{2+}]^2}{K + [Ca^{2+}]^2}, \tag{3.21}$$

where the maximal rate α is implicitly dependent on both the cytosolic ATP level and the abundance of pump proteins.

Exercise 3.4.2 Derive formula (3.21) (compare with exercise 3.3.4). ☐

3.5* Generalized Mass Action and S-System Modeling

The law of mass action provides an excellent description of reaction rates in ideal conditions: dilute molecular solutions at high molecule counts in a well-stirred reaction vessel. The rates of chemical processes occurring within the cell—a highly concentrated and inhomogeneous environment—often differ from mass action. These differences can sometimes be accounted for by describing reaction rates as functions of chemical *activities*, rather than concentrations. Chemical activities correspond to effective concentrations: they are dependent on the composition of the solution in which the reaction occurs. In some cases, the activity, a, is related to the concentration by a power-law; for example, $a_X = [X]^\gamma$, for some exponent γ. Applying the law of mass action to activities (instead of concentrations) can give rate laws with non-integer powers of concentrations; for example,

$$X \xrightarrow{k_1} Y \qquad \text{reaction rate} : k_1 a_X = k_1 [X]^\gamma$$

$$X + X \xrightarrow{k_2} Z \qquad \text{reaction rate} : k_2 (a_X)^2 = k_2 ([X]^\gamma)^2 = k_2 [X]^{2\gamma}.$$

These are referred to as *generalized mass action* (GMA) rate laws.

Generalizing mass action in this way allows simple approximations of complex rate laws. As an example, figure 3.13 compares the GMA rate law $v = s^{0.4}$ to the Michaelis–Menten rate law $v = 2s/(1+s)$. The two curves agree quite well over the range shown, although they diverge for larger values of the substrate concentration. (The GMA rate law does not saturate.) In this simple case, we generally prefer the Michaelis–Menten formulation because it was derived from the reaction mechanism, whereas the GMA rate law can only be used as an empirical fit. However, when more complex rate laws are considered—especially those involving multiple substrates and regulators—the GMA approximation becomes more attractive. The number of parameters in a Michaelis–Menten rate law increases rapidly as more substrates or regulators are included in the reaction mechanism. In contrast, a GMA formulation involves about one parameter per species and so

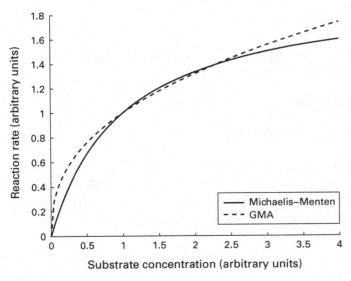

Figure 3.13
Comparison of generalized mass action and Michaelis–Menten rate laws. The GMA rate law ($v = s^{0.4}$) provides a good approximation to the Michaelis–Menten rate law ($v = 2s/(1+s)$) over the range shown but diverges as the substrate concentration increases.

provides a formulation that is much more tractable in terms of fitting to experimental data.

S-System Modeling   In the late 1960s, Michael Savageau proposed a novel modeling framework that takes the simplification inherent in a GMA formulation one step further. Savageau's approach condenses each differential equation in a model by lumping all production terms into a single GMA expression and all consumption terms into another. The result is a simple model that contains relatively few parameters and yet is able to describe a wide range of nonlinear behaviors. Models of this type have come to be called *S-system models*. (The "S" stands for synergism and saturation. S-system modeling provides a simple framework for approximating these nonlinear effects.)

One significant advantage of the S-system framework is that the steady-state species concentrations can always be determined analytically—after a transformation, the steady-state solution can be found by solving a linear system of equations. In contrast, Michaelis–Menten formulations rarely give rise to systems for which the steady state can be described explicitly.

To illustrate the S-system modeling framework, consider the reaction scheme shown in figure 3.14, in which species S_2 allosterically inhibits its own production. The concentrations s_1 and s_2 of species S_1 and S_2 satisfy

$$\frac{d}{dt} s_1(t) = v_1 - v_2$$

$$\frac{d}{dt} s_2(t) = v_2 - v_3 - v_4. \tag{3.22}$$

To construct an S-system model, we make the following power-law approximations:

$$v_1 = \alpha_1 \qquad v_2 = \alpha_2 s_1^{g_1} s_2^{g_2} \qquad v_3 + v_4 = \alpha_3 s_2^{g_3} . \tag{3.23}$$

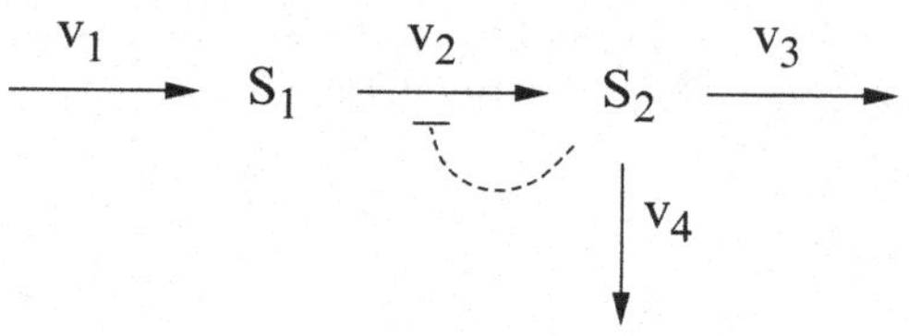

Figure 3.14
Reaction scheme for S-system model analysis. The labels v_i indicate the rates of the corresponding reactions. (They are not mass-action rate constants.) The blunted dashed arrow indicates that species S_2 allosterically inhibits its own production.

The parameters α_i are rate constants; the g_i are kinetic orders. Because S_2 acts as an allosteric inhibitor, g_2 will be negative.

The steady-state species concentrations satisfy

$$0 = \alpha_1 - \alpha_2 s_1^{g_1} s_2^{g_2}$$

$$0 = \alpha_2 s_1^{g_1} s_2^{g_2} - \alpha_3 s_2^{g_3}.$$

These equations can be rewritten as

$$\alpha_1 = \alpha_2 s_1^{g_1} s_2^{g_2}$$

$$\alpha_2 s_1^{g_1} s_2^{g_2} = \alpha_3 s_2^{g_3}.$$

The terms in these equations are all positive, allowing us to take logarithms. The logarithm of the first equation gives

$$\log \alpha_1 = \log(\alpha_2 s_1^{g_1} s_2^{g_2}) = \log \alpha_2 + \log s_1^{g_1} + \log s_2^{g_2} = \log \alpha_2 + g_1 \log s_1 + g_2 \log s_2.$$

The second equation yields

$$\log \alpha_2 + g_1 \log s_1 + g_2 \log s_2 = \log \alpha_3 + g_3 \log s_2.$$

This is a pair of linear equations in $\log s_1$ and $\log s_2$; it can be solved to yield

$$\log s_1 = \frac{\log \alpha_1 - \log \alpha_2 - \dfrac{g_2}{g_3}(\log \alpha_1 - \log \alpha_3)}{g_1} \tag{3.24}$$

$$\log s_2 = \frac{\log \alpha_1 - \log \alpha_3}{g_3}.$$

Taking exponentials on both sides gives the steady-state concentrations. This example will be revisited in problem 4.8.15, which illustrates how further insight can be drawn directly from these explicit descriptions.

Exercise 3.5.1 Consider the irreversible three-step reaction chain $\to S_1 \to S_2 \to$. Formulate an S-system description of the system, presuming a constant input rate α_0 and GMA reaction rates $\alpha_1[S_1]^{g_1}$ and $\alpha_2[S_2]^{g_2}$. Derive explicit formulas for the steady-state concentrations of s_1 and s_2. $\qquad\qquad\square$

3.6 Suggestions for Further Reading

• **Enzyme Kinetics** A concise introduction to enzyme kinetics can be found in *Enzyme Kinetics* (Cornish-Bowden and Wharton, 1988). A more complete description is provided in *Fundamentals of Enzyme Kinetics* (Cornish-Bowden, 1979). More recent treatments are *Enzyme Kinetics and Mechanism* (Cook and Cleland, 2007) and *Enzyme Kinetics for Systems Biology* (Sauro, 2011).

• **Compartmental Modeling** An extensive treatment of compartmental modeling is provided in *Compartmental Analysis in Biology and Medicine* (Jacquez, 1985).

• **S-System Modeling** S-system modeling is introduced by Michael Savageau in his book *Biochemical Systems Analysis: A Study of Function and Design in Molecular Biology* (Savageau, 1976). An updated treatment is given in *Computational Analysis of Biochemical Systems: A Practical Guide for Biochemists and Molecular Biologists* (Voit, 2000).

3.7 Problem Set

3.7.1 Michaelis–Menten Kinetics: Estimation of Parameters

(a) How would you estimate the Michaelis–Menten parameters V_{max} and K_M from a *Lineweaver–Burk plot*: a linear plot of $1/v$ against $1/s$ (where v is the reaction rate, or velocity).

(b) An alternative formulation, suggested by Hanes and Woolf, is to rearrange the kinetic description to yield a linear equation for s/V as a function of s. Derive this formula.

3.7.2 Michaelis–Menten Kinetics: Separation of Timescales

The separation of timescales that was used to derive the Michaelis–Menten rate law (3.5) from reaction scheme (3.2) depends on the substrate being much more abundant than the catalyzing enzyme. This separation of timescales can be formalized as follows.

(a) Rescale the variables in system (3.3) by defining $C = c/e_T$ and $S = s/s(0)$, where e_T is the total enzyme concentration, and $s(0)$ is the initial concentration of substrate. These new variables are dimensionless ratios. Show that these scaled concentrations S and C satisfy

$$\frac{s(0)}{e_T}\frac{d}{dt}S(t) = -k_1 S(t)s(0)(1-C(t)) + k_{-1}C(t)$$

$$\frac{d}{dt}C(t) = k_1 S(t)s(0)(1-C(t)) - (k_{-1}+k_2)C(t).$$

(b) Explain why this set of equations exhibits a difference in timescales when $s(0)/e_T$ is large. Hint: The terms describing the dynamics on the right-hand side of these equations are of roughly the same size (they are virtually the same, except for the sign). How, then, will the size of dS/dt compare to the size of dC/dt?

3.7.3 Reversible Michaelis–Menten Kinetics

(a) Derive the reversible Michaelis–Menten rate law (equation 3.9) as follows. Apply a quasi-steady-state assumption to the complex C in reaction scheme (3.8) to arrive at a description of its concentration:

$$c^{qss} = \frac{k_1 e_T s + k_{-2} e_T p}{k_1 s + k_{-2} p + k_{-1} + k_2}.$$

Next, confirm that the reaction rate is

$$v = \frac{d}{dt} p(t) = \frac{k_1 k_2 e_T s - k_{-1} k_{-2} e_T p}{k_1 s + k_{-2} p + k_{-1} + k_2}.$$

(b) Express the parameters in equation (3.9) (i.e., V_f, V_r, K_S, K_P) in terms of the rate constants k_1, k_2, k_{-1}, and k_{-2}.

(c) Confirm that when k_{-2} is zero, the irreversible Michaelis–Menten rate law is recovered.

3.7.4 Product Inhibition

Many enzymatic reactions that are irreversible are nevertheless subject to *product inhibition*, meaning that the product readily rebinds the free enzyme. To describe product inhibition, consider the scheme:

$$S + E \underset{k_{-1}}{\overset{k_1}{\rightleftharpoons}} C_1 \overset{k_r}{\longrightarrow} C_2 \underset{k_{-2}}{\overset{k_2}{\rightleftharpoons}} P + E,$$

which is equivalent to scheme (3.1), except the conversion step is irreversible. From this reaction scheme, derive the rate law

$$\frac{d}{dt} p(t) = v = \frac{V_{max} s}{s + K_M \left(1 + \dfrac{p}{K_P} \right)}.$$

3.7.5 Michaelis–Menten Kinetics: First-Order Approximation

Consider the reaction chain

$$\overset{v_0}{\longrightarrow} S_1 \overset{v_1}{\longrightarrow} S_2 \overset{v_2}{\longrightarrow} S_3 \overset{v_3}{\longrightarrow}$$

in which the v_i are labels for the reaction rates (not mass-action constants). Take the rate v_0 as fixed and presume the other reactions follow Michaelis–Menten kinetics, with

$$v_i = \frac{V_{max}^i s_i}{K_{Mi} + s_i},$$

where $s_i = [S_i]$. Take parameter values (in mM/min) $v_0 = 2$, $V_{max}^1 = 9$, $V_{max}^2 = 12$, $V_{max}^3 = 15$; (in mM) $K_{M1} = 1$, $K_{M2} = 0.4$, $K_{M3} = 3$.

(a) Simulate the system from initial conditions (in mM) $(s_1, s_2, s_3) = (0.3, 0.2, 0.1)$. Repeat with initial condition $(s_1, s_2, s_3) = (6, 4, 4)$.

(b) Generate an approximate model in which the rates of reactions 1, 2, and 3 follow first-order mass-action kinetics (i.e., $v_i = k_i s_i$, for $i = 1, 2, 3$). Choose values for the rate constants k_i that give a good approximation to the original nonlinear model. Explain your reasoning (hint: exercise 3.1.2(b) provides one viable approach).

(c) Simulate your simpler (mass-action-based) model from the sets of initial conditions in part (a). Comment on the fit. If the approximation is better in one case than the other, explain why.

3.7.6 Michaelis–Menten Kinetics: Double-Displacement Reactions

Recall the double-displacement (ping-pong) mechanism for an irreversible enzyme-catalyzed two-substrate reaction as described in section 3.1.2. Suppose the reaction scheme is

$$A + E \underset{k_{-1}}{\overset{k_1}{\rightleftharpoons}} C_1 \overset{k_2}{\longrightarrow} P + E^*$$

$$B + E^* \underset{k_{-3}}{\overset{k_3}{\rightleftharpoons}} C_2 \overset{k_4}{\longrightarrow} Q + E.$$

Derive a rate law as in equation (3.12) by using the conservation $e_T = [E] + [C_1] + [E^*] + [C_2]$ and applying a quasi-steady-state assumption to the substrate–enzyme complexes (C_1 and C_2) and to the modified enzyme E^*. You should find that the constant term K_{AB} in the denominator is zero.

3.7.7 Specificity Constants

Some enzymes catalyze multiple reactions. When distinct substrates compete for an enzyme's catalytic site, they act as competitive inhibitors of one another. In this context, we can define the *specificity constant* for each substrate as the ratio of the enzyme's corresponding catalytic constant, k_{cat} (k_2 in scheme 3.2), to the substrate's Michaelis constant, K_M:

Specificity constant: $k_s = \dfrac{k_{cat}}{K_M}$.

Suppose two species, S and S', are both substrates for an enzyme E. Verify that the ratio of the reaction rates for S and S' is the product of the ratio of their concentrations and the ratio of their specificity constants:

$$\frac{\text{Rate of reaction of } S}{\text{Rate of reaction of } S'} = \frac{[S]}{[S']} \cdot \frac{k_s}{k_{s'}}.$$

Hint: There is no need to construct a reaction scheme. Take the rate of reaction of S to be

$$\frac{e_T k_{cat}[S]}{K_M (1 + [S']/K'_M) + [S]},$$

and likewise for S' (i.e., the Michaelis constant for S' is equal to its inhibition constant with respect to S, and vice versa).

3.7.8 Allosteric Activation

Consider an allosteric *activation* scheme in which an allosteric activator must be bound before an enzyme can bind substrate. This is called *compulsory activation*. The reaction scheme resembles a two-substrate reaction, but the enzyme–activator complex stays intact after the product dissociates:

$$R + E \underset{k_{-1}}{\overset{k_1}{\rightleftharpoons}} ER$$

$$ER + S \underset{k_{-2}}{\overset{k_2}{\rightleftharpoons}} ERS \xrightarrow{k_3} P + ER,$$

where R is the allosteric activator (regulator).

(a) Apply a quasi-steady-state assumption to the two complexes ER and ERS (and use enzyme conservation) to verify that the rate law takes the form

$$v = \frac{srk_3 e_T}{r\dfrac{k_{-2} + k_3}{k_2} + \dfrac{k_{-1}(k_{-2} + k_3)}{k_1 k_2} + sr} = \frac{V_{max} sr}{K_1 r + K_2 + rs},$$

where r is the regulator concentration, and s is the substrate concentration.

(b) Next, consider the case in which catalysis can only occur after n regulator molecules have bound. Assuming that the binding involves strong cooperativity, we can approximate the regulator-binding events by

$$nR + E \underset{k_{-1}}{\overset{k_1}{\rightleftharpoons}} ER_n.$$

Verify that in this case, the rate law takes the form

$$v = \frac{V_{max} sr^n}{K_1 r^n + K_2 + r^n s}.$$

(c) Confirm that when regulator and substrate are at very low concentration, the rate law in part (b) can be approximated as

$$v = \frac{V_{max}}{K_2} sr^n.$$

3.7.9 Noncooperative Multisite Binding

A protein with multiple binding sites that are independent (no cooperativity) cannot exhibit a sigmoidal binding curve, even when the binding sites have distinct affinities. Consider a protein with two ligand binding sites of different affinities. Show that in this case, the fractional saturation is simply the sum of two hyperbolic relations:

$$Y = \frac{[X]/K_1}{2(1+[X]/K_1)} + \frac{[X]/K_2}{2(1+[X]/K_2)},$$

where K_1 and K_2 are the dissociation constants for the two sites. Plot this relation for various values of K_1 and K_2 to confirm that it describes a hyperbolic binding curve.

3.7.10 Negative Cooperativity

The discussion of cooperative binding in section 3.3 focused on *positive cooperativity*, in which the substrate binding affinity increases as substrates bind. Some proteins, such as the enzyme glyceraldehyde-3-phosphate dehydrogenase, exhibit *negative cooperativity*—substrate affinity drops as substrates bind.

(a) Consider the Adair equation for two sites in exercise 3.3.1. Plot the curve for a negative cooperative case (e.g., for $K_2 > K_1$). Is the curve sigmoidal?

(b) An extreme case of negative cooperativity is known as *half-of-the-sites reactivity*, in which the affinity drops to zero once half of the binding sites are occupied. Referring again to the Adair equation for two sites, what is the form of the binding curve in this extreme case?

3.7.11 The Concerted Model of Cooperativity

In 1965, Jacques Monod, Jeffries Wyman, and Jean-Pierre Changeux proposed a mechanistic model of cooperativity (Monod et al., 1965). Their model addresses a multimeric protein composed of identical subunits, each with one ligand binding site. They supposed that each subunit could transition between two conformations: a tensed state T and a relaxed state R. For each protein molecule, all of the subunits are presumed to have the same conformation at any given time: transitions between the relaxed and tense states are *concerted*.

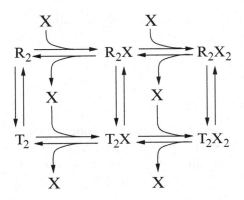

Figure 3.15
Binding scheme for concerted model of cooperativity (problem 3.7.11).

In the absence of ligand, the tensed state is more stable than the relaxed state. The relaxed state, however, has a higher affinity for ligand. Thus, at a sufficiently high ligand concentration, ligand binding causes the protein to adopt the relaxed state. This increases the protein's affinity for ligand, triggering a positive feedback, and resulting in a sigmoidal binding curve.

This mechanism is called the *MWC model*, or the *concerted model*. The ligand-binding scheme for a dimer is shown in figure 3.15, where R_2 is the dimer of two relaxed monomers, and T_2 is the dimer of two tensed monomers.

(a) Let K be the equilibrium constant for the $R_2 \leftrightarrow T_2$ conversion (i.e., $K = [T_2]/[R_2]$ at steady state). Suppose that the dissociation constant for ligand binding to R_2 is K_R, and the dissociation constant for ligand binding to T_2 is K_T. (The dissociation constants for the first and second binding events are the same, but the association/dissociation rates will depend on stoichiometric factors that reflect the number of sites.)

Confirm that in steady state, the concentrations satisfy

$$[R_2] = \frac{[T_2]}{K} \qquad [R_2X_1] = \frac{2[X][R_2]}{K_R} \qquad [R_2X_2] = \frac{[X][R_2X_1]}{2K_R}$$

$$[T_2X_1] = \frac{2[X][T_2]}{K_T} \qquad\qquad [T_2X_2] = \frac{[X][T_2X_1]}{2K_T}.$$

(The stoichiometric prefactors reflect the availability of binding sites.) Use these equilibrium conditions to verify that in steady state, the fractional saturation is given by

$$Y = \frac{K\frac{[X]}{K_T}\left(1+\frac{[X]}{K_T}\right)+\frac{[X]}{K_R}\left(1+\frac{[X]}{K_R}\right)}{K\left(1+\frac{[X]}{K_T}\right)^2+\left(1+\frac{[X]}{K_R}\right)^2}.$$

(3.25)

Plot the corresponding binding curves for $K_T = 1000$, $K_R = 1$, and $K = 500$, 1000, and 2000. Verify that although this is not a Hill function, the curves are nevertheless sigmoidal.

(b) Consider the special case of the concerted mechanism in which $K = 0$. Interpret the resulting binding mechanism and use formula (3.25) to verify that the resulting binding curve is hyperbolic. Repeat for the case when $K_R = K_T$.

(c) Verify that when the concerted model is applied to a tetramer (such as hemoglobin), the resulting fractional saturation is

$$Y = \frac{K\frac{[X]}{K_T}\left(1+\frac{[X]}{K_T}\right)^3+\frac{[X]}{K_R}\left(1+\frac{[X]}{K_R}\right)^3}{K\left(1+\frac{[X]}{K_T}\right)^4+\left(1+\frac{[X]}{K_R}\right)^4}.$$

3.7.12 Compartmental Modeling: Diffusion

Consider a system composed of three compartments: compartment 1, the nucleus, with volume V_1; compartment 2, the cytosol, with volume V_2; and compartment 3, the surrounding extracellular space, with volume V_3.

(a) Suppose that a molecular species S can diffuse across the nuclear membrane and across the cellular membrane. Confirm that the species concentrations s_i in compartment i satisfy

$$\frac{d}{dt}s_1(t) = \frac{D_1}{V_1}(s_2(t) - s_1(t))$$

$$\frac{d}{dt}s_2(t) = \frac{D_1}{V_2}(s_1(t) - s_2(t)) + \frac{D_2}{V_2}(s_3(t) - s_2(t))$$

$$\frac{d}{dt}s_3(t) = \frac{D_2}{V_3}(s_2(t) - s_3(t)),$$

where D_1 and D_2 characterize diffusion across the nuclear and cell membrane, respectively.

(b) Suppose that the initial concentrations are s_1^0, s_2^0, and s_3^0. Verify that at steady state,

$$s_1^{ss} = s_2^{ss} = s_3^{ss} = \frac{s_1^0 V_1 + s_2^0 V_2 + s_3^0 V_3}{V_1 + V_2 + V_3}.$$

(c) Suppose now that the concentration of S in the extracellular space is buffered, so that $s_3 = \bar{s}_3$ (constant). Verify that in this case, the steady state is given by

$$s_1^{ss} = s_2^{ss} = \bar{s}_3.$$

(d) Continuing with the buffered concentration assumption in part (c), suppose that species S is produced in the nucleus at rate k; that is,

$$\frac{d}{dt} s_1(t) = \frac{D_1}{V_1}(s_2(t) - s_1(t)) + k.$$

Determine the steady-state concentrations of s_1^{ss} and s_2^{ss} in this case. How does the difference $s_1^{ss} - s_2^{ss}$ depend on the model parameters? Interpret your results.

3.7.13* S-System Modeling

Consider the reaction network in figure 3.16. Formulate an S-system model of the network using reaction rates $v_1 = \alpha_1[S_1]^{g_1}$, $v_2 = \alpha_2$, $v_3 = \alpha_3[S_1]^{g_2}[S_2]^{g_3}$, $v_4 = \alpha_4[S_3]^{g_4}$. Derive explicit formulas for the steady-state concentrations in terms of the model parameters.

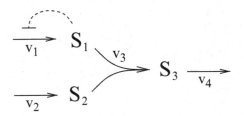

Figure 3.16
Reaction network for problem 3.7.13. The blunted dashed arrow indicates that species S_1 allosterically inhibits its own production.

4 Analysis of Dynamic Mathematical Models

A system is anything that talks to itself. All living systems and organisms ultimately reduce to a bunch of regulators — chemical pathways and neuron circuits — having conversations as dumb as "I want, I want, I want; no, you can't, you can't, you can't."
— Kevin Kelly, *Out of Control*

In the preceding chapters, we made the implicit assumption that, in the long term, the concentrations of species in a chemical reaction network will settle to a (unique) steady-state profile, regardless of the initial conditions. This is almost always a safe assumption in dealing with closed chemical reaction networks. However, when considering open networks, more interesting behaviors can occur. In this chapter, we introduce mathematical techniques that can be used to explore these dynamic behaviors.

4.1 Phase Plane Analysis

In chapter 2, we represented the dynamic behavior of reaction networks by plotting the concentrations of the reactant species as functions of time (in analogy to experimental time courses). An alternative approach to visualization is to plot concentrations against *one another*.

To provide a concrete example of this technique, consider the biochemical network shown in figure 4.1, which involves two species, S_1 and S_2. To keep the analysis simple, we suppose that all reaction rates follow mass action (or, equivalently, Michaelis–Menten kinetics with all enzymes operating in their first-order regime). The allosteric inhibition of v_1 will be modeled by presuming strong cooperative binding of n molecules of S_2. We can then write

$$v_1 = \frac{k_1}{1 + (s_2 / K)^n}, \qquad v_2 = k_2, \qquad v_3 = k_3 s_1, \qquad v_4 = k_4 s_2, \qquad v_5 = k_5 s_1,$$

where $s_1 = [S_1]$ and $s_2 = [S_2]$, so that

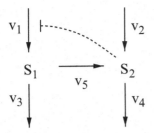

Figure 4.1
Biochemical reaction network. The production of S_1 is allosterically inhibited by S_2 (blunted dashed arrow). The labels v_i indicate the rates of the corresponding reactions. (They are not mass-action rate constants.)

$$\frac{d}{dt}s_1(t) = \frac{k_1}{1+(s_2(t)/K)^n} - k_3 s_1(t) - k_5 s_1(t)$$

$$\frac{d}{dt}s_2(t) = k_2 + k_5 s_1(t) - k_4 s_2(t). \tag{4.1}$$

Figure 4.2A shows a simulation of the system starting at initial concentrations of zero. In figure 4.2B, the same simulation is displayed by plotting the concentration s_2 against the concentration s_1 in what is called the system's **phase plane** (i.e., the s_1–s_2 plane). The phase plane plot (also called a *phase portrait*) shows the concentrations starting at the initial state $(s_1, s_2) = (0, 0)$ and converging to the steady state. This curve is called a *trajectory*. Comparing with figure 4.2A, the phase plot emphasizes the time-varying relationship between the two variables but de-emphasizes the relationship with time itself. Indeed, the direction of motion is not explicitly indicated by the curve, and although each point $(s_1(t), s_2(t))$ corresponds to a particular time instant t, the only time points that can easily be identified are at $t = 0$ (where the curve starts) and the long-time behavior ($t \to \infty$, where the curve ends).

The phase portrait allows multiple time-courses (trajectories) to be usefully described in a single plot. This is illustrated in figure 4.3A, which shows the time course in figure 4.2A along with a number of other time courses, each starting from a different initial condition. All of the trajectories reach the same steady state, but the transient behavior is a meaningless jumble. These same simulations are shown as trajectories on a phase plane in figure 4.3B. Here, the overall system behavior is clear: the trajectories follow a slow spiral as they approach the steady state.

Because phase portraits are two-dimensional, they cannot capture system behavior when more than two species are involved in the network. However, we will find that insights gained from application of phase plane analysis to these low-dimensional systems will be directly applicable to larger and more complex models.

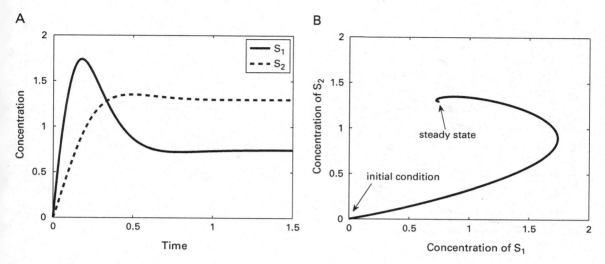

Figure 4.2
Simulation of model (4.1). (A) Concentrations plotted against time. Both $[S_1]$ and $[S_2]$ overshoot their steady-state values before coming to rest. (B) Concentration $[S_1]$ plotted against concentration $[S_2]$ in the phase plane. Parameter values: (in concentration · time^{-1}) $k_1 = 20$ and $k_2 = 5$; (in concentration) $K = 1$; (in time^{-1}) $k_3 = 5$, $k_4 = 5$, and $k_5 = 2$; and $n = 4$. Units are arbitrary.

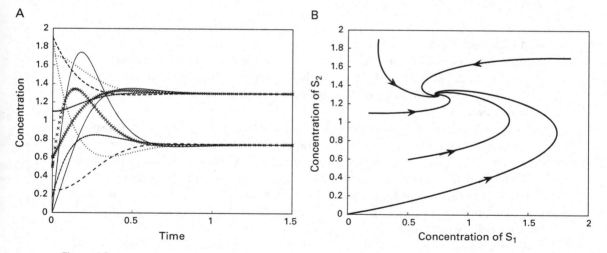

Figure 4.3
Simulations of model (4.1). (A) Multiple time courses confirm the steady-state concentrations, but the transient behavior cannot be usefully resolved. (B) On the phase plane, the individual trajectories provide a unified picture of the dynamic behavior of the system. Parameters are as for figure 4.2.

A

B

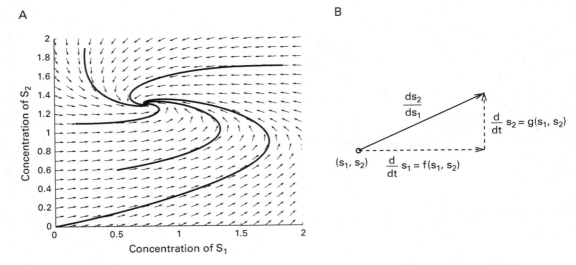

Figure 4.4
Direction field for model (4.1). (A) The field of arrows indicates the direction of motion at each point.
Trajectories are curves that lie tangent to the arrows; they follow the flow. (B) Arrows in the direction
field can be generated directly from the model—no simulation is needed. At each point (s_1, s_2), the direction of the arrow is the slope of s_2 with respect to s_1 (ds_2/ds_1). This slope can be determined from the
model dynamics, which specify the rates of change of s_2 and s_1. In (A), these vectors have been normalized to display a field of arrows of equal length.

4.1.1 Direction Fields

A phase portrait can become crowded as more and more trajectories are added. An
alternative to drawing all of these curves is to use short arrows to indicate the direction and speed of motion at each point on the phase plane. The resulting plot is
called a **direction field**.

Figure 4.4A shows the phase portrait from figure 4.3B along with the corresponding direction field. The trajectories lie parallel to (i.e., tangent to) the vector field at
each point. Additional trajectories can be sketched by simply following the arrows.
An analogy can be made to fluid flow, as follows. Imagine a two-dimensional flow
(on, for instance, a water table). The vector field describes, at each point, the direction of motion of a particle suspended in the fluid. The trajectories are the paths
such particles would traverse as they follow the flow.

A direction field is, in a sense, easier to construct than a phase portrait. To plot
trajectories in the phase portrait, simulations of the differential equation model
must be carried out. In contrast, the direction field can be determined directly from
the differential equation model, as follows. For a general system involving two
species concentrations s_1 and s_2,

$$\frac{d}{dt} s_1(t) = f(s_1(t), s_2(t))$$

$$\frac{d}{dt} s_2(t) = g(s_1(t), s_2(t)),$$

the motion in the phase plane at any given point (s_1, s_2) is given by the vector $(f(s_1, s_2), g(s_1, s_2))$, as indicated in figure 4.4B. The direction field can be constructed by selecting a mesh of points (s_1, s_2) in the phase plane and, at each point, drawing an arrow in the appropriate direction.

Exercise 4.1.1   Consider the system

$$\frac{d}{dt} x(t) = -y(t) \qquad\qquad \frac{d}{dt} y(t) = x(t).$$

Sketch the direction field by drawing the direction vectors at the following points in the x–y phase plane: $(1, 0)$, $(1, 1)$, $(0, 1)$, $(-1, 1)$, $(-1, 0)$, $(-1, -1)$, $(0, -1)$, $(1, -1)$. (Place each vector as an arrow with its tail at the corresponding point in the phase plane.) Can you infer the overall behavior of the system around its steady state at $(0, 0)$? You may want to draw a few more arrows to confirm your conjecture.   □

Exercise 4.1.2   Explain why trajectories in the phase portrait cannot cross one another. Hint: Consider the direction of motion at the intersection point.   □

4.1.2   Nullclines

A key feature of a system's phase portrait are points at which the trajectories "turn around"—that is, points at which trajectories change their direction with respect to one of the axes. These are the points at which one of the two variables $s_1(t)$ or $s_2(t)$ reaches a local maximum or local minimum.

On the phase plane, these turning points occur whenever a trajectory is directed either vertically (the direction arrow points straight up or down) or horizontally (the arrow points directly left or right). Rather than identify these points by examining the phase portrait, we can determine them directly from the model (because the direction of motion is specified by the model, as in figure 4.4B). These "turning points" constitute the system *nullclines*:

The set of points (s_1, s_2) where $d/dt(s_1(t)) = f(s_1, s_2) = 0$ is called the s_1-**nullcline**. Likewise, the set of points where $d/dt(s_2(t)) = g(s_1, s_2) = 0$ is called the s_2-**nullcline**.

Referring to figure 4.4B, we confirm that points on the s_1-nullcline have direction arrows with no horizontal component (so are oriented vertically), whereas points

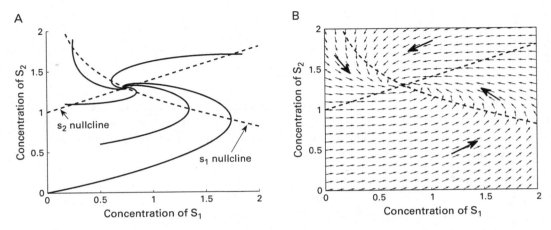

Figure 4.5
Nullclines for model (4.1). (A) The trajectories intersect the nullclines at turning points. The steady state occurs at the intersection of the nullclines. (B) The nullclines divide the phase plane into four regions. Because the direction arrows only "flip" directions at the nullclines, each region is characterized by motion in a particular direction (up/down, left/right), as indicated.

on the s_2-nullcline have direction arrows with no vertical component (so are oriented horizontally).

Figure 4.5A shows the phase portrait from figure 4.3B along with the nullclines. The trajectories cross the s_1-nullcline when they are oriented vertically and cross the s_2-nullcline when they are oriented horizontally. The nullclines intersect at the steady state, because at that point $ds_1/dt = f(s_1, s_2) = 0$ and $ds_2/dt = g(s_1, s_2) = 0$.

The nullclines are shown together with the direction field in figure 4.5B. The nullclines separate the phase plane into four regions; in each region, the direction arrows all have the same up-or-down and left-or-right orientation (as the arrows change these directions only when a nullcline is crossed). Thus, as shown in the figure, a rough picture of system behavior can be generated by specifying the direction of motion in each of the regions.

The nullclines, like the direction field, can be determined directly from the model—without running simulations. However, the equations $f(s_1, s_2) = 0$ and $g(s_1, s_2) = 0$ are typically nonlinear and so may not be solvable except via computational software.

Exercise 4.1.3 For model (4.1), the nullclines can be determined analytically. Verify that the s_1-nullcline is given by

$$s_1 = \frac{k_1}{(k_3 + k_5)(1 + (s_2 / K)^n)}$$

whereas the s_2-nullcline is the line

$$s_2 = \frac{k_2 + k_5 s_1}{k_4}.$$ □

4.2 Stability

The long-time (i.e., asymptotic) behavior of biochemical and genetic networks will be either

- convergence to a steady state; or
- convergence to a sustained periodic oscillation, referred to as limit-cycle oscillation.

Other dynamic behaviors (divergence and chaos, for example) do not often occur in systems biology models.

For the network studied in the previous section (model 4.1), we saw that all trajectories converge to a unique steady state. To explore an alternative asymptotic behavior, we next consider the network in figure 4.6. This reaction scheme is symmetric—each species allosterically inhibits production of the other, resulting in a mutual antagonism.

With cooperative inhibition and first-order consumption rates, the model is

$$\frac{d}{dt}s_1(t) = \frac{k_1}{1 + (s_2(t)/K_2)^{n_1}} - k_3 s_1(t)$$

$$\frac{d}{dt}s_2(t) = \frac{k_2}{1 + (s_1(t)/K_1)^{n_2}} - k_4 s_2(t). \tag{4.2}$$

We first consider an asymmetric model parameterization in which $n_1 > n_2$. In this case, the inhibition by S_2 is more effective than the inhibition by S_1. If the other

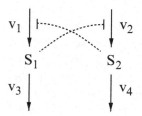

Figure 4.6
Symmetric biochemical network. Each species allosterically inhibits production of the other (blunted dashed arrows).

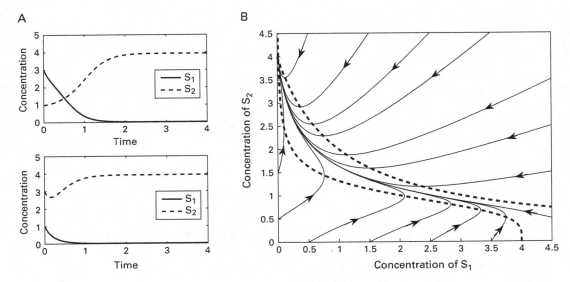

Figure 4.7
Model (4.2) with imbalanced inhibition strength. (A) Time-series plots show that regardless of the initial condition, the system settles to a steady state with high S_2 concentration and low S_1 concentration. (B) This phase portrait confirms that all trajectories approach the high-$[S_2]$, low-$[S_1]$ steady state at which the nullclines (dashed lines) intersect. Parameter values: $k_1 = k_2 = 20$ (concentration $\cdot$ time^{-1}), $K_1 = K_2 = 1$ (concentration), $k_3 = k_4 = 5$ (time^{-1}), $n_1 = 4$, and $n_2 = 1$. Units are arbitrary.

parameters are symmetric ($k_1 = k_2, K_1 = K_2, k_3 = k_4$), we should expect the model to exhibit a steady state in which the concentration of S_1 is low and the concentration of S_2 is high (the mutual antagonism "competition" will be won by S_2). This intuition is confirmed by Figure 4.7A, which shows two time courses starting from different initial conditions. Regardless of whether S_1 or S_2 is initially more abundant, the imbalance in inhibition strength leads to the same steady state (low $[S_1]$, high $[S_2]$). The phase portrait in figure 4.7B confirms this finding. All trajectories converge to the steady state at the intersection of the s_1- and s_2-nullclines, at which S_2 dominates.

We next consider the symmetric case. Changing the Hill coefficients so that $n_1 = n_2$, the two species S_1 and S_2 are perfectly balanced. Because neither species has an advantage over the other, we should expect the system to exhibit symmetric behavior: the two concentrations might converge to the same value (a "tie game") or one species will maintain dominance over the other (and emerge as the "winner"). The system's steady-state behaviors are illustrated in figure 4.8. Panel A shows that the long-time behavior depends on the initial conditions—whichever species is initially more abundant maintains its dominance. The phase portrait in panel B shows a symmetric phase plane. Trajectories are attracted to whichever steady state is closer.

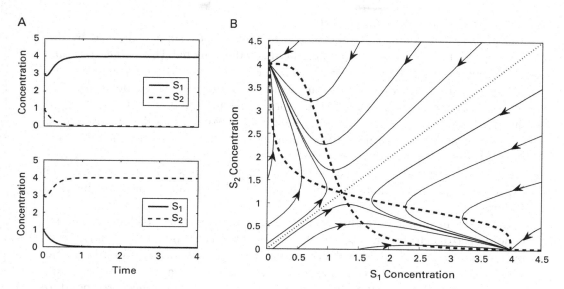

Figure 4.8
Model (4.2) with balanced inhibition strength. (A) Time-series plots show that the steady-state behavior depends on the initial conditions. Either species can dominate over the other if its initial concentration is larger. (B) The phase portrait confirms the presence of two steady states at which the nullclines (dashed lines) intersect. Each trajectory converges to the closer steady state. The two basins of attraction are separated by the diagonal (dotted line). Parameter values: $k_1 = k_2 = 20$ (concentration · time^{-1}), $K_1 = K_2 = 1$ (concentration), $k_3 = k_4 = 5$ (time^{-1}), $n_1 = n_2 = 4$. Units are arbitrary.

The region of the phase plane from which trajectories converge to each steady state is called the *basin of attraction* of that steady state. Curves that separate basins of attraction are called *separatrices*. In this perfectly symmetric case, the separatrix is the diagonal.

A system that exhibits two distinct steady states is called **bistable**. (In contrast, a system with a single steady state is called *monostable*.) Bistability provides a system with a type of memory—the system's long-term behavior reflects its past condition. Biological implications of bistability will be discussed in later chapters.

There are two essential ingredients to bistability: positive feedback and nonlinearity. In the model considered here, the positive feedback is implemented in a double-negative feedback loop: each species inhibits production of the other and thus inhibits the inhibition of itself. Thus, each species acts to enhance its own production —a positive feedback. (Double-negative feedback is sometimes called *derepression*.) Nonlinearity is provided by the cooperative inhibition mechanism.

These two ingredients are necessary for bistability, but they do not guarantee it: the model structure and parameter values must also be properly aligned.

4.2.1 Stable and Unstable Steady States

The nullclines in figure 4.8B intersect at the two steady states exhibited by the system, and they also intersect at a third point, on the diagonal. This symmetric point is a steady state for the system, but it will not be observed as a long-time state of the system. The behavior of the trajectories near this symmetric steady state is shown more clearly in figure 4.9. Trajectories near this state are repelled from it: they tend toward one of the other two steady states. We say that this steady state is **unstable** because nearby trajectories diverge away from it. In contrast, the other two steady states in figure 4.8B—which attract nearby trajectories—are called **stable**. This system thus has two stable steady states and one unstable steady state. The existence of an intermediate unstable steady state is a defining feature of bistable systems.

In theory, the unstable steady state can be maintained by perfectly balanced initial conditions (trajectories balanced on the diagonal will converge to this point). However, any deviation from this balance causes the trajectory to tend toward a stable steady state. This behavior is illustrated by a standard metaphor for stability: a ball rolling on an undulating slope (figure 4.10), in which valley bottoms correspond to stable steady states and hilltops correspond to unstable steady states. A ball within a valley is attracted to the valley bottom and settles to rest. A ball balanced on a

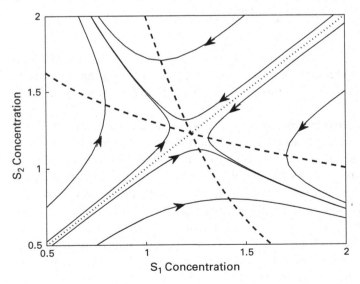

Figure 4.9
Model (4.2) with balanced inhibition strength: unstable steady state. Figure 4.8B shows the nullclines intersecting three times. This close-up of the middle intersection shows that trajectories are repelled from this unstable steady state. The dashed lines are the nullclines. The dotted line (on the diagonal) is the separatrix that forms the boundary between the two basins of attraction.

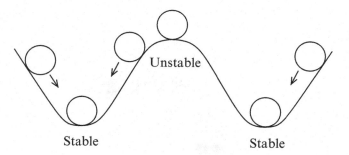

Figure 4.10
Stability and instability. In this analogy of a ball rolling on an undulating slope, the valley bottoms correspond to stable steady states: balls in each valley are attracted to the valley bottom, where they settle to rest. The hilltops correspond to unstable states: a ball perfectly balanced on a hilltop will stay there, but any deviation will topple the ball toward one of the valley bottoms. The slope shown here corresponds to a bistable system, with an unstable steady state separating the basins of attractions (the valleys) of the two stable steady states.

hilltop will theoretically remain in that position, but the slightest nudge in any direction will send it rolling toward a valley bottom.

By extending this analogy to three dimensions (figure 4.11), we can illustrate the stability behaviors of the phase plots in figures 4.7B and 4.8B. The single valley in figure 4.11A corresponds to a monostable system. In figure 4.11B, the unstable point attracts trajectories that are perfectly balanced on the ridge, but any imbalance causes trajectories to fall to one of the valley bottoms. Because of the shape of the surface, this unstable point is called a *saddle*. The valleys are the two basins of attraction; the ridge is the separatrix between them.

Exercise 4.2.1 Consider the simple reaction chain

$$\xrightarrow{v_1} S \xrightarrow{v_2} .$$

Let $s = [S]$. If $v_1 = k_1$ (constant) and $v_2 = k_2 s$, then there is no feedback, and the system will be monostable, with steady state $s = k_1/k_2$.

(a) Alternatively, if $v_1 = k_1 s$, then S enhances its own production in a positive feedback. Take the consumption rate to be nonlinear: $v_2 = k_2 s^2$. Verify that in this case, the system exhibits two steady states: one with $s = 0$ and one with $s = k_1/k_2$. In this one-dimensional case, the stability of these steady states can be determined by evaluating the rate of change of s near each point. For instance, when s is near zero, $s^2 \ll s$, so $k_1 s^2$ will be small compared to $k_2 s$. The rate of change $d/dt(s(t))$ will then be positive, so $s(t)$ increases away from the steady state at $s = 0$; this steady state is unstable. Verify that the steady state $s = k_1/k_2$ is stable by determining the sign of $d/dt(s(t))$ for s-values above and below k_1/k_2.

A   B

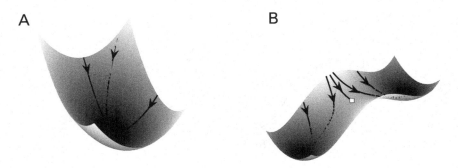

Figure 4.11
Monostability and bistability. (A) For a monostable system, there is a single valley bottom, representing a unique steady state to which all trajectories converge. (B) A bistable system corresponds to a pair of valleys, separated by a ridge. The low point of the ridge (white box) is the unstable steady state (a saddle point). Most trajectories settle to one of the valley bottoms, but trajectories that remain perfectly balanced on the ridge will settle to the unstable saddle point.

(b) Next, consider the case in which $v_1 = k_0 + k_1 s^2 / (k_2 + s^2)$ and $v_2 = k_3 s$; then the system exhibits positive feedback and significant nonlinearity. In this case, the system is bistable for appropriate values of the parameters. Take $k_0 = 6/11$, $k_1 = 60/11$, $k_2 = 11$, and $k_3 = 1$ and verify that $s = 1$, $s = 2$, and $s = 3$ are all steady states. Evaluate the rate of change $d/dt\,(s(t))$ around these points to verify that $s = 1$ and $s = 3$ are stable, whereas $s = 2$ is unstable. □

So far, we have determined stability from system phase portraits. We next introduce a technique for stability analysis that does not rely on graphical representations and is not restricted to two-species networks.

4.2.2 Linearized Stability Analysis

The behavior of trajectories near a steady state is called *local* behavior. As we shall see in this section, the local behavior of any nonlinear system can be approximated by a linear system. This approximation, called the *linearization*, can be used to test for stability of steady states.

Linearization   As reviewed in appendix B, we can approximate any function $f(s)$ near a particular point $s = \bar{s}$ by the tangent line centered at $\bar{s}$, as illustrated in figure 4.12:

$$f(s) \approx f(\bar{s}) + \frac{df}{ds}(\bar{s}) \cdot (s - \bar{s}).$$

The tangent line is called the *linearization* (or *linear approximation*) of $f(s)$ at $s = \bar{s}$.

Exercise 4.2.2   Consider the Michaelis–Menten rate law

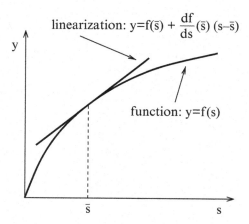

Figure 4.12
Linear approximation of a function of a single variable. The tangent line centered at $s = \bar{s}$ provides a good approximation to the function $f(s)$ when the argument s is near $\bar{s}$.

$$f(s) = \frac{V_{\max} s}{K_M + s}.$$

(a) Determine the linear approximation of $f(s)$ at an arbitrary point $\bar{s}$.

(b) Verify that the linear approximation centered at $\bar{s} = 0$ has the form of a (first-order) mass-action rate law.

(c) Verify that when $\bar{s}$ is large compared to K_M (so that $K_M + \bar{s} \approx \bar{s}$), the linearization is almost horizontal (i.e., it approximates a zero-order rate law). □

Linear approximations can also be constructed for functions of more than one variable. For a function of two variables, $f(s_1, s_2)$, the linearization centered at a point $(s_1, s_2) = (\bar{s}_1, \bar{s}_2)$ is

$$f(s_1, s_2) \approx f(\bar{s}_1, \bar{s}_2) + \frac{\partial f}{\partial s_1}(\bar{s}_1, \bar{s}_2) \cdot (s_1 - \bar{s}_1) + \frac{\partial f}{\partial s_2}(\bar{s}_1, \bar{s}_2) \cdot (s_2 - \bar{s}_2). \tag{4.3}$$

(Readers unfamiliar with partial derivatives may wish to consult appendix B.) This approximation is valid for arguments (s_1, s_2) near the point $(\bar{s}_1, \bar{s}_2)$. This linearization corresponds to a tangent plane approximating the surface $z = f(s_1, s_2)$, as in figure 4.13.

Exercise 4.2.3 Consider the rate law for a competitively inhibited enzyme:

$$f(s, i) = \frac{V_{\max} s}{K_M (1 + i / K_i) + s}.$$

Determine the linear approximation centered at $(s, i) = (1, 0)$. □

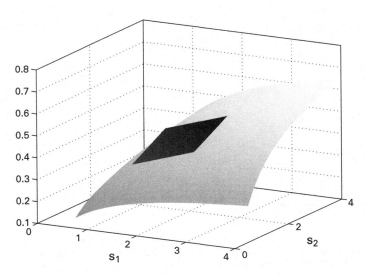

Figure 4.13
Linear approximation of a function of two variables. The tangent plane (black) centered at $(s_1, s_2) = (\overline{s}_1, \overline{s}_2)$ provides a good approximation to the function $f(s_1, s_2)$ (gray surface) when the argument (s_1, s_2) is near $(\overline{s}_1, \overline{s}_2)$.

Now, consider the general two-species system introduced in section 4.1.1:

$$\frac{d}{dt} s_1(t) = f(s_1(t), s_2(t))$$

$$\frac{d}{dt} s_2(t) = g(s_1(t), s_2(t)).$$

We will construct linear approximations to $f(s_1, s_2)$ and $g(s_1, s_2)$. By centering these linearizations at a steady state $(\overline{s}_1, \overline{s}_2)$, we have (because $f(\overline{s}_1, \overline{s}_2) = 0$ and $g(\overline{s}_1, \overline{s}_2) = 0$)

$$\frac{d}{dt} s_1(t) = f(s_1(t), s_2(t)) \approx \frac{\partial f}{\partial s_1}(\overline{s}_1, \overline{s}_2) \cdot (s_1(t) - \overline{s}_1) + \frac{\partial f}{\partial s_2}(\overline{s}_1, \overline{s}_2) \cdot (s_2(t) - \overline{s}_2)$$

$$\tag{4.4}$$

$$\frac{d}{dt} s_2(t) = g(s_1(t), s_2(t)) \approx \frac{\partial g}{\partial s_1}(\overline{s}_1, \overline{s}_2) \cdot (s_1(t) - \overline{s}_1) + \frac{\partial g}{\partial s_2}(\overline{s}_1, \overline{s}_2) \cdot (s_2(t) - \overline{s}_2).$$

Because we will be applying this linearization to address stability, we are particularly interested in the behavior of trajectories whose initial conditions are near a steady state. If the steady state is stable, these trajectories will converge to the steady state; if it is unstable, they will diverge. We can think of these trajectories as small displacements from the steady state that either shrink or grow with time. To focus on these displacements, we introduce a change of variables that describes these deviations explicitly:

$$x_1(t) = s_1(t) - \bar{s}_1 \qquad x_2(t) = s_2(t) - \bar{s}_2.$$

A small displacement from the steady state $(\bar{s}_1, \bar{s}_2)$ thus corresponds to a value of (x_1, x_2) near $(0, 0)$. The dynamics of these displacement variables is easily determined. Observing that

$$\frac{d}{dt}x_1(t) = \frac{d}{dt}s_1(t) + 0 \qquad \text{and} \qquad \frac{d}{dt}x_2(t) = \frac{d}{dt}s_2(t) + 0,$$

we have the approximate dynamics (from equation 4.4):

$$\frac{d}{dt}x_1(t) = \frac{\partial f}{\partial s_1}(\bar{s}_1, \bar{s}_2) \cdot x_1(t) + \frac{\partial f}{\partial s_2}(\bar{s}_1, \bar{s}_2) \cdot x_2(t)$$

$$\frac{d}{dt}x_2(t) = \frac{\partial g}{\partial s_1}(\bar{s}_1, \bar{s}_2) \cdot x_1(t) + \frac{\partial g}{\partial s_2}(\bar{s}_1, \bar{s}_2) \cdot x_2(t),$$

which is valid for (x_1, x_2) near $(0, 0)$. This set of differential equations forms a **linear system**. The steady state $(x_1, x_2) = (0, 0)$ of this linear system corresponds to the steady state $(\bar{s}_1, \bar{s}_2)$ of the original nonlinear system. We next present a procedure for testing stability for this linear system: this will determine stability for the nonlinear system as well.

Stability Analysis for Linear Systems   The general form for a two-state linear system is

$$\frac{d}{dt}x_1(t) = ax_1(t) + bx_2(t)$$

$$\frac{d}{dt}x_2(t) = cx_1(t) + dx_2(t). \tag{4.5}$$

This system can be solved explicitly. Solutions take the form

$$x_1(t) = c_{11}e^{\lambda_1 t} + c_{12}e^{\lambda_2 t}$$

$$x_2(t) = c_{21}e^{\lambda_1 t} + c_{22}e^{\lambda_2 t}. \tag{4.6}$$

The constants c_{ij} depend on the initial conditions, but the values of λ_1 and λ_2 are inherent to the system. They are the *eigenvalues* of the system's **Jacobian**, which is constructed from system (4.5) as the matrix

$$\text{Jacobian: } \mathbf{J} = \begin{bmatrix} a & b \\ c & d \end{bmatrix}.$$

The eigenvalues of this matrix are the roots of the quadratic equation

$$\lambda^2 - (a + d)\lambda + (ad - bc) = 0.$$

Applying the quadratic formula gives

$$\lambda_1 = \frac{(a+d)+\sqrt{(a+d)^2 - 4(ad-bc)}}{2}, \qquad \lambda_2 = \frac{(a+d)-\sqrt{(a+d)^2 - 4(ad-bc)}}{2}. \qquad (4.7)$$

Depending on the sign of the discriminant $(a + d)^2 - 4(ad - bc)$, these eigenvalues may be real-valued or complex-valued.[1]

Exercise 4.2.4 Verify that if either of the off-diagonal terms b or c is zero, then the eigenvalues of the Jacobian are the diagonal entries of the matrix: $\lambda_1 = a$ and $\lambda_2 = d$. ☐

Exercise 4.2.5 Consider the system

$$\frac{d}{dt}x_1(t) = -\frac{5}{3}x_1(t) + \frac{1}{3}x_2(t)$$

$$\frac{d}{dt}x_2(t) = \frac{2}{3}x_1(t) - \frac{4}{3}x_2(t). \qquad (4.8)$$

Find the eigenvalues of the system's Jacobian matrix

$$J = \begin{bmatrix} -5/3 & 1/3 \\ 2/3 & -4/3 \end{bmatrix}.$$

Next, use formula (4.6) to determine the solution of system (4.8) that satisfies initial conditions $x_1(0) = 1/3$, $x_2(0) = 5/3$. Hint: The initial conditions provide two constraints on the unknowns c_{ij}. Two more constraints can be determined by evaluating the time-derivative of the solutions (4.6) at $t = 0$ and substituting those derivatives and the initial conditions into system (4.8). ☐

The general behavior of the solutions (4.6) depends on the nature of the exponential functions $e^{\lambda_1 t}$ and $e^{\lambda_2 t}$. To classify the behavior of these functions, we consider two cases.

Case I The discriminant is nonnegative, so that λ_1 and λ_2 are real numbers. We note that:

---

1. Complex numbers (e.g., $3 + 4i$) involve the square root of -1, denoted i. The generic complex number $x + yi$ has x as its *real part* and y as its *imaginary part*. For the complex number $3 + 4i$, the real part is 3 and the imaginary part is 4 (not $4i$).

(i) If both eigenvalues are negative (e.g., $\lambda_1 = -3$, $\lambda_2 = -1$), then both solutions $x_1(t)$ and $x_2(t)$ tend to zero (regardless of the values of the constants c_{ij}). Because (x_1, x_2) is the displacement of (s_1, s_2) from $(\bar{s}_1, \bar{s}_2)$, the steady state $(\bar{s}_1, \bar{s}_2)$ is stable. In this case, the steady state is called a *stable node*.

(ii) If *either* eigenvalue is positive (e.g., $\lambda_1 = -3$, $\lambda_2 = 1$), then most solutions diverge (because one of the exponentials grows indefinitely). Thus, the displacement of (s_1, s_2) from $(\bar{s}_1, \bar{s}_2)$ grows; the steady state is unstable. If both eigenvalues are positive, then all trajectories diverge, and the steady state is called an *unstable node*. If one eigenvalue is negative and the other is positive, then the steady state is called a *saddle point*. (The unstable steady state in figure 4.8 is a saddle point. The negative eigenvalue causes perfectly balanced trajectories to approach the saddle point along the diagonal, as in figure 4.11B.)

Case II The discriminant $(a + d)^2 - 4(ad - bc)$ is negative, so the eigenvalues are complex-valued. In this case, referring back to equation (4.7), let

$$\alpha = \frac{a+d}{2} \quad \text{and} \quad \beta = \frac{\sqrt{-((a+d)^2 - 4(ad-bc))}}{2}.$$

We can then write the eigenvalues as[2]

$$\lambda_1 = \alpha + \beta i \text{ and } \lambda_2 = \alpha - \beta i.$$

Because the solutions (4.6) involve the terms $e^{\lambda_1 t}$ and $e^{\lambda_2 t}$, we will need to evaluate the exponential of complex numbers. *Euler's formula* states that

$$e^{\lambda_1 t} = e^{(\alpha+\beta i)t} = e^{\alpha t}(\cos(\beta t) + i \sin(\beta t))$$

$$e^{\lambda_2 t} = e^{(\alpha-\beta i)t} = e^{\alpha t}(\cos(-\beta t) + i \sin(-\beta t)) = e^{\alpha t}(\cos(\beta t) - i \sin(\beta t)).$$

Substituting these expressions into the solution formulas (4.6), we find

$$x_1(t) = c_{11}e^{\alpha t}(\cos(\beta t) + i \sin(\beta t)) + c_{12}e^{\alpha t}(\cos(\beta t) - i \sin(\beta t))$$

$$x_2(t) = c_{21}e^{\alpha t}(\cos(\beta t) + i \sin(\beta t)) + c_{22}e^{\alpha t}(\cos(\beta t) - i \sin(\beta t)).$$

(4.9)

These expressions may raise some eyebrows. If $x_1(t)$ and $x_2(t)$ describe the displacement of species concentrations from steady state, how can they possibly take complex values? This issue is resolved by a special property of these formulas: if the initial displacements $x_1(0)$ and $x_2(0)$ are specified as real numbers, then the corresponding constants c_{ij} guarantee that the formulas in (4.9) evaluate to real numbers.

---

2. For a negative number y, we have $\sqrt{y} = (\sqrt{-y})(\sqrt{-1}) = (\sqrt{-y})i$.

(The imaginary parts of these expressions will cancel to zero. See problem 4.8.2 for an example.)

Having resolved the unsettling appearance of the solution formulas (4.9), we next consider their behavior. In each formula, the exponential e^{α} is multiplied by terms involving cosines and sines. These sines and cosines contribute an oscillatory component to the trajectories, but they have no influence over whether the solutions diverge or converge. The long-term behavior is determined solely by the exponential term e^{α}. We note that:

(i) If α, the real part of the eigenvalues, is negative, then the solutions converge to zero. In this case, the steady state is stable. It is called a *stable spiral point*, or *focus*. Solutions will exhibit damped oscillations as they converge.

(ii) If α, the real part of the eigenvalues, is positive, then the solutions diverge. The steady state is unstable and is called an *unstable spiral point*.

Fortunately, the conclusions about stability in the real-valued case and the complex-valued case are consistent. Because the real part of a real number is simply the number itself, we arrive at the following general statement.

Linearized Stability Criterion

(i) If both eigenvalues of the Jacobian have negative real part, then the steady state is *stable*.

(ii) If *either* eigenvalue has positive real part, then the steady state is *unstable*.

We have not addressed the case of eigenvalues with zero real part. This occurs only for systems that exhibit certain symmetries and is rarely encountered in models of biochemical and genetic networks. (When both eigenvalues have zero real part, the trajectories are periodic: they follow circular arcs around the steady state, which is then called a *center*.)

We derived the linearized stability criterion for systems involving two species. It can be shown that this eigenvalue-based criterion applies to systems of any size. (In particular, a steady state s^{ss} of a one-dimensional system $d/dt\,(x(t)) = f(x(t))$ is stable if the Jacobian, which is simply $d/dx\,(f(x))$, is negative at x^{ss}. This Jacobian is a single number, which is also the eigenvalue.)

In summary, to apply the linearized stability criterion to a nonlinear model:

1. Identify a steady state of interest.

2. Construct the system Jacobian at that point (by taking the appropriate partial derivatives).

3. Evaluate the eigenvalues of the Jacobian.

4. Check the sign of their real parts.

To illustrate, consider model (4.2), with symmetric parameter values as in figure 4.8, except $n_1 = n_2 = 2$:

$$\frac{d}{dt}s_1(t) = \frac{20}{1 + s_2^2(t)} - 5s_1(t) \qquad\qquad \frac{d}{dt}s_2(t) = \frac{20}{1 + s_1^2(t)} - 5s_2(t).$$

Defining

$$f(s_1, s_2) = \frac{20}{1 + s_2^2} - 5s_1 \qquad \text{and} \qquad g(s_1, s_2) = \frac{20}{1 + s_1^2} - 5s_2,$$

we construct the system Jacobian as

$$\mathbf{J}(s_1, s_2) = \begin{bmatrix} \dfrac{\partial f}{\partial s_1} & \dfrac{\partial f}{\partial s_2} \\[2mm] \dfrac{\partial g}{\partial s_1} & \dfrac{\partial g}{\partial s_2} \end{bmatrix} = \begin{bmatrix} -5 & -\dfrac{20}{(1 + s_2^2)^2}2s_2 \\[3mm] -\dfrac{20}{(1 + s_1^2)^2}2s_1 & -5 \end{bmatrix}.$$

To determine the steady-state concentration profiles, we must solve the system of equations

$$\frac{20}{1 + s_2^2} - 5s_1 = 0 \qquad\qquad \frac{20}{1 + s_1^2} - 5s_2 = 0.$$

The three solutions (the system is bistable) can be found numerically as

$$(\bar{s}_1, \bar{s}_2) = (3.73, 0.268), \quad (1.38, 1.38), \quad (0.268, 3.73).$$

Evaluating the Jacobian at each of these three points and determining the corresponding eigenvalues, we have

$$\mathbf{J}(3.73, 0.268) = \begin{bmatrix} -5 & -9.33 \\ -0.671 & -5 \end{bmatrix}, \lambda_1 = -2.50, \lambda_2 = -7.50$$

$$\mathbf{J}(1.38, 1.38) = \begin{bmatrix} -5 & -6.54 \\ -6.54 & -5 \end{bmatrix}, \lambda_1 = -11.5, \lambda_2 = 1.55$$

$$\mathbf{J}(0.268, 3.73) = \begin{bmatrix} -5 & -0.671 \\ -9.33 & -5 \end{bmatrix}, \lambda_1 = -7.50, \lambda_2 = -2.50$$

We thus confirm that the balanced steady state (1.38, 1.38) is a saddle point, whereas the other two steady states are stable nodes.

Exercise 4.2.6 Perform a linearized stability analysis for the model (4.2) with unbalanced inhibition as specified by the parameter values in figure 4.7. The steady state is $(\bar{s}_1, \bar{s}_2) = (0.0166, 3.94)$. $\qquad\square$

Exercise 4.2.7   Solve for the steady state of the system

$$\frac{d}{dt}s_1(t) = V_0 - k_1 s_1(t)$$

$$\frac{d}{dt}s_2(t) = k_1 s_1(t) - \frac{V_2 s_2(t)}{K_M + s_2(t)}.$$

Assume that $V_2 > V_0$. Use linearized stability analysis to verify that this steady state is stable for all (nonnegative) values of the model parameters. $\square$

4.3   Limit-Cycle Oscillations

So far, our analysis of long-term behavior has been restricted to steady states. We next consider systems whose long-term behavior is sustained oscillation.

We make a distinction between *damped* oscillations, which display ever-decreasing amplitude and converge eventually to a steady state, and *persistent* (or *sustained*) oscillations, which are periodic and continue indefinitely. These two behaviors look very different in the phase plane, but they can look similar in time series unless simulations are run for sufficiently long time periods.

As an example of a system that displays persistent oscillatory behavior, we consider the network in figure 4.14. In this scheme, species S_2 allosterically activates its own production. This sort of positive feedback, called *autocatalysis*, is common in biology. We model the network as

$$\frac{d}{dt}s_1(t) = k_0 - k_1(1 + (s_2(t)/K)^n)s_1(t)$$

$$\text{(4.10)}$$

$$\frac{d}{dt}s_2(t) = k_1(1 + (s_2(t)/K)^n)s_1(t) - k_2 s_2.$$

The allosteric activation is presumed to be strongly cooperative.

Intuition suggests that this system might be prone to oscillations: the positive feedback will cause a continual increase in the rate of S_2 production until the pool of S_1 is depleted. The S_2 concentration will then crash and stay low until more S_1 is available, at which point the cycle can repeat.

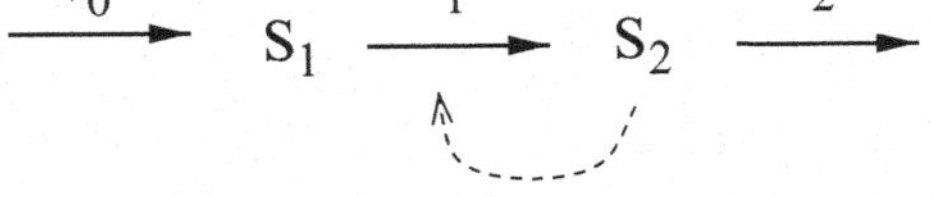

Figure 4.14
Autocatalytic biochemical reaction network. Species S_2 activates its own production (dashed arrow).

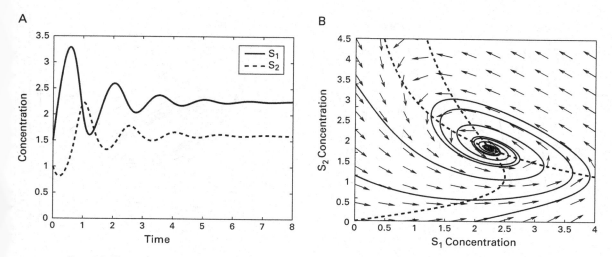

Figure 4.15
Model (4.10) with moderate nonlinearity. (A) Time series. The species concentrations exhibit damped oscillations as they converge to steady state. (B) Phase plane. Damped oscillations correspond to trajectories (solid curves) spiraling toward the steady state at the intersection of the nullclines (dashed curves). Parameter values: $k_0 = 8$ (concentration $\cdot$ time^{-1}), $k_1 = 1$ (time^{-1}), $K = 1$ (concentration), $k_2 = 5$ (time^{-1}), and $n = 2$. Units are arbitrary.

Figure 4.15 shows the system's behavior when the Hill coefficient has value $n = 2$. The model exhibits a single steady state—a stable spiral point. Damped oscillations of the species concentrations are evident in both the times series (panel A) and the phase portrait (panel B). These damped oscillations suggest that the system may be "close" to displaying persistent periodic behavior.

Next, consider the behavior when the Hill coefficient is raised to $n = 2.5$, shown in figure 4.16. In the times series (panel A), we see a short transient followed by sustained periodic behavior. The phase portrait (panel B) shows a cyclic track, called a **limit cycle**, to which all trajectories are attracted. Comparing with the phase portrait in figure 4.15B, the nullcline structure has not changed significantly; what has changed is the stability of the steady state. In figure 4.15B, the steady state is a stable spiral point. In contrast, figure 4.16B reveals an unstable spiral point in the center of the limit cycle. The close-up in figure 4.17 shows how trajectories are repelled from the unstable steady state and converge toward the limit cycle from the inside.

In the following chapters, we will see a variety of oscillatory phenomenon. Sustained limit-cycle behaviors are generated from two necessary ingredients: negative feedback and nonlinearity. Many biological oscillators can be classified as either *delay oscillators* or *relaxation oscillators*. The periodic behavior of delay oscillators is caused by a lag in a negative feedback loop, which causes repeated rounds of

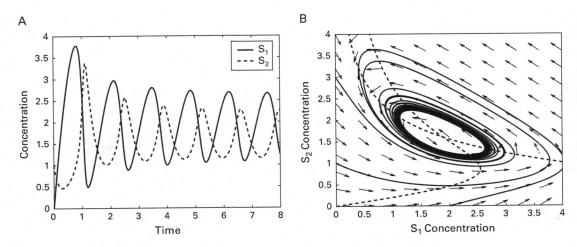

A

B

Figure 4.16
Model (4.10) with strong nonlinearity. (A) This time series shows convergence to sustained periodic behavior. (B) In this phase portrait, all trajectories converge to a cyclic track called a limit cycle. The steady state at the intersection of the nullclines is an unstable spiral point. Parameter values are as for figure 4.15 except $n = 2.5$.

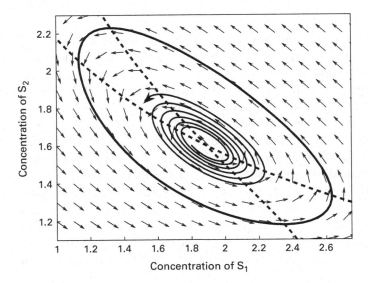

Figure 4.17
Model (4.10) with strong nonlinearity: unstable steady state. Trajectories that start near the unstable steady state spiral away from it, eventually converging to the limit cycle. Parameter values are as for figure 4.16.

accumulated activity. Relaxation oscillators exhibit an interplay of positive and negative feedback. Positive feedback causes these system to display near-bistable behavior; the negative feedback causes cyclic transfers between the two quasi-stable conditions. Relaxation oscillators exhibit behavior at different timescales, with slow negative feedback operating over fast positive feedback. This results in sudden switches from one state to the other, leading to sharp "pulse-like" oscillations. Delay oscillators, in contrast, tend to exhibit smoothly varying time courses.

Model (4.10) is best described as a relaxation oscillator. The allosteric activation introduces both positive feedback (increasing the rate of S_2 production) and negative feedback (depleting the pool of S_1, and so eventually stifling S_2 production). As the degree of cooperativity n is increased, the pulse-like nature of the oscillations becomes more pronounced, with spike-like rise-and-crash behaviors followed by longer intervals in the depleted state.

It is usually difficult to infer the existence of limit-cycle oscillations directly from model structure. In the special case of two-species models, a result called the Poincaré–Bendixson theorem can be used. This theorem states that if all trajectories are bounded (i.e., do not escape by diverging to infinity) and the system exhibits no stable steady states, then there must be a limit cycle. The intuition is that the trajectories have to settle somewhere: as they cannot diverge, or settle to a steady state, the only remaining option is convergence to a limit cycle.

Exercise 4.3.1 The Brusselator is a theoretical model of an oscillatory chemical reaction network (developed at the Free University of Brussels). The network is

$$\xrightarrow{\;k_1\;} X$$

$$X \xrightarrow{\;k_2\;} Y$$

$$2X + Y \xrightarrow{\;k_3\;} 3X$$

$$X \xrightarrow{\;k_4\;}$$

(a) Verify that the model is

$$\frac{d}{dt} x(t) = k_1 - k_2 x(t) + k_3 x(t)^2 y(t) - k_4 x(t)$$

$$\frac{d}{dt} y(t) = k_2 x(t) - k_3 x(t)^2 y(t).$$

(b) Find the steady state of the system.

(c) Take $k_2 = 2$ (time^{-1}), $k_3 = 1/2$ (time^{-1} · concentration^{-1}), and $k_4 = 1$ (time^{-1}). Use linearized stability analysis to verify that the steady state is unstable when $0 < k_1 < \sqrt{2}$.

It can be shown that trajectories do not diverge. The Poincaré–Bendixson theorem thus indicates that the system exhibits limit-cycle oscillations for these k_1 values.   □

In our discussions of bistability and limit-cycle oscillations, we saw that a model's qualitative behavior can change as parameter values vary. We will next introduce an analytic approach that provides deeper insight into these changes in model behavior.

4.4 Bifurcation Analysis

In most cases, the position of a model's steady state shifts if the model parameters are changed. For instance, figure 4.18 shows the steady-state concentration of S_1 in model (4.1) as the parameter k_1 varies. In an experimental context, this plot would be called a *dose-response curve*; when it is constructed from a model, it is called a *continuation diagram*.

As we have seen, variation in parameter values can cause qualitative changes in long-term system behavior (e.g., changes in the number of steady states or in their stability properties). Parameter values at which such changes occur are called *bifurcation points*; a continuation diagram on which bifurcation points appear is called a **bifurcation diagram**.

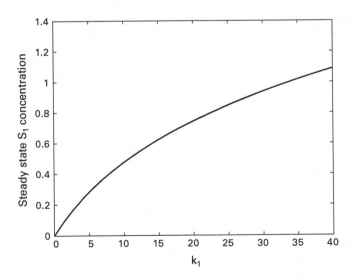

Figure 4.18
Continuation diagram. The steady state of species S_1, from model (4.1), is shown as a function of the value of parameter k_1. In an experimental context, this would be called a dose-response curve. Other parameter values are as for figure 4.2.

Exercise 4.4.1 Consider the differential equation

$$\frac{d}{dt}x(t) = (a-1)x(t).$$

By determining the sign of the rate of change dx/dt for positive and negative values of x, verify that the steady state at $x = 0$ is stable if $a < 1$ and unstable if $a > 1$. The parameter value $a = 1$ is thus a bifurcation point for this system. □

Figure 4.19 shows a bifurcation diagram for the symmetric reaction network modeled by (4.2). The phase plots in panel A show the nullclines at four different values of parameter k_1. As k_1 varies, the s_1-nullcline (gray curve) shifts, changing the number of points at which the nullclines intersect. The bifurcation diagram in panel B shows the steady-state behavior of $[S_1]$ as k_1 varies. The points corresponding to each subplot in panel A are marked. This S-shaped bifurcation curve is characteristic of bistable systems. The points where the bifurcations occur (at $k_1 = 16.1$ and $k_1 = 29.0$) are called *saddle-node* bifurcations (because they occur when an unstable

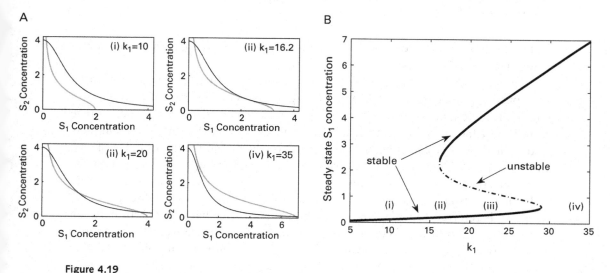

Figure 4.19
Bifurcation diagram for model (4.2). (A) Nullclines at various values of k_1. As k_1 increases, the s_1-nullcline (gray curve) shifts: (i) at low k_1 there is a single steady state (low $[S_1]$, high $[S_2]$); (ii) at a higher value of k_1, a new steady state appears when a new intersection appears; (iii) at still higher values, three intersection points are exhibited—the system is bistable; (iv) finally, at high k_1 values, there is again a single intersection point (high $[S_1]$, low $[S_2]$). (B) Bifurcation diagram showing the S_1 steady-state concentration as a function of the value of parameter k_1. The k_1 values represented in (A) are indicated. At low and high k_1 values, the system is monostable and exhibits a single stable steady state (solid curves). Over a mid-range interval, the two stable steady states coexist, separated by an unstable steady state (dashed curve). The k_1 values at which steady states appear or disappear are saddle-node bifurcations. Parameter values: $k_2 = 20$ (concentration · time^{-1}), $K_1 = K_2 = 1$ (concentration), $k_3 = k_4 = 5$ (time^{-1}), $n_1 = n_2 = 2$. Units are arbitrary.

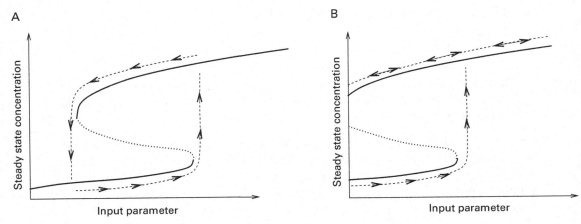

Figure 4.20
Switching in bistable systems. (A) Hysteresis loop. Changes in the input parameter can push the system from one steady state to the other (dashed arrows). Over the intermediate bistable region, the state depends on the recent past. Transitions between the two states occur abruptly at the bifurcation points. (B) Irreversible switching. If one of the two bifurcation points is inaccessible, the system can become trapped in one of the steady states.

saddle point and a stable node come together). Between these bifurcation points, three steady states coexist.

Figure 4.19B reflects the ability of this bistable system to act as a switch: an input that pushes parameter k_1 back and forth past the saddle-node bifurcations will toggle the system between low- and high-[S_1] states. The intermediate bistable range introduces a lag into this switching action: over this interval, the state is not uniquely determined by the value of k_1 but also depends on the previous condition. The sketch in figure 4.20A illustrates this behavior. If the bistable range is entered from the high state, then the system remains in the high state over this interval. The opposite holds if the bistable region is entered from the low state. This "path-dependent" property is referred to as *hysteresis*. As the system cycles back and forth between the two states, it follows a *hysteresis loop*, in which transitions between the two states occur at two separate threshold values (i.e., at the two bifurcation points). Bistable switches can be irreversible. As shown in figure 4.20B, if one of the two saddle-node bifurcations is outside the range of relevant parameter values, then the system executes a one-way transition between the two states.

Next, we turn to the oscillatory model (4.10). Recall that for this model, oscillatory behavior is dependent on the degree of cooperativity n. A bifurcation diagram for this system is shown in figure 4.21. For small values of n, a single stable steady state is shown. At $n = 2.4$, a change occurs—the steady state becomes unstable, and a limit cycle appears. The bifurcation diagram shows both the change in stability and the upper and lower bounds of the limit-cycle oscillations.

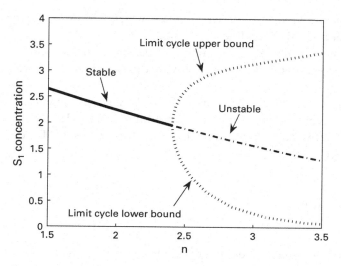

Figure 4.21
Bifurcation diagram for the autocatalytic model (4.10). For small n values, a stable steady state is shown. At higher n values, this steady state is unstable. At the bifurcation point ($n = 2.4$), a stable limit cycle is born; the two dotted curves show the maximal and minimal concentrations reached by the limit cycle. The transition point is called a Hopf bifurcation. Parameter values are as for figure 4.15.

The bifurcation in figure 4.21 occurs when the stability of the steady state changes. From our discussion of linearized stability analysis (section 4.2.2), we know that this change occurs when eigenvalues of the Jacobian at the steady state transition from having negative real part (stable) to positive real part (unstable). The steady state in model (4.10) is a spiral point, and so the eigenvalues are complex numbers. The bifurcation in figure 4.21, in which a pair of complex-valued eigenvalues transition between negative and positive real part, is called a *Hopf bifurcation*.

Bifurcation diagrams provide insight into the *robustness* of system behavior. A behavior is called robust if it is not significantly affected by disturbances. Robustness is indicated by a bifurcation diagram: if a system is operating far from any bifurcation points, then perturbations are unlikely to result in a qualitative change in system behavior; alternatively, the behavior of a system operating near a bifurcation point may change dramatically in response to a disturbance. In the next section, we will address another tool for analysis of robustness: parametric sensitivity analysis.

4.5 Sensitivity Analysis

Continuation and bifurcation diagrams illustrate how model behavior depends on parameter values. The general study of this dependence is called **(parametric) sensitivity analysis**. It can be divided into *global sensitivity analysis*, which addresses wide variations in parameter values, and *local sensitivity analysis*, which addresses small variations around a nominal operating condition.

Global sensitivity analysis typically involves sampling the space of parameter values and determining the corresponding system behavior. Statistical methods are usually used to analyze the results.

In this section, we will address local sensitivity analysis, which makes use of linearized model approximations. There is a long tradition of applications of local sensitivity analysis to biochemical networks (through metabolic control analysis (MCA) and biochemical systems theory (BST)).

4.5.1 Local Sensitivity Analysis

Consider the simple reaction scheme:

$$\to S \to.$$

Suppose the rate of production is maintained at a constant rate V_0, and the rate of consumption is described by Michaelis–Menten kinetics as $V_{max}s/(K_M + s)$, where $s = [S]$. The steady-state concentration s^{ss} is characterized by

$$V_0 = \frac{V_{max}s^{ss}}{K_M + s^{ss}}.$$

Solving for s^{ss}, we find

$$s^{ss} = \frac{V_0 K_M}{V_{max} - V_0}. \tag{4.11}$$

Now, suppose that V_{max} is varied while V and K_M are held fixed. Equation (4.11) then defines the steady-state concentration as a function of V_{max}. For concreteness, we take $V_0 = 2$ mM/min and $K_M = 1.5$ mM. In that case, equation (4.11) becomes

$$s^{ss} = \frac{3}{V_{max} - 2}. \tag{4.12}$$

This relationship is plotted in figure 4.22 (a continuation diagram).

Because most models do not admit explicit steady-state formulas, construction of continuation curves generally requires significant computational effort. As an alternative, parametric sensitivities provide an easily calculated description of the continuation curve near a nominal parameter value.

The **absolute local sensitivity** of a steady state s^{ss} with respect to a variable p is defined as the rate of change of s^{ss} with respect to p, that is, as ds^{ss}/dp. This is the slope of the tangent to the continuation curve (figure 4.22). This sensitivity coefficient can be used to predict the effect of small perturbations Δp at the parameter value $p = p_0$ through the linearization formula

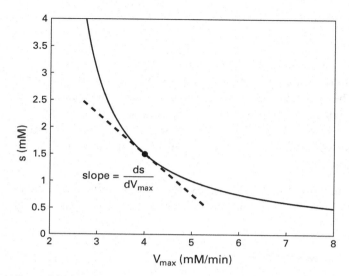

Figure 4.22
Local sensitivity. The solid continuation curve shows the steady state as a function of the parameter V_{max}. The dashed line is the tangent at $V_{max} = 4$ mM/min. The slope of this tangent is the absolute local sensitivity coefficient, which describes the effect of small changes in V_{max} on the steady state.

$$s^{ss}(p_0 + \Delta p) \approx s^{ss}(p_0) + \Delta p \, \frac{ds^{ss}}{dp}. \tag{4.13}$$

Returning to our example, from the explicit formula for steady state in equation (4.12), the absolute local sensitivity coefficient with respect to V_{max} can be calculated directly:

$$\frac{ds^{ss}}{dV_{max}} = \frac{d}{dV_{max}} \left(\frac{V_0 K_M}{V_{max} - V_0} \right) = \frac{-3}{(V_{max} - 2)^2}. \tag{4.14}$$

Choosing a nominal value of $V_{max} = 4$ mM/min, we find a sensitivity coefficient of $ds^{ss}/dV_{max} = -0.75$ min (figure 4.22). Thus, an increase of, say, 0.1 mM/min in V_{max} leads to a 0.75(0.1) = 0.075 mM decrease in s^{ss}, for V_{max} near 4 mM/min (by applying equation 4.13 with $p_0 = 4$ mM/min and $\Delta p = 0.1$ mM/min).

Although this sensitivity coefficient can be used to make predictions, it is not usually used directly. An improved sensitivity measure describes the *relative* effect of perturbations. We define the **relative sensitivity** as

$$\frac{ds^{ss} / s^{ss}}{dp / p} = \frac{p}{s^{ss}} \frac{ds^{ss}}{dp}.$$

The relative sensitivity relates the size of a relative perturbation in p to a relative change in s^{ss}. Referring back to equation (4.12), we find that at $V_{max} = 4$ mM/min, the relative sensitivity coefficient of s^{ss} with respect to V_{max} is

$$\frac{V_{max}}{s^{ss}} \frac{ds^{ss}}{dV_{max}} = \left(\frac{4}{1.5}\right)(-0.75) = -2.$$

Thus, a 1% increase in V_{max} results in a 2% decrease in s^{ss}.

Relative sensitivity coefficients are frequently used to provide a concise description of model behavior. These coefficients provide insight into robustness: if the system shows a small sensitivity coefficient with respect to a parameter, then behavior is robust with respect to perturbations of that parameter. In contrast, large sensitivity coefficients (positive or negative) suggest "control points" at which interventions will have significant effects.

Exercise 4.5.1 Starting from equation (4.11), verify that the relative sensitivity coefficients of s^{ss} with respect to K_M is equal to one, regardless of the parameter values.   ☐

4.5.2 Determination of Local Sensitivity Coefficients

Numerical Approximation Local sensitivity coefficients are typically determined by simulation. The sensitivity coefficient at a parameter value $p = p_0$ can be determined by simulating the model at $p = p_0$ and at another nearby value $p = p_0 + \Delta p_0$, where Δp_0 should normally be chosen less than a 5% deviation from p_0. The derivative ds^{ss}/dp at $p = p_0$ can then be approximated by

$$\frac{ds^{ss}}{dp} \approx \frac{s^{ss}(p_0 + \Delta p_0) - s^{ss}(p_0)}{\Delta p_0}. \tag{4.15}$$

This ratio can then be scaled by p_0/s^{ss} to arrive at the relative sensitivity. When using this approximation, care must be taken to avoid the significant round-off errors that can occur when calculating the ratio of two small numbers. (To ensure accuracy, the approximation can be calculated for a handful of Δp_0 values; e.g., at 1%, 3%, and 5% displacements from p_0. If these approximations do not agree, then a more accurate simulation procedure may be needed.)

Exercise 4.5.2 Verify the accuracy of the finite difference approach by using equation (4.15) to approximate the absolute sensitivity coefficient in equation (4.14) (at the nominal value of $V_{max} = 4$ mM/min). Calculate approximations with $\Delta p_0 = 0.2$ (5% deviation) and $\Delta p_0 = 0.04$ (1% deviation).   ☐

Implicit Differentiation When an explicit formula for steady state is available (as in equation 4.12), sensitivity coefficients can be determined by direct differentiation. For most models, no such steady-state formula is available. Nevertheless, sensitivity coefficients can be derived by implicit differentiation of the differential equation model, as in the following exercise.

Exercise 4.5.3   Consider a species S that is consumed at rate $k_2[S]$ and inhibits its own production, so that it is produced at rate $k_1/(1 + [S]^n)$. The steady-state concentration s is then given by

$$0 = k_1/(1 + s^n) - k_2 s.$$

Use implicit differentiation (reviewed in appendix B) to determine the absolute sensitivity coefficient ds/dk_1 and verify that it is positive for all values of k_1, k_2, and n.   □

4.6*   Parameter Fitting

Chapters 2 and 3 addressed techniques for model construction but gave no indication of how the values of model parameters should be chosen. The task of finding appropriate parameter values is called *model calibration* or *parameter fitting*.

Some parameters can be measured directly. For instance, rates of degradation can be determined from observations of half-lives, and specialized enzymological assays have been developed to determine the kinetic parameters of enzyme catalysis. However, in construction of models for systems biology, most model parameters are not measured directly. Instead, parameter values are assigned by fitting model behavior to observations of system behavior. We will next outline the most commonly used parameter calibration approach, called *least-squares fitting*.

Suppose observations of a biological system have been collected, and a model structure (i.e., a network with reaction kinetics) has been chosen. To be concrete, suppose that the model involves three species concentrations s_1, s_2, s_3 and depends on two parameters p_1 and p_2:

$$\frac{d}{dt} s_1(t) = f_1(s_1(t), s_2(t), s_3(t), p_1, p_2)$$

$$\frac{d}{dt} s_2(t) = f_2(s_1(t), s_2(t), s_3(t), p_1, p_2)$$

$$\frac{d}{dt} s_3(t) = f_3(s_1(t), s_2(t), s_3(t), p_1, p_2).$$

Depending on the experimental observations that have been made, corresponding simulations of the model can be carried out. For instance, observations of the steady-state concentrations correspond directly to the model's steady state: $s_1^{ss}, s_2^{ss}, s_3^{ss}$. Time-course data can be compared to time-points along a model trajectory.

The goal of parameter fitting is to determine the parameter values for which model simulation best matches experimental data. The accuracy of the model can be assessed by comparing the model predictions to each of the experimental observations. This collection of comparisons can be combined into a single measure of the quality of fit. For the model described in this section, if steady-state observations of the concentrations are available (denoted s_i^{obs}) the *sum of squared errors* is defined by

$$\text{SSE}(p_1, p_2) = \left[s_1^{\text{ss}}(p_1, p_2) - s_1^{\text{obs}} \right]^2 + \left[s_2^{\text{ss}}(p_1, p_2) - s_2^{\text{obs}} \right]^2 + \left[s_3^{\text{ss}}(p_1, p_2) - s_3^{\text{obs}} \right]^2.$$

(The errors are squared to avoid cancellation between terms of opposite sign. A relative SSE can also be used, in which the error terms are scaled by the observations.)

When replicate observations are available, the predictions are usually compared to the mean of the replicates. The error terms can then be inversely weighted by the variance in the replicates, so that observations with high variability (in which we have less confidence) make a reduced contribution to the total error.

The *least-squares fit* corresponds to the parameter values that minimize the sum of squared errors. This parameter set can be found by numerical function-minimization techniques. Although fitting to data almost always demands numerical calculations, we can illustrate the general principles with a simple example, as follows. Consider the reaction chain

$$\xrightarrow{k_1} S_1 \xrightarrow{k_2} S_2 \xrightarrow{k_3} \qquad\qquad (4.16)$$

where the parameters k_1, k_2, and k_3 are mass-action rate constants. The reaction rates are then $k_1, k_2[S_1]$, and $k_3[S_2]$. We will illustrate least-squares fitting of the model parameters in three separate scenarios.

Case I Suppose the consumption rate of S_2 has been measured directly, $k_3 = 4$ mM/min, and that steady-state measurements have been made: $s_1^{\text{obs}} = 5$ mM, $s_2^{\text{obs}} = 2$ mM. In this case, an exact model fit can be found. We begin by solving for the model steady-state concentrations:

$$s_1^{\text{ss}} = \frac{k_1}{k_2} \qquad s_2^{\text{ss}} = \frac{k_2 s_1^{\text{ss}}}{k_3} = \frac{k_1}{k_3}.$$

The sum of squared errors is then

$$\text{SSE} = \left[s_1^{\text{ss}}(k_1, k_2, k_3) - s_1^{\text{obs}} \right]^2 + \left[s_2^{\text{ss}}(k_1, k_2, k_3) - s_2^{\text{obs}} \right]^2$$

$$= \left(\frac{k_1}{k_2} - 5 \right)^2 + \left(\frac{k_1}{k_3} - 2 \right)^2.$$

This error takes its minimum value (of zero) when

$$\frac{k_1}{k_2} = 5 \text{ mM} \qquad \text{and} \qquad \frac{k_1}{k_3} = 2 \text{ mM} . \qquad\qquad (4.17)$$

Because we know $k_3 = 4$ mM/min, we can solve for

$$k_1 = 8/\text{min} \quad \text{and} \quad k_2 = \frac{8}{5}/\text{min}.$$

Case II Suppose now that the same steady-state measurements have been made ($s_1^{\text{obs}} = 5$ mM, $s_2^{\text{obs}} = 2$ mM), but k_3 is unknown. In this case, we have the same error function:

$$\text{SSE} = \left(\frac{k_1}{k_2} - 5\right)^2 + \left(\frac{k_1}{k_3} - 2\right)^2,$$

but we cannot determine a unique parameter set that minimizes this function. Solving equations (4.17) indicates that the error will be zero whenever

$$k_2 = \frac{k_1}{5} \quad \text{and} \quad k_3 = \frac{k_1}{2},$$

regardless of the value of k_1. In this case, the fitting problem is called *underdetermined*, as there are multiple equivalently good solutions. Unfortunately, parameter calibration of system biology models is often underdetermined because it is a challenge to collect the experimental data needed to fit dynamic models. In such cases, model reduction techniques can sometimes be used to reduce the number of parameters in the model.

Case III Suppose that the value $k_3 = 4$ mM/min is known and that steady-state observations have been made in two conditions: in the control condition, $s_1^{\text{obs}} = 5$ mM, $s_2^{\text{obs}} = 2$ mM, whereas in the experimental condition, the (unknown) production rate k_1 has been reduced by 90% and measurements have been made of $s_1^{\text{obs}} = 0.5$ mM, $s_2^{\text{obs}} = 0.3$ mM.

In this case, there are four terms in the sum of squared errors. The steady states in the experimental condition are

$$s_2^{\text{ss}} = \frac{k_1/10}{k_2} \qquad s_2^{\text{ss}} = \frac{k_1/10}{k_3},$$

so we have

$$\text{SSE} = \left(\frac{k_1}{k_2} - 5\right)^2 + \left(\frac{k_1}{4} - 2\right)^2 + \left(\frac{k_1/10}{k_2} - 0.5\right)^2 + \left(\frac{k_1/10}{4} - 0.3\right)^2.$$

There are *no* choices of k_1 and k_2 that will make this error equal to zero: this fitting problem is *overdetermined*. Overdetermined fits are caused by inaccuracies in model formulation and errors in experimental measurements. The "solution" to an overdetermined fitting problem is a compromise parameter set that minimizes the resulting error.

Exercise 4.6.1   Consider again the network (4.16). Suppose that the degradation rate $k_3 = 4$ mM/min has been measured directly and that observations are made in two conditions, but only the pooled concentration of S_1 and S_2 can be measured. Consider two cases:

(i) Suppose that in the control condition, $s_1^{obs} + s_2^{obs} = 6$ mM, whereas in the experimental condition, the production rate k_1 has been reduced by 90% and the resulting observation is $s_1^{obs} + s_2^{obs} = 0.6$ mM. Perform a least-squares fit.

(ii) Suppose that in the control condition, $s_1^{obs} + s_2^{obs} = 6$ mM, whereas in the experimental condition, the rate constant k_2 has been reduced by 90% and the resulting observation is $s_1^{obs} + s_2^{obs} = 18$ mM. Perform a least-squares fit.

How does your analysis in cases (i) and (ii) compare? Explain the difference. $\square$

4.7 Suggestions for Further Reading

• **Nonlinear Dynamics** An accessible introduction to bistability, oscillations, and bifurcations can be found in the book *Nonlinear Dynamics and Chaos* (Strogatz, 2001). More formal treatments of nonlinear dynamics appear in many texts on differential equations, including *Elementary Differential Equations and Boundary Value Problems* (Boyce and DiPrima, 2008).

• **Sensitivity Analysis** Techniques for global sensitivity analysis are introduced in *Global Sensitivity Analysis: The Primer* (Saltelli et al., 2008). Local sensitivity analysis for chemical systems is addressed in *Parametric Sensitivity in Chemical Systems* (Varma et al., 2005). Several applications of local sensitivity analysis to systems biology are reviewed in Ingalls (2008).

• **Parameter Fitting** An introduction to parameter fitting in systems biology, including coverage of a range of computational function-minimization techniques, can be found in *Systems Biology: A Textbook* (Klipp et al., 2009).

4.8 Problem Set

4.8.1 Phase Line Analysis
Phase analysis can be applied to systems of a single dimension. The phase portrait of a one-dimensional system lies on a *phase line*. For example, the phase portrait of the system

$$\frac{dx(t)}{dt} = x^2(t) - x(t) = x(t)(x(t) - 1)$$

Figure 4.23
Phase line for problem 4.8.1.

is shown in figure 4.23, with the open circle indicating an unstable steady state, the closed circle indicating a stable steady state, and the arrows indicating the direction of motion.

(a) Sketch the phase lines of the following one-dimensional systems.

(i) $\dfrac{d}{dt}V(t) = V^3(t) - V(t)$

(ii) $\dfrac{d}{dt}r(t) = r^4(t) - 3r^2(t) + 2$

(iii) $\dfrac{d}{dt}w(t) = \sin(w(t))$

(iv) $\dfrac{d}{dt}p(t) = p^3(t) - 2p^2(t) + p(t).$

(Note that case (iv) involves a "semistable" steady state.)

(b) Use a phase line argument to confirm that a one-dimensional system can never display oscillatory behavior.

(c) Consider the simple model

$$\frac{d}{dt}s(t) = k - \frac{V_{\max}s}{K_M + s}$$

in which species s is produced at a fixed rate and consumed via Michaelis–Menten kinetics. Sketch a phase line for this system. Verify that the steady state is stable for any nonnegative parameter values, provided $V_{\max} > k$.

4.8.2 Linear System: Complex Eigenvalues
Consider the system

$$\frac{d}{dt}x_1(t) = -x_1(t) + x_2(t)$$

$$\frac{d}{dt}x_2(t) = -x_1(t) - x_2(t).$$

Find the eigenvalues of the Jacobian matrix. Determine the solution $(x_1(t), x_2(t))$ satisfying initial condition $(x_1(0), x_2(0)) = (1, 1)$ by substituting the general form of the solution (4.6) into the system of equations and solving for the parameters c_{ij}.

4.8.3 Linear System: Dynamics
Consider the general linear system

$$\frac{d}{dt}x(t) = ax(t) + by(t)$$

$$\frac{d}{dt}y(t) = cx(t) + dy(t).$$

Note that the steady state is $(x, y) = (0, 0)$. Choose six sets of parameter values (a, b, c, d) that yield the following behaviors:

(i) stable node (real negative eigenvalues)

(ii) stable spiral point (complex eigenvalues with negative real part)

(iii) center (purely imaginary eigenvalues)

(iv) unstable spiral point (complex eigenvalues with positive real part)

(v) unstable node (real positive eigenvalues)

(vi) saddle point (real eigenvalues of different sign).

In each case, prepare a phase portrait of the system, including the x and y nullclines, a direction field, and a few representative trajectories. Hint: Recall from exercise 4.2.4 that if either of the off-diagonal entries in the Jacobian matrix are zero, then the eigenvalues are simply the entries on the diagonal.

4.8.4 Phase Portrait
Consider the nonlinear system

$$\frac{d}{dt}x(t) = \mu x(t) - x^3(t)$$

$$\frac{d}{dt}y(t) = -y(t).$$

(a) Take $\mu = -1$. Show that the system has a single steady state and characterize its stability by finding the eigenvalues of the Jacobian matrix at this point. Confirm your results by producing a phase portrait.

(b) Repeat with $\mu = 1$. In this case, there are three steady states.

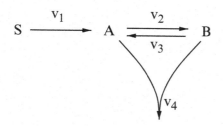

Figure 4.24
Reaction network for problem 4.8.5.

4.8.5 Linearized Stability Analysis

Consider the chemical reaction network in figure 4.24, with reaction rates v_i as labeled. Suppose the concentration of S is fixed at 1 mM and that the reaction rates are given by mass action as $v_1 = k_1[S]$, $v_2 = k_2[A]$, $v_3 = k_3[B]$, and $v_4 = k_4[A][B]$.

(a) Write a pair of differential equations that describe the concentrations of A and B.

(b) Presuming that $k_1 = 1$/min, $k_2 = 2$/min, $k_3 = 0.5$/min, and $k_4 = 1$/mM/min, determine the steady-state concentrations of A and B.

(c) Evaluate the system Jacobian at the steady state found in (b) and verify that this steady state is stable.

4.8.6 Global Dynamics from Local Stability Analysis

(a) Consider the chemical reaction network with mass-action kinetics:

$$A + X \xrightarrow{k_1} 2X$$

$$X + X \xrightarrow{k_2} Y$$

$$Y \xrightarrow{k_3} B.$$

Assume that $[A]$ and $[B]$ are held constant.

(i) Write a differential equation model describing the concentrations of X and Y.

(ii) Verify that the system has two steady states.

(iii) Determine the system Jacobian at the steady states and characterize the local behavior of the system near these points.

(iv) By referring to the network, provide an intuitive description of the system behavior starting from any initial condition for which $[X] = 0$.

(v) Sketch a phase portrait for the system that is consistent with your conclusions from (iii) and (iv).

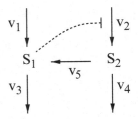

Figure 4.25
Reaction network for problem 4.8.7. The production of S_2 is allosterically inhibited by S_1 (blunted dashed arrow).

(b) Repeat for the system

$$A + X \xrightarrow{k_1} 2X$$

$$X + Y \xrightarrow{k_2} 2Y$$

$$Y \xrightarrow{k_3} B.$$

In this case, you'll find that the nonzero steady state is a center: it is surrounded by concentric periodic trajectories.

4.8.7 Nullcline Analysis

Consider the network in figure 4.25. Suppose the reaction rates are given by

$$v_1 = V \qquad v_2 = f(s_1) \qquad v_3 = k_3 s_1$$

$$v_4 = k_4 s_2 \qquad v_5 = k_5 s_2.$$

Suppose that the parameters V, k_3, k_4, and k_5 are positive constants and that $f(s_1)$ takes positive values and is a *decreasing* function of s_1 (i.e., as the values of s_1 increase, the values of $f(s_1)$ decrease). By sketching the nullclines, demonstrate that this system cannot exhibit bistability.

4.8.8 Linearization

Consider the simple reaction system $\rightarrow S \rightarrow$, where the reaction rates are

Production: V_0 $\qquad\qquad$ Consumption: $\dfrac{V_{max}[S]}{K_M + [S]}$.

(a) Write the differential equation that describes the dynamics in $s=[S]$. Find the steady state. Next, approximate the original system by linearizing the dynamics around the steady state. This approximation takes the form of a linear differential equation in the new variable $x(t) = s(t) - s^{ss}$.

(b) Take parameter values $V_0 = 2$, $V_{max} = 3$, and $K_M = 1$ and run simulations of the nonlinear and linearized systems starting at initial conditions $[S] = 2.1$, $[S] = 3$, and $[S] = 12$. Comment on the discrepancy between the linear approximation and the original nonlinear model.

4.8.9 Saddle-Node Bifurcation

Consider the system

$$\frac{d}{dt} x(t) = \mu - x^2(t).$$

Draw phase lines (as in problem 4.8.1) for $\mu = -1, 0, 1$. For this system, $\mu = 0$ is a bifurcation value. Use your one-dimensional phase portraits to sketch a bifurcation diagram for the system showing steady states of x against μ. Be sure to indicate the stability of each branch of steady states.

4.8.10 Pitchfork Bifurcation

Recall from problem 4.8.4 that the nonlinear system

$$\frac{d}{dt} x(t) = \mu x(t) - x^3(t)$$

$$\frac{d}{dt} y(t) = -y(t)$$

exhibits different steady-state profiles depending on the value of the parameter μ. Sketch a bifurcation diagram showing steady states of x against μ. Your diagram should make it clear why the point $\mu = 0$ is referred to as a pitchfork bifurcation.

4.8.11 Bifurcation Diagram: Bistability

Consider model (4.2) with parameter values $k_1 = k_2 = 20$ (concentration $\cdot$ time^{-1}), $K_1 = K_2 = 1$ (concentration), $k_3 = k_4 = 5$ (time^{-1}), and $n_2 = 2$. Use a software package to generate a bifurcation diagram showing the steady-state concentration of S_1 as a function of the parameter n_1. The system is bistable when $n_1 = 2$. Does the system become monostable at high n_1, low n_1, or both?

4.8.12 Bifurcation Diagram: Limit-Cycle Oscillations

Consider the autocatalytic model (4.10), with parameter values $k_0 = 8$, $k_1 = 1$, $K = 1$, and $n = 2.5$. Use a software package to generate a bifurcation diagram showing the steady-state concentration of S_1 as a function of the parameter k_2. The system exhibits limit-cycle oscillations when $k_2 = 5$. Are the oscillations lost at high k_2, low k_2, or both?

4.8.13 Sensitivity Analysis: Reversible Reaction

Consider the reversible reaction

$$A \underset{k_2}{\overset{k_1}{\rightleftharpoons}} A^*$$

with mass-action rate constants as shown. Let T be the total concentration of A and A^*.

(a) Solve for the steady-state concentration of A^* and verify that an increase in k_1 leads to an increase in $[A^*]^{ss}$.

(b) Use parametric sensitivity analysis to determine whether the steady-state concentration of A^* is more sensitive to a 1% increase in T or a 1% increase in k_1. Does the answer depend on the values of the parameters?

4.8.14 Sensitivity Analysis: Branched Network

Consider the branched network in figure 4.26. Suppose the reaction rates are $v_0 = V$, $v_1 = k_1[S]$, $v_2 = k_2[S]$, with V, k_1, and k_2 constant. Suppose that $k_1 > k_2$. Use sensitivity analysis to determine whether the steady state of $[S]$ is more sensitive to a 1% increase in k_1 or a 1% increase in k_2.

4.8.15* Sensitivity Coefficients in S-System Models

The relative sensitivity coefficients defined in section 4.5 can be formulated as ratios of changes in the logarithms of concentrations and parameter values:

$$\frac{p}{s^{ss}} \frac{ds^{ss}}{dp} = \frac{d\log(s^{ss})}{d\log(p)}.$$

For this reason, sensitivity coefficients are sometimes called *logarithmic gains*. Because S-system models (see section 3.5) involve linear relationships among logarithmic quantities, relative sensitivities reveal themselves immediately in these model.

Refer to the S-system model (3.22)–(3.23) in chapter 3. Verify that relative sensitivities with respect to the rate constants α_i appear as coefficients in equations (3.24).

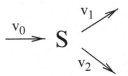

Figure 4.26
Reaction network for problem 4.8.14.

4.8.16* Model Fitting: Time-Series Data

Consider the network

$$\xrightarrow{k_1} S \xrightarrow{k_2}$$

with mass-action rate constants, as shown. Suppose that the following time-series data are available for model fitting (in mM): $s(0) = 0, s(1) = 1, s(2) = 1.5$. Determine the corresponding model parameters. (The data can be fit exactly.) Recall from equation (2.6) of chapter 2 that the predicted time-course takes the form

$$s(t) = \left(s(0) - \frac{k_1}{k_2} \right) e^{-k_2 t} + \frac{k_1}{k_2}.$$

Hint: You will need to solve an equation of the form $ae^{-2k_2} + be^{-k_2} + c = 0$. This is a quadratic equation in $x = e^{-k_2}$.

5 Metabolic Networks

[I]t would be a delusion ... to imagine that the detailed behaviour of systems as complex as metabolic pathways can be predicted with the aid of a few qualitative, verbal principles ....
—David Fell, *Understanding the Control of Metabolism*

A cell's metabolism is the network of enzyme-catalyzed reactions in which sources of energy and materials are broken down and cellular components (e.g., amino acids, lipids) are produced. The reactants and products in metabolic reactions are referred to as *metabolites*: they typically are small molecules.

A cell's complete metabolic network is organized roughly in a "bow-tie" structure, in which a wide range of substrates are broken down into a much smaller number of intermediates, from which a large number of distinct biomolecules are formed (figure 5.1). The structure of metabolic networks varies from species to species, but there are many aspects of "core" metabolism that are conserved across all organisms. Metabolic networks for a wide range of species are archived in online databases such as KEGG and MetaCyc.[1]

Reactions that break down sources of energy and materials are called *catabolic*. Well-studied catabolic pathways include *glycolysis*, which converts glucose to pyruvate (and produces adenosine triphosphate, ATP), and the *tricarboxylic acid (TCA) cycle* (also called the *citric acid cycle* or the *Krebs cycle*), which consumes pyruvate and produces substrates for biosynthetic and energy-producing pathways.

Metabolic reactions that build up cellular components are called *anabolic*. Standard examples of anabolism include the branched pathways leading to amino acid production and the pathways responsible for generation of lipids and nucleic acids.

Visualizations of a cell's complete metabolic network typically appear as unintelligible "spaghetti diagrams" involving multiple branches, cycles, hubs, and pathways. However, careful study of these networks reveals that they are organized into well-defined *subnetworks*, each responsible for a particular function: these subnetworks are interconnected by the flow of material and energy.

---

1. KEGG: www.genome.jp/kegg. MetaCyc: www.metacyc.org.

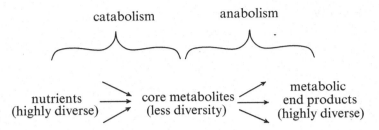

Figure 5.1
Bow-tie structure of metabolic networks. A wide variety of nutrients is funneled through a network of enzyme-catalyzed reactions into a smaller set of core metabolites. This set of metabolites is then used to generate the wide range of end-products that is needed for growth and other cellular activities.

The energy needs of the cell are met by a small number of universal cofactors. The most important of these is ATP—the primary energy carrier within the cell. In modeling metabolism, cofactors such as ATP are often considered to be part of the cellular "background" in which reactions take place (unless the model explicitly describes their dynamics). The concentration of each cofactor can be presumed constant: their influence on reaction rates is then implicitly incorporated into the values of kinetic parameters.

In addition to catabolic and anabolic reactions, metabolic networks also involve allosteric regulatory interactions: these ensure that the metabolic needs of the cell are met under a range of conditions. In this chapter, we will explore models that provide insight into the function and regulation of metabolic networks.

5.1   Modeling of Metabolism

Metabolic networks are biochemical reaction networks. To construct differential-equation models of these networks, we will follow the procedure developed in chapter 2 and make use of the rate laws described in chapter 3.

5.1.1   Example: A Pathway Model
As a first example, consider the reaction network shown in figure 5.2. Presuming that the concentrations of both the substrate S_0 and product P are held fixed, the time-varying metabolite concentrations are described by

$$\frac{d}{dt}s_1(t) = v_1(t) - v_2(t) - v_3(t)$$

$$\frac{d}{dt}s_2(t) = v_2(t) + v_4(t) - v_5(t)$$

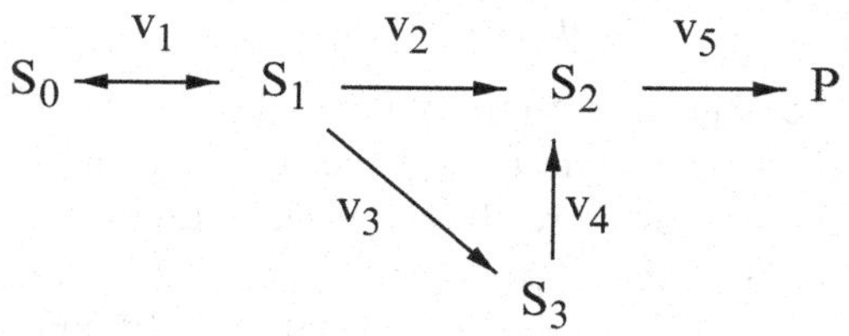

Figure 5.2
Simple metabolic pathway. Except for the production of S_1, all reactions are presumed irreversible. The concentrations of the pathway substrate, S_0, and product, P, are held fixed. The reaction labels v_i indicate reaction rates. (They are not mass-action rate constants.)

$$\frac{d}{dt} s_3(t) = v_3(t) - v_4(t),$$

where $s_i = [S_i]$. We treat all but the first reaction as irreversible. Suppose that each reaction is catalyzed by an enzyme, E_i, with concentration e_i. To keep our analysis simple, we presume that all enzymes are operating in their first-order regimes. The reaction rates are then given by

$$v_1 = e_1(k_0[S_0] - k_1 s_1) \qquad v_2 = e_2 k_2 s_1 \qquad v_3 = e_3 k_3 s_1$$

$$v_4 = e_4 k_4 s_3 \qquad v_5 = e_5 k_5 s_2.$$

(Recall from chapter 3 that reaction rates are proportional to enzyme concentration.) For simplicity, we take the constants $k_i = 1$. The model can then be written as

$$\frac{d}{dt} s_1(t) = e_1([S_0] - s_1(t)) - (e_2 + e_3) s_1(t)$$

$$\frac{d}{dt} s_2(t) = e_2 s_1(t) + e_4 s_3(t) - e_5 s_2(t) \tag{5.1}$$

$$\frac{d}{dt} s_3(t) = e_3 s_1(t) - e_4 s_3(t).$$

The steady-state concentrations of the metabolites can be found by setting the time rates of change to zero, resulting in

$$s_1^{ss} = \frac{e_1[S_0]}{e_1 + e_2 + e_3}$$

$$s_2^{ss} = \frac{(e_2 + e_3) e_1 [S_0]}{e_5 (e_1 + e_2 + e_3)} \tag{5.2}$$

$$s_3^{ss} = \frac{e_3 e_1 [S_0]}{e_4 (e_1 + e_2 + e_3)}.$$

Exercise 5.1.1 Derive the steady-state concentrations in (5.2). ☐

The **flux** through the pathway is a measure of network activity. Flux—the flow-rate of material—is a time-varying quantity. The steady-state rate of the reactions in an unbranched pathway is called the *pathway flux*. Because the network in figure 5.2 has a single substrate and a single product, we can define the pathway flux as the steady-state rate of production of P (which is equal to the steady-state rate of consumption of S_0). We will use the notation J_k to indicate the steady-state rate of reaction k and simply use J for the pathway flux. For this example:

$$J = J_5 = v_5^{ss} = J_1 = v_1^{ss}.$$

From equation (5.2), we can write the pathway flux J in terms of the concentration of the substrate S_0 and the enzyme concentrations:

$$J = v_5^{ss} = e_5 s_2^{ss} = \frac{(e_2 + e_3)e_1[S_0]}{(e_1 + e_2 + e_3)}. \tag{5.3}$$

Exercise 5.1.2 Verify that for the network in figure 5.2, the flux satisfies $J = v_2^{ss} + v_3^{ss} = v_2^{ss} + v_4^{ss}$. ☐

Exercise 5.1.3 Why does the pathway flux J in equation (5.3) not depend on the enzyme concentrations e_4 and e_5? Provide an intuitive explanation in terms of the network structure. ☐

5.1.2 Sensitivity Analysis of Metabolic Networks: Metabolic Control Analysis

Because experimental observations of metabolism are often carried out in steady state, modeling efforts often focus on steady-state behavior. Parametric sensitivity analysis (see section 4.5) plays a key role in these studies. The variables of primary interest in metabolic systems are the steady-state metabolite concentrations and the steady-state reaction fluxes. The parameters of primary interest are the enzyme concentrations.

Researchers working in theoretical biochemistry use a sensitivity approach called **metabolic control analysis** (MCA). Within MCA, specialized terminology and notation are used to describe the effects of changes in enzyme activity on metabolite concentrations and reaction fluxes. The relative sensitivities of metabolite concentrations are called *concentration control coefficients*, defined by (as in section 4.5):

$$C_{e_j}^{s_i} = \frac{e_j}{s_i} \frac{ds_i}{de_j},$$

where s_i is the steady-state concentration of species i, and e_j is the abundance of enzyme j. Likewise, the *flux control coefficients* are given by

$$C_{e_j}^{J_k} = \frac{e_j}{J_k} \frac{dJ_k}{de_j}, \tag{5.4}$$

where J_k is the flux through reaction k (i.e., the steady-state reaction rate v_k).

The rate of a metabolic reaction is proportional to the abundance of the catalyzing enzyme (see section 3.1.1). Thus, a relative change in enzyme concentration results in an equivalent relative change in the corresponding reaction rate (i.e., a 2% increase in e_k leads to a 2% increase in rate v_k). In the typical case that no enzyme in the network catalyzes multiple reactions or interacts directly with another enzyme, each flux control coefficient indicates how changes in the rate of a particular reaction affect the reaction fluxes throughout the network. In particular, these coefficients indicate how much control each reaction exerts over the overall pathway flux and so can provide significant insight into the behavior and regulation of metabolic activity.

Sensitivity analysis can be used to predict the response of a system to perturbations. Perturbations of particular interest in metabolism are those that result from the action of drugs or from modification of the organism (by, e.g., genetic modification). In both cases, the perturbation causes a change in the concentration e_j of active enzyme. (Drugs typically cause such changes biochemically; genetic modification usually leads to changes in enzyme abundance.)

For the model (5.1), we can calculate flux control coefficients for the pathway flux J from the explicit description of flux in equation (5.3):

$$C_{e_1}^{J} = \frac{e_1}{J} \frac{dJ}{de_1} = \frac{e_2 + e_3}{e_1 + e_2 + e_3}. \tag{5.5}$$

Thus, a 1% change in the concentration of enzyme e_1 leads to an $(e_2 + e_3)/(e_1 + e_2 + e_3)$% change in the steady-state pathway flux. Further calculation yields

$$C_{e_2}^{J} = \frac{e_2}{J} \frac{dJ}{de_2} = \frac{e_1 e_2}{(e_2 + e_3)(e_1 + e_2 + e_3)} \quad \text{and} \quad C_{e_3}^{J} = \frac{e_3}{J} \frac{dJ}{de_3} = \frac{e_1 e_3}{(e_2 + e_3)(e_1 + e_2 + e_3)}. \tag{5.6}$$

Exercise 5.1.4

(a) Verify equations (5.5) and (5.6).

(b) The flux control coefficients for e_4 and e_5 in model (5.1) are zero. Why?

(c) Verify that for this network

$$C_{e_1}^{J} + C_{e_2}^{J} + C_{e_3}^{J} + C_{e_4}^{J} + C_{e_5}^{J} = 1.$$

This follows from the summation theorem of metabolic control analysis (see section 5.2). ☐

Exercise 5.1.5

(i) Verify the following concentration control coefficients for model (5.1):

$$C_{e1}^{s1} = \frac{e_2 + e_3}{e_1 + e_2 + e_3}, \quad C_{e2}^{s1} = -\frac{e_2}{e_1 + e_2 + e_3}, \quad C_{e3}^{s1} = -\frac{e_3}{e_1 + e_2 + e_3}, \quad C_{e4}^{s1} = C_{e5}^{s1} = 0. \qquad \square$$

We next address the behavior and regulation of metabolic pathways, starting with unbranched reaction chains.

5.2 Metabolic Pathways

The simplest metabolic network is an unbranched (so-called "linear") chain of reactions, as in figure 5.3. Truly unbranched metabolic pathways are rare, and they typically consist of only a few reactions. Nevertheless, by neglecting side reactions and reaction cofactors, models of metabolism frequently make use of this unbranched structure—it is the simplest way to describe a path from substrate to product.

5.2.1 Flux Control of Unbranched Pathways

In an unbranched pathway, all reactions rates must be equal at steady state, and so the pathway flux is the steady-state rate of every reaction.

Chains consisting of irreversible reactions display a simple behavior: the first reaction has a fixed rate and so dictates the pathway flux. In this case, all of the reactions except the first have flux control coefficients of zero, whereas the first reaction—which exerts total control over the pathway flux—has a coefficient of one.

To address reversible reaction chains, we consider, for concreteness, the three-reaction chain shown in figure 5.3. Suppose that all enzymes are operating in their first-order regimes. In that case we can write, for $i = 1, 2, 3$:

$$v_i = e_i(k_i s_{i-1} - k_{-i} s_i),$$

where e_i are enzyme concentrations, $s_i = [S_i]$, and $P = S_3$. To simplify the notation, we write each reaction rate in terms of the corresponding equilibrium constant: $q_i = k_i / k_{-i}$, so that

$$v_i = e_i k_i \left(s_{i-1} - \frac{s_i}{q_i} \right).$$

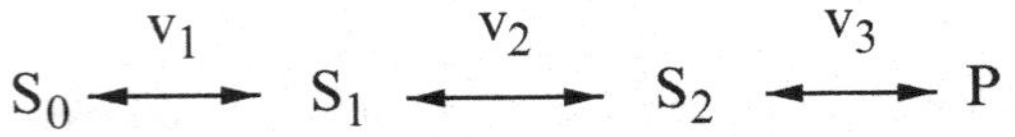

$$S_0 \xrightarrow{\quad v_1 \quad} S_1 \xrightarrow{\quad v_2 \quad} S_2 \xrightarrow{\quad v_3 \quad} P$$

Figure 5.3
Unbranched metabolic chain. The substrate S_0 and the product P are held at fixed concentrations.

The steady-state pathway flux can be written explicitly as

$$J = \frac{[S_0]q_1q_2q_3 - [P]}{\dfrac{q_1q_2q_3}{e_1k_1} + \dfrac{q_2q_3}{e_2k_2} + \dfrac{q_3}{e_3k_3}}.$$

(5.7)

Differentiation and scaling of equation (5.7) gives the flux control coefficients as

$$C_{e_1}^J = \frac{\dfrac{q_1q_2q_3}{e_1k_1}}{\dfrac{q_1q_2q_3}{e_1k_1} + \dfrac{q_2q_3}{e_2k_2} + \dfrac{q_3}{e_3k_3}}, \quad C_{e_2}^J = \frac{\dfrac{q_2q_3}{e_2k_2}}{\dfrac{q_1q_2q_3}{e_1k_1} + \dfrac{q_2q_3}{e_2k_2} + \dfrac{q_3}{e_3k_3}}, \quad C_{e_3}^J = \frac{\dfrac{q_3}{e_3k_3}}{\dfrac{q_1q_2q_3}{e_1k_1} + \dfrac{q_2q_3}{e_2k_2} + \dfrac{q_3}{e_3k_3}}.$$

(5.8)

In each case, the flux control coefficient $C_{e_i}^J$ varies inversely with the abundance e_i of the corresponding enzyme, so that as enzyme activity increases, the control coefficient decreases—the reaction loses "control" over the pathway flux. This fact provides a lesson for anyone seeking to increase the rate of pathway flux by increasing enzyme abundance (e.g., for metabolic engineering purposes). A law of diminishing returns limits the effect that changes in individual enzyme concentrations will have on pathway flux.

Exercise 5.2.1 Derive the flux control coefficients (5.8). ☐

The Summation Theorem Note that the flux control coefficients in (5.8) sum to one:

$$C_{e_1}^J + C_{e_2}^J + C_{e_3}^J = 1.$$

(5.9)

This is a consequence of the *summation theorem* of metabolic control analysis, which states that the flux control coefficients in any reaction chain will sum to one, regardless of the form of the kinetics (or any allosteric regulation). The summation theorem describes how the control of flux is shared among the steps in a pathway. The distribution of sensitivity dictated by this result invalidates the misconception that pathway flux is routinely controlled by a single "rate-limiting step" (which would have flux control coefficient equal to one). The summation theorem reveals that a rate-limiting step can only be present when all other reactions have virtually no influence over the pathway flux, which is rarely the case.[2]

---

2. An intuitive proof of the summation theorem follows from considering a perturbation that simultaneously increases all enzyme abundances by a factor α. The net effect of these changes on pathway flux is given by $\alpha C_{e_1}^J + \alpha C_{e_2}^J + \alpha C_{e_3}^J$. However, because all reaction rates increase by identical amounts, there is no impact on species concentrations, so the pathway flux increases by the factor α.

5.2.2   Regulation of Unbranched Pathways

We can think of an unbranched pathway as an assembly line that generates a product. In many cases, cellular activities require a steady rate of production—the pathway flux should be robust to perturbations. In other cases, pathways need to be responsive—to generate product when it is called for, and otherwise keep the production rate low. Sensitivity analysis can be used to address these aspects of pathway performance: robustness corresponds to insensitivity to perturbations, whereas responsiveness is characterized by high sensitivity with respect to the appropriate input signals.

Reversibility of reactions provides an inherent robustness to pathway flux. Robustness can also be introduced through allosteric regulation. The most common strategy for regulation of pathway flux is *end-product inhibition*—a feedback mechanism whereby increases in product concentration reduce the production rate.

Strategies for Regulation   In an end-product inhibition scheme, the pathway's product allosterically inhibits the enzyme that catalyzes the first reaction in the pathway, as in figure 5.4A. In a series of papers published in the 1970s, Michael Savageau used dynamic modeling to justify the prevalence of this inhibition scheme (Savageau, 1976). He compared a range of feedback strategies, such as those shown in panels B and C of figure 5.4. Savageau's analysis indicated that end-product inhibition exhibits better performance than alternative feedback architectures, where performance includes (i) robustness to disturbances in demand for product P, (ii) robustness to perturbations in enzyme activity levels, and (iii) responsiveness to changes in the availability of substrate S. (See problem 5.6.5 for Savageau's approach.)

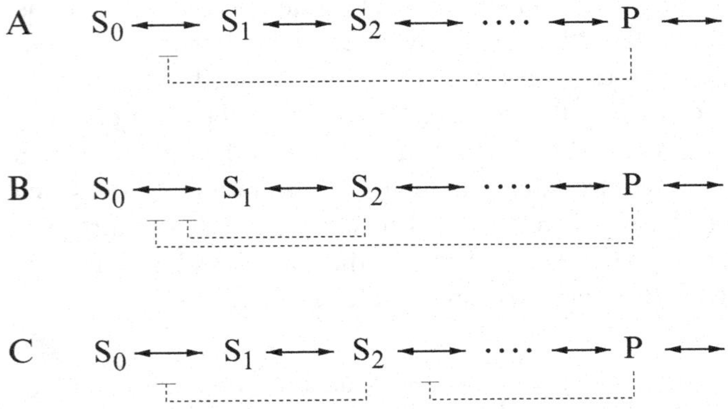

Figure 5.4
Strategies for regulation of unbranched metabolic chains. (A) End-product inhibition is the most commonly observed regulatory scheme. (B) A nested feedback scheme. (C) A sequential feedback scheme. The end-product inhibition scheme regulates flux better than the alternatives but is prone to instability. Adapted from figure 10.1 of Savageau (1976).

Savageau also identified a defect in the end-product inhibition motif—the potential for instability. A strong negative feedback, coupled with the delay inherent in passing material through the pathway, can lead to oscillatory behavior. (This is a delay oscillation, as discussed in section 4.3.) Such unstable behavior becomes more likely with increased time delay and thus with increased pathway length (see problem 5.6.2).

Metabolic Control Analysis of End-Product Inhibition In the 1973 paper that set up the MCA framework, Heinrich Kascer and Jim Burns addressed how end-product inhibition redistributes control of flux (Kascer et al., 1995). Before summarizing their findings, we need to introduce an additional piece of MCA terminology.

The relative rate of change of the reaction rate v_k with respect to metabolite S_j is called the **elasticity** of reaction k with respect to species j and is denoted $\varepsilon_{S_j}^k$:

$$\varepsilon_{S_j}^k = \frac{s_j}{v_k} \frac{\partial v_k}{\partial s_j}. \tag{5.10}$$

The elasticities are fundamentally different from the control coefficients. As partial derivatives of the rate laws, elasticities are properties of individual (isolated) reactions. Control coefficients, in contrast, reflect the role of the reaction *within* the network, as they describe the change in reaction flux at the system steady state. If S_j is a reactant in reaction k, then the elasticity of S_j coincides with its kinetic order (as defined in section 3.1.1). Species that allosterically regulate an enzyme's activity will have nonzero elasticities with respect to the corresponding reaction.

Exercise 5.2.2 Find the elasticity of the allosterically inhibited rate law

$$v = \frac{V_{\max}}{1+[I]/K_i} \frac{[S]}{K_M+[S]}$$

with respect to the substrate S and the inhibitor I. □

Referring to the end-product–inhibited chain in figure 5.5, the control coefficients for the flux J can be expressed in terms of the reaction elasticities as follows (see problem 5.6.4):

$$C_{e_1}^J = \frac{\varepsilon_{S_1}^2 \varepsilon_{S_2}^3}{\varepsilon_{S_1}^2 \varepsilon_{S_2}^3 - \varepsilon_{S_1}^1 \varepsilon_{S_2}^3 + \varepsilon_{S_1}^1 \varepsilon_{S_2}^2 - \varepsilon_{S_1}^2 \varepsilon_{S_2}^1}$$

$$C_{e_2}^J = \frac{-\varepsilon_{S_1}^1 \varepsilon_{S_2}^3}{\varepsilon_{S_1}^2 \varepsilon_{S_2}^3 - \varepsilon_{S_1}^1 \varepsilon_{S_2}^3 + \varepsilon_{S_1}^1 \varepsilon_{S_2}^2 - \varepsilon_{S_1}^2 \varepsilon_{S_2}^1} \tag{5.11}$$

$$C_{e_3}^J = \frac{\varepsilon_{S_1}^1 \varepsilon_{S_2}^2 - \varepsilon_{S_1}^2 \varepsilon_{S_2}^1}{\varepsilon_{S_1}^2 \varepsilon_{S_2}^3 - \varepsilon_{S_1}^1 \varepsilon_{S_2}^3 + \varepsilon_{S_1}^1 \varepsilon_{S_2}^2 - \varepsilon_{S_1}^2 \varepsilon_{S_2}^1}.$$

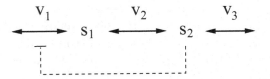

Figure 5.5
End-product–inhibited three-reaction chain.

The feedback effect (characterized by $\varepsilon_{S_2}^1$) appears in the numerator of $C_{e_3}^J$ and in the final term of the (identical) denominators. Because S_2 inhibits the first reaction, $\varepsilon_{S_2}^1$ is negative. If the inhibition were absent, $\varepsilon_{S_2}^1$ would be zero. Comparing the flux control coefficients in the presence or absence of end-product inhibition, we note that the control coefficients for enzymes E_1 and E_2 are decreased by the inhibitory feedback (the numerators are unchanged by the feedback, whereas the denominators include the additional positive term $-\varepsilon_{S_1}^2 \varepsilon_{S_2}^1$). Thus, the inhibitory feedback reduces the sensitivity of the pathway flux J to changes in enzyme activities e_1 and e_2: the regulation insulates the flux against these perturbations. Intuitively, if the activity of either of these enzymes is increased, there will be an increase in the concentration s_2, which will partially counteract any increase in the rates of the first two reactions.

We next consider the effect of end-product inhibition on the flux control coefficient for enzyme E_3. Recall that from the summation theorem, the control coefficients must sum to one—regardless of the feedback structure. Consequently, the inhibitory feedback, which decreases $C_{e_1}^J$ and $C_{e_2}^J$, must *increase* the value of $C_{e_3}^J$ by the same amount. The inhibition thus increases the sensitivity of the flux to perturbations of enzyme E_3 (which is outside of the feedback loop). This result surprised some members of the biochemistry community when it was published. Many had believed that the best target for manipulation of the flux of this pathway would be enzyme E_1—the regulated reaction. Initially, there was some resistance to the notion that end-product inhibition draws sensitivity away from enzymes within the feedback loop, but this theoretical prediction has since been confirmed by experimental evidence showing the lack of effect when such regulated enzymes are perturbed. Modern attempts at flux modification make use of the fact that a perturbation outside of a feedback loop is more likely to yield a successful intervention.

Exercise 5.2.3 Verify that every term in the denominator of the control coefficients in equations (5.11),

$$\varepsilon_{S_1}^2 \varepsilon_{S_2}^3, \ -\varepsilon_{S_1}^1 \varepsilon_{S_2}^3, \ \varepsilon_{S_1}^1 \varepsilon_{S_2}^2, \text{ and } -\varepsilon_{S_1}^2 \varepsilon_{S_2}^1,$$

is nonnegative.   □

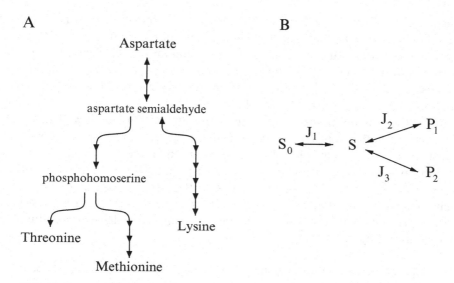

Figure 5.6
(A) Branched metabolic pathway: biosynthesis of lysine, threonine, and methionine in *E. coli*. Compound arrows indicate multistep reaction chains. (B) Pathway branch point.

5.2.3 Branched Pathways

Models of unbranched metabolic pathways are typically simplifications: most networks are highly branched, with metabolites being shared among multiple pathways. As an example, figure 5.6A shows the biosynthetic pathways that lead to production of the amino acids lysine, threonine, and methionine in the bacterium *Escherichia coli*.

We next discuss branched pathways, focusing on the flux control at a branch point (figure 5.6B).

Control of Flux at a Branch Point The flux at the branch point in figure 5.6B consists of three unequal flows, labeled J_1, J_2, and J_3. These steady-state fluxes satisfy $J_1 = J_2 + J_3$. The flux control coefficients depend on the *split ratio*—the proportion of flux that passes through each branch. When the split is equal, control is symmetric between the two balanced branches. To gain insight into the behavior of imbalanced split ratios, we will consider the extreme case in which the flow is carried almost exclusively by J_2, so that $J_2 \approx J_1$. As discussed by Herbert Sauro (Sauro, 2009), the effects of changes in enzyme E_3, which carries a negligible flux, are described by

$$C_{e3}^{J_2} \approx 0 \quad \text{and} \quad C_{e3}^{J_3} \approx 1.$$

Thus, when the flux through J_3 is minimal, changes in enzyme E_3 have very little effect on the larger flux J_2 and have complete influence over the smaller flux J_3. This makes intuitive sense: the branch carrying J_3 has minimal impact on what is otherwise a linear pathway through reactions 1 and 2. Furthermore, because changes in E_3 have almost no impact on $[S]$, there is no systemic response to these perturbations, so the resulting changes are felt directly in J_3.

Recall that in the case of an unbranched chain, the summation theorem dictates that if one reaction has flux control coefficient near one, the remaining control coefficients must have values near zero. The summation theorem also holds for a branched pathway—the flux control coefficients sum to one. However, in this case the control coefficients may take negative values. In the limiting case that $J_2 \approx J_1$, the summation theorem dictates that $C_{e_1}^{J_3} + C_{e_2}^{J_3} \approx 0$ (because $C_{e_3}^{J_3} \approx 1$). That is, these flux control coefficients, which dictate the effects of e_1 and e_2 on the minimal flux J_3, are of equal magnitude and of opposite signs.

Regulation of Branch Point Flux   Biosynthetic pathways generate a wide range of end-products from a relatively small pool of precursors. This allows for an efficient use of materials, but the resulting branching structure poses a challenge for regulation: what feedback strategies best allow the production rates to be continually adjusted to meet cellular demand? A wide variety of regulatory architectures have been discovered in branching pathways. These can be broadly classified as *sequential feedback* or *nested feedback*.

In a sequential feedback strategy (figure 5.7A), the end-product of each branch inhibits flux through the respective branch, whereas consumption of the initial substrate is only inhibited by common intermediates. Michael Savageau analyzed this

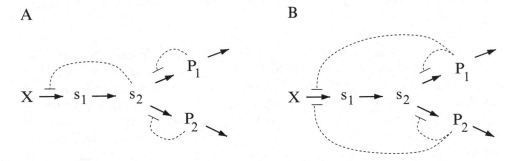

Figure 5.7
Regulation of branched biosynthesis. (A) Sequential feedback inhibition. (B) Nested feedback inhibition. Adapted from figure 12.1 of Savageau (1976).

feedback strategy and found that it provides for efficient independent regulation of the two end-products (Savageau, 1976).

However, the more commonly occurring regulatory architecture is nested feedback (figure 5.7B), in which the end-products directly inhibit one or more of the shared steps in the pathway. Savageau's analysis revealed that this nested feedback structure is susceptible to significant inefficiency. In particular, if the shared pathway is inhibited too strongly, an increase in the availability of one end-product can inappropriately suppress production of the other. This dysfunctional regulation has been observed experimentally: some microbial strains, when provided with an abundant external supply of a particular amino acid, fail to generate sufficient supplies of closely related amino acids, and thus exhibit stunted growth.

Inefficiency in nested feedback can be tempered by a number of mechanisms, such as *enzyme multiplicity*, in which multiple enzymes catalyze a single reaction. Each enzyme can be inhibited independently, thus allowing each end-product to have influence over a portion of the shared flux. An example is provided by the production of the amino acids lysine, threonine, and methionine in *E. coli* (figure 5.6A). In that branched pathway, each end-product inhibits both the first dedicated step in its own branch and one of three distinct enzymes (called aspartate kinase I, II, and III) responsible for catalyzing the first step in the pathway (not shown in the figure). (An example of a detailed study addressing regulation of amino acid production, including the roles of enzyme multiplicity, is Curien et al. (2009).)

5.3 Modeling of Metabolic Networks

5.3.1 Model Construction
As described in section 5.1, the development of metabolic models is straightforward once the kinetics of the enzyme-catalyzed reactions have been characterized and the corresponding parameter values have been determined.

Decades of effort have gone into the investigation of enzymatic mechanisms, and there is an extensive literature on enzyme kinetics. These studies focus on reaction mechanisms: they rarely address the in vivo role of enzymes. As a consequence, most metabolic pathways can only be modeled after in vivo experimental data have been collected.

Constructing Models from the Literature   Without targeted laboratory experiments, the potential for building accurate models of metabolic networks is limited, but in some cases there is sufficient data available in the literature to support a plausible model. Knowledge of reaction networks is often readily available: the major metabolic pathways of most organisms follow a common framework, and atlases of "standard" metabolism have been compiled. Species-specific reaction networks are archived in online databases (such as KEGG and MetaCyc), which are continually updated with new information.[3]

Much of the published enzymological data are available in online databases such as BRENDA and SABIO-RK.[4] Unfortunately, construction of accurate metabolic models from the information in these databases is not straightforward, for two reasons.

First, in attempting to determine the specifics of kinetic mechanisms, enzymologists usually study enzyme activity in simplified conditions. The enzymes are often purified, and reaction rates are typically measured in the absence of reaction products, in the presence of only one regulator at a time, and at convenient or revealing temperature and pH. These experiments successfully demonstrate reaction mechanism but do not always provide characterizations of the in vivo activity of the enzyme. Second, enzymological studies focus on the role of enzymes as catalysts, not as components of cellular networks. In particular, they do not usually address in vivo enzyme abundance.

Enzymological databases typically list a Michaelis constant (K_M) for each enzyme substrate and an inhibition constant (K_i) for each inhibitor. Rather than report a V_{max} value, these studies measure the catalytic constant (or *turnover number*) of the enzyme. The catalytic constant, k_{cat}, is related to V_{max} by $V_{max} = k_{cat}e_T$, where e_T is the enzyme concentration. (Some studies report the *specific activity*, which is a measure of catalytic activity per milligram of enzyme. This is related to k_{cat} by the enzyme's molecular mass.) To determine a V_{max} value for use in a metabolic model, the in vivo enzyme abundance must be determined. This in vivo data is specific to cell type and condition and is not yet well-archived online (although it appears in some databases, such as the *E. coli*–focused GENOBASE).[5]

5.3.2   Case Study: Modeling the Regulation of the Methionine Cycle

In the year 2000, Michael Martinov and his colleagues published a model describing aspects of methionine metabolism in mammalian lever cells (Martinov et al., 2000). The network of interest, shown in figure 5.8, starts with methionine combining with ATP to form *S*-adenosylmethionine (referred to as AdoMet). Cells use AdoMet as a methyl (CH_3) donor in a wide range of reactions. When donating a methyl group in one of these *transmethylation* reactions, AdoMet is converted to *S*-adenosylhomocysteine (referred to as AdoHcy). AdoHcy is then converted, reversibly, to homo-

---

3. KEGG: www.genome.jp/kegg. MetaCyc: www.metacyc.org.

4. BRENDA: www.brenda-enzymes.org. SABIO-RK: http://sabio.villa-bosch.de.

5. GENOBASE: http://ecoli.naist.jp/GB8.

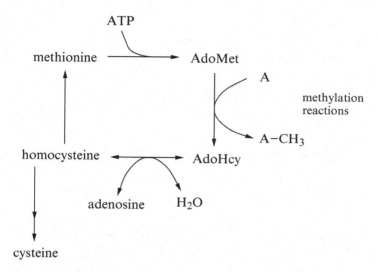

Figure 5.8
The methionine cycle. Methionine combines with ATP to form S-adenosylmethionine (AdoMet). When AdoMet donates a methyl group in any of a range of methylation reactions, it is converted to S-adenosylhomocysteine (AdoHcy). AdoHcy releases the adenosine moiety of the previously bound ATP molecule to become homocysteine, which can be converted to methionine or cysteine. Adapted from figure 5.1 of Martinov et al. (2000).

cysteine (Hcy), releasing the adenosine moiety of the previously bound ATP molecule in the process. Homocysteine can then follow one of two paths. It can be converted to methionine, hence completing the *methionine cycle*, or it can be converted to cysteine.

Martinov and colleagues constructed their model to address the following issues:

• The enzyme that produces AdoMet from methionine is called methionine adenosyl transferase (MAT). In liver cells, there are two distinct forms of MAT, called MATI and MATIII. (Another form, MATII, occurs in non-liver cells.) AdoMet interacts with both MATI and MATIII, causing inhibition of MATI and activation of MATIII. Both MATI and MATIII were included in the model in an attempt to determine their distinct roles in the pathway.

• In all cell types, there are various methyltransferase enzymes whose action results in the conversion of AdoMet to AdoHcy. In liver cells, there is an additional enzyme, glycine N-methyltransferase (GNMT), that catalyzes this conversion. The presence of GNMT is somewhat puzzling. It methylates glycine (an amino acid) to form sarcosine, but sarcosine has no apparent cellular function—it can be converted back to glycine or is exported from the cell.

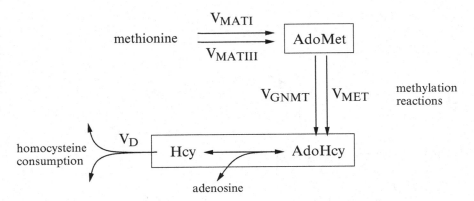

Figure 5.9
Martinov model of methionine metabolism in mammalian liver cells. The production of AdoMet is catalyzed by two distinct enzymes: methionine adenosyl transferase I (MATI) and methionine adenosyl
transferase III (MATIII). Except for the reaction catalyzed by glycine *N*-methyltransferase (GNMT), all
other methylation processes are described by a single lumped reaction. Likewise, consumption of homocysteine (Hcy) is lumped into a single consumption reaction. The pool of AdoHcy and Hcy is presumed
to be in rapid equilibrium. Adapted from figure 5.2 of Martinov et al. (2000).

• Experimental evidence has shown that as dietary methionine levels rise 10-fold,
the methionine concentration in liver cells shows only a modest increase. The intracellular abundance of AdoMet and AdoHcy, in contrast, show a different behavior.
These concentrations increase slowly over a low range of methionine levels and then
jump abruptly (a fourfold increase) at a threshold concentration of methionine.

   Figure 5.9 shows the model network. In constructing the model, Martinov and
colleagues made the following simplifying assumptions: (i) the methionine concentration is held constant, as are the concentrations of the reaction cofactors: ATP,
adenosine, H_2O, glycine, and the methyl acceptors (lumped into a single species A);
and (ii) the interconversion of AdoHcy and Hcy is in rapid equilibrium. The latter
assumption is justified by the experimental observation that adenosyl homocyste-
inase—the enzyme responsible for catalyzing this reaction—has an activity level at
least ten times higher than the other enzymes in the model. This rapid equilibrium
assumption leads to the condition

$$K_{AHC} = \frac{[\text{Adenosine}][\text{Hcy}]}{[\text{AdoHcy}]}, \tag{5.12}$$

where K_{AHC} is the equilibrium constant for the reaction.

   The rates of the reactions in the model are labeled by the names of the corresponding enzymes: V_{MATI}, V_{MATIII}, V_{GNMT}. Except for the reaction catalyzed by GNMT,
all other methylation reactions are lumped into a single process with rate V_{MET}.

Likewise, the two reactions consuming homocysteine are lumped into a single consumption reaction with rate V_D. The model describes the concentration of AdoMet and the equilibrated pool of AdoHcy and Hcy by

$$\frac{d}{dt}[\text{AdoMet}] = V_{\text{MATI}} + V_{\text{MATIII}} - V_{\text{GNMT}} - V_{\text{MET}} \tag{5.13}$$

$$\frac{d}{dt}[\text{AdoHcy} - \text{Hcy pool}] = V_{\text{GNMT}} + V_{\text{MET}} - V_{\text{D}}. \tag{5.14}$$

To reformulate the model in terms of the metabolite AdoHcy, we describe the concentration of AdoHcy as a fraction of the combined pool. From equation (5.12):

$$[\text{AdoHcy}] = \left(\frac{1}{1 + \dfrac{K_{\text{AHC}}}{[\text{Adenosine}]}} \right) [\text{AdoHcy} - \text{Hcy pool}]. \tag{5.15}$$

We can then use equation (5.14) to write

$$\frac{d}{dt}[\text{AdoHcy}] = \frac{1}{1 + \dfrac{K_{\text{AHC}}}{[\text{Adenosine}]}} (V_{\text{GNMT}} + V_{\text{MET}} - V_{\text{D}}). \tag{5.16}$$

The reaction rates are described as follows. They are specified in terms of Michaelis–Menten kinetics, incorporating allosteric regulation and cooperative effects (as developed in chapter 3).

Methionine Adenosyl Transferase I (MATI) The substrate for this reaction is methionine (Met). MATI is inhibited by its product AdoMet[6]:

$$V_{\text{MATI}} = \frac{V_{\text{max}}^{\text{MATI}}}{1 + \dfrac{K_{\text{m}}^{\text{MATI}}}{[\text{Met}]}} \left(\frac{1}{1 + \dfrac{[\text{AdoMet}]}{K_{\text{i}}^{\text{MATI}}}} \right).$$

Methionine Adenosyl Transferase III (MATIII) This reaction shows a sigmoidal dependence on its substrate Met, suggesting a cooperative mode of action. Furthermore, this enzyme is allosterically activated by its product AdoMet:

6. Note that the standard Michaelis–Menten formula can be written as

$$\frac{V_{\text{max}}}{1 + \dfrac{K_{\text{M}}}{[S]}}.$$

$$V_{\text{MATIII}} = \frac{V_{\max}^{\text{MATIII}}}{1 + \dfrac{K_{m1}^{\text{MATIII}} K_{m2}^{\text{MATIII}}}{[\text{Met}]^2 + [\text{Met}] K_{m2}^{\text{MATIII}}}}, \quad \text{where} \quad K_{m1}^{\text{MATIII}} = \frac{20{,}000}{1 + 5.7 \left(\dfrac{[\text{AdoMet}]}{[\text{AdoMet}] + 600} \right)^2}$$

Glycine N-Methyltransferase (GNMT) This reaction shows a sigmoidal dependence on its substrate, AdoMet, and is inhibited by its product AdoHcy:

$$V_{\text{GNMT}} = \left(\frac{V_{\max}^{\text{GNMT}}}{1 + \left(\dfrac{K_m^{\text{GNMT}}}{[\text{AdoMet}]} \right)^{2.3}} \right) \left(\frac{1}{1 + \dfrac{[\text{AdoHcy}]}{K_i^{\text{GNMT}}}} \right).$$

Methylation Reactions (MET) Methylation reactions other than GNMT are described by a single lumped reaction. These reactions are inhibited by AdoHcy. The rate law uses a single Michaelis constant for the methyl-accepting substrates A:

$$V_{\text{MET}} = \frac{V_{\max}^{\text{MET}}}{1 + \dfrac{K_{m1}^{\text{MET}}}{[\text{AdoMet}]} + \left(\dfrac{K_{m2}^{\text{MET}}}{[\text{A}]} \right) + \left(\dfrac{K_{m2}^{\text{MET}}}{[\text{A}]} \dfrac{K_{m1}^{\text{MET}}}{[\text{AdoMet}]} \right)} \quad \text{where}$$

$$K_{m1}^{\text{MET}} = 10 \left(1 + \frac{[\text{AdoHcy}]}{4} \right).$$

Homocysteine Consumption (D) A mass-action rate is used for the lumped reaction consuming homocysteine:

$$V_{\text{D}} = \alpha_d [\text{Hcy}] = \alpha_d [\text{AdoHcy}] \frac{K_{\text{AHC}}}{[\text{Adenosine}]}.$$

Martinov and his colleagues determined the values of the corresponding kinetic parameters (maximal rates $V_{\max}$, Michaelis constants K_m, and inhibition constants K_i) from previously reported studies (figure 5.10).

Model Behavior Figure 5.10 shows a simulation of the model at a low methionine level. There is a separation of timescales between the two state variables. The AdoHcy concentration drops within minutes and remains steady, whereas the AdoMet concentration relaxes over several hours.

This separation of timescales is also apparent in the phase portraits in figure 5.11. All trajectories converge directly to the AdoHcy nullcline and then follow it to a steady state. Figure 5.11A corresponds to the low methionine concentration in figure 5.10. Trajectories converge to the single steady state, at the intersection of the

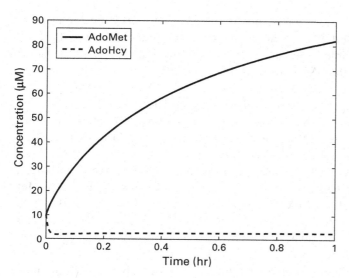

Figure 5.10
Simulation of the Martinov methionine metabolism model. Initial concentrations are [AdoMet] = 10 μM, AdoHcy = 10 μM. A separation of timescales is apparent: the AdoHcy level drops rapidly to its low steady value, whereas [AdoMet] climbs to steady state over a longer time interval. (Steady state is reached after about 5 hours.) The methionine concentration is 48.5 μM. Other parameter values are (in μM hr^{-1}): $V_{\max}^{\text{MATI}} = 561$, $V_{\max}^{\text{MATIII}} = 22870$, $V_{\max}^{\text{GNMT}} = 10600$, $V_{\max}^{\text{MET}} = 4544$; (in hr^{-1}): $\alpha_d = 1333$; (in μM): $K_{\text{AHC}} = 0.1$, [Adenosine] = 1, $K_{\text{m}}^{\text{MATI}} = 41$, $K_{\text{i}}^{\text{MATI}} = 50$, $K_{\text{m2}}^{\text{MATIII}} = 21.1$, $K_{\text{m}}^{\text{GNMT}} = 4500$, $K_{\text{i}}^{\text{GNMT}} = 20$; $K_{\text{m2}}^{\text{MET}}/[\text{A}] = 10$.

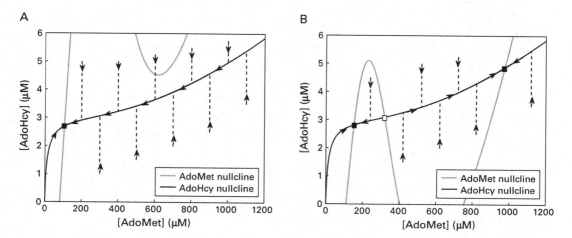

Figure 5.11
Phase portraits of the Martinov methionine metabolism model. A separation of timescales is apparent. The trajectories (dashed lines) converge rapidly to the AdoHcy nullcline and then follow it to steady state. (A) At a methionine concentration of 48.5 μM, the nullclines intersect once: the system is mono-stable. (B) At a methionine concentration of 51 μM, the nullclines intersect three times, at two stable steady states (filled boxes) and one unstable steady state (empty box). Other parameter values are as for figure 5.10. Adapted from figure 5.3 of Martinov et al. (2000).

nullclines. Figure 5.11B shows the phase portrait at a higher methionine concentration. The nullclines have shifted: they now intersect three times, indicating that the system is bistable. One stable steady state exhibits a low AdoMet level; the other corresponds to a much higher AdoMet concentration.

The system's bistability is caused by the activity of MATIII and GNMT, both of which show nonlinear dependence on AdoMet levels. The positive feedback on MATIII (whereby AdoMet enhances its own production) generates the high-AdoMet steady state. The AdoMet concentration that is attained in this elevated state is set by the GNMT-specific AdoMet consumption rate.

At the low AdoMet state, the bulk of the pathway flux is through MATI and MET. This is the standard operating behavior for the cycle. In contrast, at the high AdoMet state, the positive feedback on MATIII increases the flux through MATIII while the negative feedback on MATI decreases flux through that reaction. Thus, MATIII carries the bulk of AdoMet production. This high level of AdoMet induces a significant increase in GNMT flux.

The network uses bistability to keep the methionine level steady: when methionine levels rise, the system switches to the high-AdoMet state in which the MATIII–GNMT "shunt" is active, shuttling extra methionine directly to AdoHcy without affecting the activity of cellular methylases. This is a valuable safety measure: methylation reactions are crucial to a wide range of cellular processes, and perturbations to the rates of these reactions might have wide-reaching consequences. The presence of GNMT as a safety valve insulates methylation rates from changes in methionine level.

Further insight into system behavior comes from the bifurcation diagram in figure 5.12, which reveals the system's response to changing methionine levels. As noted earlier, experimental observations show that increases in methionine over a low range cause only a small increase in the AdoMet concentration. Then, at a threshold methionine concentration, the AdoMet level increases dramatically. In the model, this threshold corresponds to the narrow region of bistability, above which only the high AdoMet state is present. (The model's prediction of a sharp rise in both AdoMet concentration and methionine consumption rate was later experimentally validated (Korendyaseva et al., 2008).)

This model cannot be used to explore the pathway's ability to regulate methionine levels because it treats the methionine concentration [Met] as a fixed input. A follow-up study appeared in 2004, in which Michael Reed and colleagues presented an expanded model that takes methionine production rate—rather than methionine concentration—as input (Reed et al., 2004; see also Martinov et al., 2010).

5.4* Stoichiometric Network Analysis

As we have seen, the construction of dynamic models begins with a description of reaction kinetics. In this section, we address an approach that is used to study reaction

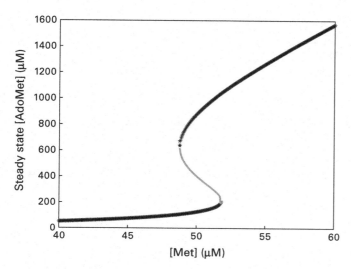

Figure 5.12
Bifurcation diagram for the Martinov methionine metabolism model. As the methionine concentration rises, the system passes through a narrow region of bistability. At low methionine levels, the AdoMet concentration is nearly constant. Beyond the region of bistability, [AdoMet] increases rapidly with methionine concentration. Parameter values are as for figure 5.10. Adapted from figure 5.6 of Martinov et al. (2000).

networks for which the kinetics are unknown: stoichiometric network analysis. Stoichiometric techniques focus on steady-state behavior: they rely primarily on network architecture. The material in this section makes use of vectors and matrices (i.e., linear algebra). Readers unfamiliar with this material may wish to consult appendix B.

As mentioned in section 5.3, there are a great many metabolic systems for which the reaction network is known, but the associated kinetic mechanisms and parameters are yet to be determined. This is especially true for large-scale networks that have been reconstructed from genomic data (Palsson, 2006). These networks regularly contain thousands of species and reactions.

The Stoichiometry Matrix For a reaction network composed of n species involved in m reactions, the **stoichiometry matrix**, denoted $\mathbf{N}$, is an n-by-m matrix whose ij-th element is the net number of molecules of species i involved in reaction j. As an example, consider the network shown in figure 5.13. There are two species, S_1 and S_2, involved in four reactions, with rates v_1, v_2, v_3, and v_4. The stoichiometry matrix is

$$\mathbf{N} = \begin{bmatrix} 1 & -1 & 0 & 0 \\ 0 & 1 & -1 & -1 \end{bmatrix} \begin{matrix} \leftarrow S_1 \\ \leftarrow S_2 \end{matrix}$$
$$\begin{matrix} \uparrow & \uparrow & \uparrow & \uparrow \\ v_1 & v_2 & v_3 & v_4 \end{matrix}$$

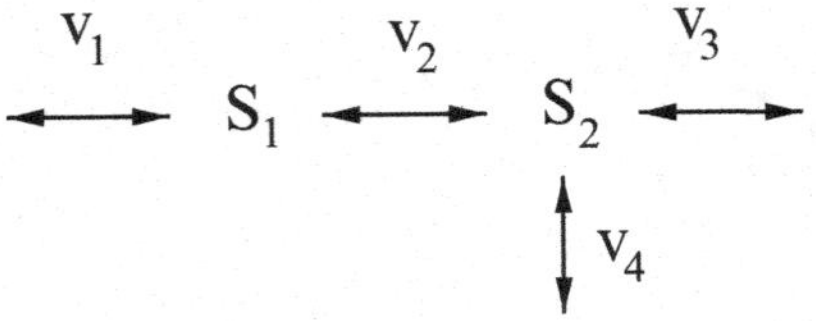

Figure 5.13
Branched metabolic network. The reaction rates are denoted v_i. The forward direction for each reaction is rightward (downward for v_4).

If the species concentrations and reaction rates are organized into vectors $\mathbf{s}$ and $\mathbf{v}$ (with $s_i = [S_i]$), that is,

$$\mathbf{s} = \begin{bmatrix} s_1 \\ s_2 \end{bmatrix}, \qquad \mathbf{v} = \begin{bmatrix} v_1 \\ v_2 \\ v_3 \\ v_4 \end{bmatrix},$$

then the network dynamics are described by the vector differential equation

$$\frac{d}{dt}\mathbf{s}(t) = \mathbf{N}\mathbf{v}, \tag{5.17}$$

which, in expanded form, is

$$\frac{d}{dt}\begin{bmatrix} s_1(t) \\ s_2(t) \end{bmatrix} = \begin{bmatrix} 1 & -1 & 0 & 0 \\ 0 & 1 & -1 & -1 \end{bmatrix} \begin{bmatrix} v_1 \\ v_2 \\ v_3 \\ v_4 \end{bmatrix} = \begin{bmatrix} v_1 - v_2 \\ v_2 - v_3 - v_4 \end{bmatrix}.$$

Structural Conservations   The stoichiometry matrix provides a description of all structural conservations (e.g., conserved moieties) in the network, as follows.

Let the row vector $\mathbf{N}_i$ denote the i-th row of $\mathbf{N}$, so that

$$\mathbf{N} = \begin{bmatrix} \mathbf{N}_1 \\ \mathbf{N}_2 \\ \vdots \\ \mathbf{N}_n \end{bmatrix}.$$

Then, we can expand the system dynamics in equation (5.17) as

$$\frac{d}{dt}s_1(t) = \mathbf{N}_1\mathbf{v}$$

$$\frac{d}{dt} s_2(t) = \mathbf{N}_2 \mathbf{v}$$

$$\vdots$$

$$\frac{d}{dt} s_n(t) = \mathbf{N}_n \mathbf{v}.$$

The dynamic behavior of combined quantities can be expressed in terms of the products $\mathbf{N}_i \mathbf{v}$. For instance

$$\frac{d}{dt}(s_1(t) + s_2(t)) = \mathbf{N}_1 \mathbf{v} + \mathbf{N}_2 \mathbf{v} = (\mathbf{N}_1 + \mathbf{N}_2) \mathbf{v}.$$

It follows that conserved moieties—which are combinations of species whose time derivatives sum to zero—correspond to combinations of species for which the corresponding rows $\mathbf{N}_i$ sum to zero.

As an example, consider the enzyme-catalysis network shown in figure 5.14. The stoichiometry matrix is

$$\mathbf{N} = \begin{bmatrix} \mathbf{N}_1 \\ \mathbf{N}_2 \\ \mathbf{N}_3 \\ \mathbf{N}_4 \\ \mathbf{N}_5 \end{bmatrix} = \begin{bmatrix} -1 & 0 & 0 \\ -1 & 0 & 1 \\ 1 & -1 & 0 \\ 0 & 1 & -1 \\ 0 & 0 & 1 \end{bmatrix} \begin{array}{l} \leftarrow S \\ \leftarrow E \\ \leftarrow ES \\ \leftarrow EP \\ \leftarrow P \end{array}$$
$$\begin{array}{ccc} \uparrow & \uparrow & \uparrow \\ v_1 & v_2 & v_3 \end{array}$$

In this case, there are two conservations: rows $\mathbf{N}_2$, $\mathbf{N}_3$, and $\mathbf{N}_4$ sum to zero, so $[E]$ + $[ES]$ + $[EP]$ = constant, and rows $\mathbf{N}_1$, $\mathbf{N}_3$, $\mathbf{N}_4$, and $\mathbf{N}_5$ sum to zero, so $[S]$ + $[ES]$ + $[EP]$ + $[P]$ = constant.

For simple networks, it may seem unnecessary to identify conservations from the stoichiometry matrix because they can be readily determined by inspection of the network. However, for larger networks—with hundreds or thousands of metabolites

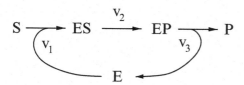

Figure 5.14
Closed enzyme-catalysis reaction network. The reaction rates are denoted v_i.

—a computational algorithm for identification of conservations is required. Each structural conservation corresponds to an independent vector in the *left nullspace* of **N** (i.e., the *nullspace*, or *kernel*, of the transpose of **N**). The elements in this nullspace can be determined by standard linear algebraic methods (e.g., Gaussian elimination).

Exercise 5.4.1   Verify that for the network in figure 5.14, the vectors $[0\ 1\ 1\ 1\ 0]^T$ and $[1\ 0\ 1\ 1\ 1]^T$, (superscript T for transpose) which correspond to the two conservations, lie in the kernel of the transpose of **N**.   □

Exercise 5.4.2   Consider an open variant of the network of figure 5.14, in which two additional reactions have been included: production of S, with rate v_0; and consumption of P, with rate v_4. Determine the stoichiometry matrix **N** for this expanded network. By addressing the rows of **N**, verify that in this case, the total enzyme is still conserved, but there is no conservation involving S or P.   □

Exercise 5.4.3   Consider the network shown in figure 5.15. Determine the stoichiometry matrix, and, by analyzing its rows, verify that the system exhibits two structural conservations, one of which cannot be written as a conserved moiety (i.e., as a conserved sum of concentrations).   □

5.4.1 Metabolic Pathway Analysis

If kinetic descriptions of the reaction rates were available, we could incorporate the kinetics into the reaction-rate vector $\mathbf{v} = \mathbf{v}(\mathbf{s})$ and write the differential equation (5.17) in terms of the concentration vector $\mathbf{s}(t)$. When kinetic descriptions are not available, we abandon the goal of describing dynamics and restrict our attention to steady state. Equation (5.17) gives a steady-state condition of

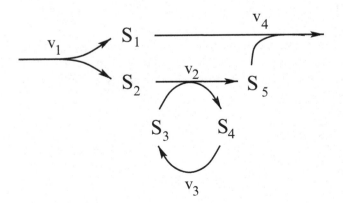

Figure 5.15
Reaction network for exercise 5.4.3. The reaction rates are denoted v_i.

$$\mathbf{0} = \mathbf{Nv}. \tag{5.18}$$

This is called the *balance equation*. Regardless of the kinetics, this equation must be satisfied by the reaction rate profile $\mathbf{v}$ at steady state. We can thus explore possible steady-state flux profiles by treating $\mathbf{v}$ as an unknown variable in the balance equation.

Equation (5.18) is a homogeneous system of linear equations. This is a standard object of study in linear algebra. (It can be solved by, for instance, Gaussian elimination.) The solutions are vectors that lie in the nullspace of $\mathbf{N}$. For metabolic networks, the balance equation does not typically have a unique solution; instead, it admits a family of solutions. As we will see, each vector $\mathbf{v}$ in this family describes a *pathway* within the network.

As an example, consider the network in figure 5.13. The corresponding balance equation $\mathbf{0} = \mathbf{Nv}$ is satisfied by the two flux profiles

$$\mathbf{v}_1 = \begin{bmatrix} 1 \\ 1 \\ 1 \\ 0 \end{bmatrix} \quad \mathbf{v}_2 = \begin{bmatrix} 1 \\ 1 \\ 0 \\ 1 \end{bmatrix}. \tag{5.19}$$

Vector $\mathbf{v}_1$ corresponds to the steady-state behavior in which there is equal flow through reactions 1, 2, and 3. Vector $\mathbf{v}_2$ corresponds to equal flow through reactions 1, 2, and 4. These are the two branches of the network. Vectors $\mathbf{v}_1$ and $\mathbf{v}_2$ are not the only solutions to the balance equation. By choosing any coefficients α_1, α_2, we can construct a new solution as the sum $\mathbf{v} = \alpha_1 \mathbf{v}_1 + \alpha_2 \mathbf{v}_2$ (called a *linear combination* of $\mathbf{v}_1$ and $\mathbf{v}_2$). In fact, for this network *every* steady-state flux profile $\mathbf{v}$ takes this form. That is, given any vector $\mathbf{v}$ satisfying $\mathbf{0} = \mathbf{Nv}$, there is some pair of numbers α_1, α_2, for which $\mathbf{v} = \alpha_1 \mathbf{v}_1 + \alpha_2 \mathbf{v}_2$. (This is described technically by saying that $\mathbf{v}_1$ and $\mathbf{v}_2$ *span* the family of solutions.) This family of linear combinations thus provides a concise description of all steady-state flux profiles for the network.

Exercise 5.4.4 For the following three vectors, verify that the corresponding flux profile satisfies the balance equation for the network in figure 5.13, and then find numbers α_1 and α_2 for which the vector can be written as $\mathbf{w}_i = \alpha_1 \mathbf{v}_1 + \alpha_2 \mathbf{v}_2$ for $\mathbf{v}_1$ and $\mathbf{v}_2$ as in equation (5.19):

$$\mathbf{w}_1 = \begin{bmatrix} 2 \\ 2 \\ 1 \\ 1 \end{bmatrix}, \quad \mathbf{w}_2 = \begin{bmatrix} 6 \\ 6 \\ 5 \\ 1 \end{bmatrix}, \quad \mathbf{w}_3 = \begin{bmatrix} 0 \\ 0 \\ -1 \\ 1 \end{bmatrix}.$$

In each case, describe the corresponding flow pattern (with reference to figure 5.13). □

Exercise 5.4.5 The choice of flux profiles $\mathbf{v}_1$ and $\mathbf{v}_2$ in (5.19) is not unique. Consider the pair

$$\hat{\mathbf{v}}_1 = \begin{bmatrix} -2 \\ -2 \\ -2 \\ 0 \end{bmatrix} \quad \hat{\mathbf{v}}_2 = \begin{bmatrix} 1 \\ 1 \\ -1 \\ 2 \end{bmatrix}.$$

For each vector $\mathbf{w}_i$ in exercise 5.4.4, find numbers α_1 and α_2 so that $\mathbf{w}_i = \alpha_1 \hat{\mathbf{v}}_1 + \alpha_2 \hat{\mathbf{v}}_1$. □

Irreversibility Constraints Some reactions within a metabolic reaction network can be readily identified as irreversible. (For example, a reaction that involves ATP hydrolysis will not proceed in the opposite direction.) Irreversibility imposes constraints on the possible steady-state flux profiles. We say that a steady-state flux profile $\mathbf{v}$ is *feasible* if it satisfies the balance equation (5.18) and does not violate any irreversibility conditions.

As an example, consider the network shown in figure 5.16. This network has the same structure as in figure 5.13, but three of the four reactions are considered irreversible, so that the rates satisfy $v_1 \geq 0$, $v_2 \geq 0$, and $v_3 \geq 0$.

In this case, the vectors

$$\tilde{\mathbf{v}}_1 = \begin{bmatrix} -1 \\ -1 \\ -1 \\ 0 \end{bmatrix} \quad \tilde{\mathbf{v}}_2 = \begin{bmatrix} 0 \\ 0 \\ -1 \\ 1 \end{bmatrix}$$

both satisfy the balance equation but are not feasible because they violate the irreversibility conditions.

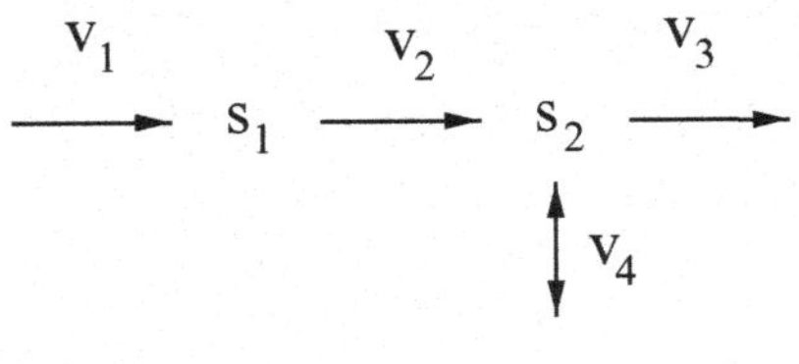

Figure 5.16
Branched metabolic network. Reactions 1, 2, and 3 are irreversible.

To characterize the set of all feasible flux profiles, we need to determine the set of solutions of the balance equation that satisfy $v_i \geq 0$ for all irreversible reactions. Methods from linear algebra are not sufficient for solving this problem: tools from convex analysis are needed. We will not delve into this theory but will briefly introduce the approaches that have become standard in dealing with metabolic networks.

To describe the set of feasible flux profiles as linear combinations of the form $\alpha_1 \mathbf{v}_1 + \alpha_2 \mathbf{v}_2 + \ldots + \alpha_k \mathbf{v}_k$, we cannot allow the coefficients α_i to take negative values, as that would correspond to a reversal of flux. For the network in figure 5.16, the flux profiles $\mathbf{v}_1$ and $\mathbf{v}_2$ from equation (5.19) have the property that any flux of the form $\alpha_1 \mathbf{v}_1 + \alpha_2 \mathbf{v}_2$ is feasible provided that $\alpha_1 \geq 0$ and $\alpha_2 \geq 0$. However, this set of linear combinations does not capture all feasible flux profiles, as the following exercise demonstrates.

Exercise 5.4.6   Verify that

$$
\mathbf{v} = \begin{bmatrix} 2 \\ 2 \\ 3 \\ -1 \end{bmatrix}
$$

is a feasible steady-state flux profile for the network in figure 5.16 but cannot be written in the form $\alpha_1 \mathbf{v}_1 + \alpha_2 \mathbf{v}_2$ with $\mathbf{v}_1$ and $\mathbf{v}_2$ from equation (5.19) and $\alpha_1 \geq 0$, $\alpha_2 \geq 0$. $\square$

We can describe all feasible flux profiles as linear combinations with nonnegative coefficients α_i if we extend the set of vectors used in the sum. For the network in figure 5.16, we can use

$$
\mathbf{v}_1 = \begin{bmatrix} 1 \\ 1 \\ 1 \\ 0 \end{bmatrix} \qquad \mathbf{v}_2 = \begin{bmatrix} 1 \\ 1 \\ 0 \\ 1 \end{bmatrix} \qquad \mathbf{v}_3 = \begin{bmatrix} 0 \\ 0 \\ 1 \\ -1 \end{bmatrix}, \tag{5.20}
$$

and then describe all steady-state feasible fluxes by

$$
\mathbf{v} = \alpha_1 \mathbf{v}_1 + \alpha_2 \mathbf{v}_2 + \alpha_3 \mathbf{v}_3, \quad \alpha_1 \geq 0, \alpha_2 \geq 0, \alpha_3 \geq 0.
$$

The strategy of adding more flux profiles to the linear combination works in this case but leads to some questions. How are we to know that we have added enough profiles to describe completely the set of feasible flux profiles? Conversely, is there some way we can tell if we have included more profiles than necessary?

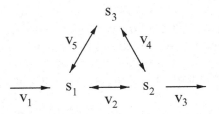

Figure 5.17
Reaction network for exercise 5.4.7. Reactions 1 and 3 are irreversible. The forward direction for all reactions is to the right.

A number of methods have been developed to address these issues. One of the most commonly used approaches is based on the work of Stefan Schuster, who introduced the term *flux mode* to refer to feasible steady-state flux profiles. If such a profile has the property that it cannot be decomposed into a collection of simpler flux modes, then it is called an *elementary flux mode*. The vectors v_1, v_2, and v_3 in equation (5.20) form a complete set of elementary flux modes for the network in figure 5.16, as can be verified by an algorithm for generation of elementary flux modes that Schuster published with Claus Hilgetag in 1994.[7]

Exercise 5.4.7 Consider the reaction network in figure 5.17. Suppose reactions 1 and 3 are assumed irreversible.

(a) Verify that

$$\mathbf{v} = \begin{bmatrix} 1 \\ 2 \\ 1 \\ -1 \\ -1 \end{bmatrix}$$

is a flux mode, but is not elementary.

(b) Find, by inspection, a complete set of elementary flux modes (there are three).   □

To illustrate pathway analysis further, consider the network in figure 5.18, which involves six species involved in ten reactions. The stoichiometry matrix is

7. Reviewed in the book *The Regulation of Cellular Systems* (Heinrich and Schuster, 1996). This algorithm is implemented in the software package METATOOL, which is freely available (http://pinguin.biologie .uni-jena.de/bioinformatik/networks).

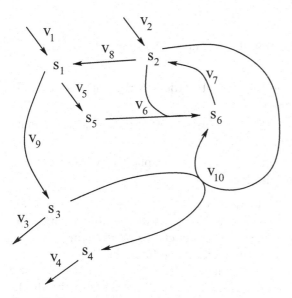

Figure 5.18
Metabolic network with six species and ten reactions. Note that reaction 10 is $S_2 + S_3 \rightarrow S_4 + S_6$.

$$
\mathbf{N} =
\begin{bmatrix}
1 & 0 & 0 & 0 & -1 & 0 & 0 & 1 & -1 & 0 \\
0 & 1 & 0 & 0 & 0 & -1 & 1 & -1 & 0 & -1 \\
0 & 0 & -1 & 0 & 0 & 0 & 0 & 0 & 1 & -1 \\
0 & 0 & 0 & -1 & 0 & 0 & 0 & 0 & 0 & 1 \\
0 & 0 & 0 & 0 & 1 & -1 & 0 & 0 & 0 & 0 \\
0 & 0 & 0 & 0 & 0 & 1 & -1 & 0 & 0 & 1
\end{bmatrix}.
$$

Suppose that all reactions are irreversible, as shown in the figure. A complete set of elementary flux modes for the network is

$$
\mathbf{v}_1 =
\begin{bmatrix} 1 \\ 0 \\ 0 \\ 0 \\ 1 \\ 1 \\ 1 \\ 0 \\ 0 \\ 0 \end{bmatrix}, \quad
\mathbf{v}_2 =
\begin{bmatrix} 1 \\ 0 \\ 1 \\ 0 \\ 0 \\ 0 \\ 0 \\ 0 \\ 1 \\ 0 \end{bmatrix}, \quad
\mathbf{v}_3 =
\begin{bmatrix} 1 \\ 0 \\ 0 \\ 1 \\ 0 \\ 0 \\ 0 \\ 1 \\ 0 \\ 1 \end{bmatrix}, \quad
\mathbf{v}_4 =
\begin{bmatrix} 0 \\ 1 \\ 0 \\ 0 \\ 1 \\ 1 \\ 1 \\ 1 \\ 0 \\ 0 \end{bmatrix}, \quad
\mathbf{v}_5 =
\begin{bmatrix} 0 \\ 1 \\ 1 \\ 0 \\ 0 \\ 0 \\ 0 \\ 0 \\ 1 \\ 0 \end{bmatrix}, \quad
\mathbf{v}_6 =
\begin{bmatrix} 0 \\ 1 \\ 0 \\ 1 \\ 0 \\ 0 \\ 0 \\ 1 \\ 1 \\ 1 \end{bmatrix}.
$$

These flux modes are illustrated in figure 5.19.

Exercise 5.4.8 Consider the network in figure 5.18 and the flux modes v_1 to v_6 listed above.

(a) Consider the case in which reaction 1 is reversible. Identify the single additional elementary flux mode exhibited by the system.

(b) Consider the case in which reaction 3 is reversible. Identify the single additional elementary flux mode exhibited by the system.

(c) Consider the case in which reaction 5 is reversible. Explain why there are no additional elementary flux modes in this case.   □

Elementary flux modes provide a valuable framework for investigation of potential network behaviors. For instance, flux mode analysis allows the identification of all possible paths from a given metabolite to a given product: this provides insight into which reactions are most important for a given metabolic function.

The elementary mode concept suffers from some deficiencies. For instance, although the set of elementary flux modes is unique (up to a scaling of the flux), the construction of flux modes from the elementary modes is non-unique. Moreover, the number of elementary flux modes expands rapidly with network size, so that for large networks the set of elementary modes may be unworkable. There are alternative notions that better handle these issues: Bruce Clarke's "extreme currents" and Bernhard Palsson's "extreme pathways" (Palsson, 2006). These alternative notions, however, do not always lend themselves as easily to biological interpretation.

In the next section, we extend our discussion of feasible steady-state flux profiles by imposing additional constraints on the reaction rates.

5.4.2 Constraint-Based Modeling: Metabolic Flux Analysis

Metabolic flux analysis (MFA) offers techniques to address networks for which specific information about the reaction rates is available. Most commonly, this information consists of measurements of the rates of exchange reactions—reactions that involve transport of metabolites across the cell membrane (i.e., uptake or secretion). We will address two MFA techniques: metabolic balancing and flux balance analysis.

Metabolic Balancing Consider a network in which some of the steady-state reaction fluxes have been measured experimentally. These data provide constraints on the feasible fluxes through the network, as follows.

To begin, we relabel the reactions so that the flux vector v can be partitioned into a vector of known fluxes v_k and a vector of unknown fluxes v_u, as:

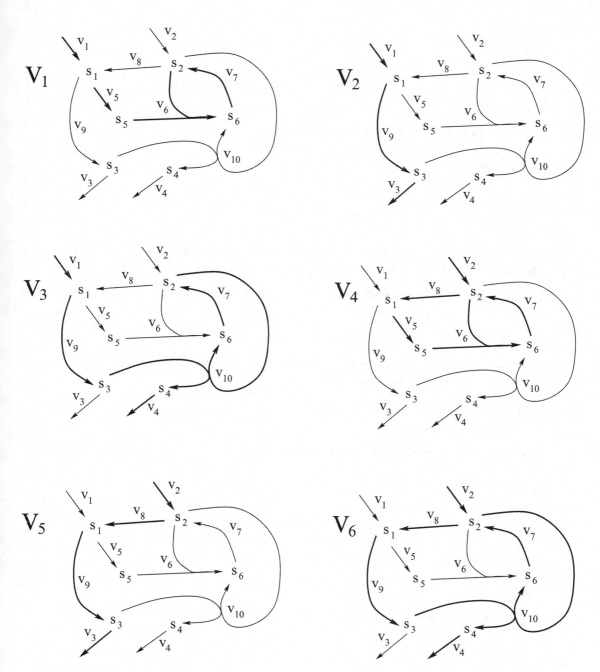

Figure 5.19
Elementary flux modes (heavy lines) for the network in figure 5.18. Note that in modes v_1 and v_4, uptake is balanced by the loop composed of reactions 6 and 7, which consumes S_5.

$$\mathbf{v} = \begin{bmatrix} \mathbf{v_k} \\ \mathbf{v_u} \end{bmatrix}.$$

We then partition the stoichiometry matrix accordingly,

$$\mathbf{N} = [\mathbf{N_k} \ \mathbf{N_u}], \tag{5.21}$$

so that we can write the balance condition as

$$\mathbf{0} = \mathbf{Nv} = [\mathbf{N_k} \ \mathbf{N_u}] \begin{bmatrix} \mathbf{v_k} \\ \mathbf{v_u} \end{bmatrix} = \mathbf{N_k v_k} + \mathbf{N_u v_u},$$

or, equivalently,

$$\mathbf{N_u v_u} = -\mathbf{N_k v_k}. \tag{5.22}$$

The unknown in this equation is $\mathbf{v_u}$. We next consider the three cases that can arise when using this equation to characterize the unknown fluxes $\mathbf{v_u}$.

Case I   Exactly Determined Systems   If $\mathbf{N_u}$ is invertible, then we can determine $\mathbf{v_u}$ directly:

$$\mathbf{v_u} = -\mathbf{N_u}^{-1}\mathbf{N_k v_k}. \tag{5.23}$$

To illustrate, consider the network in figure 5.20, which has stoichiometry matrix (note the numbering of the reactions)

$$\mathbf{N} = \begin{bmatrix} 0 & 0 & 1 & -1 \\ -1 & -1 & 0 & 1 \end{bmatrix}.$$

Suppose reaction rates v_1 and v_2 have been measured, so

$$\mathbf{v_k} = \begin{bmatrix} v_1 \\ v_2 \end{bmatrix} \quad \text{and} \quad \mathbf{v_u} = \begin{bmatrix} v_3 \\ v_4 \end{bmatrix}.$$

In this simple case, the relationship between $\mathbf{v_k}$ and $\mathbf{v_u}$ is clear from inspection of the pathway. Nevertheless, to illustrate the general method, we partition the stoichiometry matrix as in equation (5.21) with

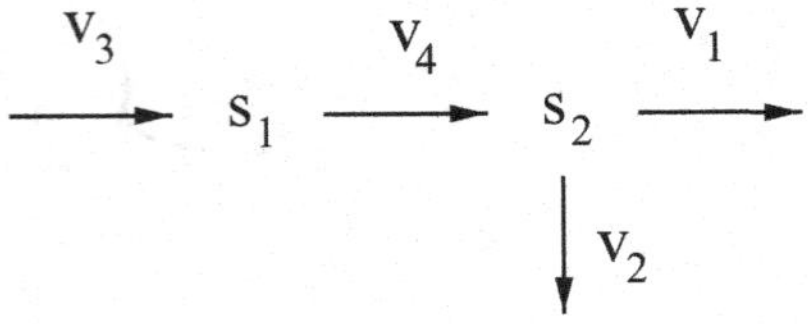

Figure 5.20
Network for metabolic balancing analysis.

$$\mathbf{N_k} = \begin{bmatrix} 0 & 0 \\ -1 & -1 \end{bmatrix} \quad \text{and} \quad \mathbf{N_u} = \begin{bmatrix} 1 & -1 \\ 0 & 1 \end{bmatrix}.$$

In this case, $\mathbf{N_u}$ is invertible:

$$\mathbf{N_u}^{-1} = \begin{bmatrix} 1 & 1 \\ 0 & 1 \end{bmatrix}.$$

We can solve for the unknown reaction rates as

$$\mathbf{v_u} = -\mathbf{N_u}^{-1}\mathbf{N_k}\mathbf{v_k} = -\begin{bmatrix} 1 & 1 \\ 0 & 1 \end{bmatrix}\begin{bmatrix} 0 & 0 \\ -1 & -1 \end{bmatrix}\begin{bmatrix} v_1 \\ v_2 \end{bmatrix} = \begin{bmatrix} 1 & 1 \\ 1 & 1 \end{bmatrix}\begin{bmatrix} v_1 \\ v_2 \end{bmatrix} = \begin{bmatrix} v_1 + v_2 \\ v_1 + v_2 \end{bmatrix}.$$

Exercise 5.4.9 Returning to the network in figure 5.18, suppose that the rate of reactions v_1, v_2, v_3, and v_4 have all been measured. Verify that in this case, the remaining fluxes are completely determined, and solve for their values in terms of the measured rates. (Note that this system is simple enough that this task can be carried out by inspection of the network—using the steady-state condition for each species.) ☐

In practice, the matrix $\mathbf{N_u}$ is almost never invertible, and so equation (5.22) cannot be solved so easily. The more commonly occurring cases are addressed next.

Case II Overdetermined Systems To illustrate the case of an overdetermined system, consider again the network in figure 5.20 and suppose that three fluxes have been measured:

$$\mathbf{v_k} = \begin{bmatrix} v_1 \\ v_2 \\ v_3 \end{bmatrix} \quad \text{and} \quad \mathbf{v_u} = [v_4]. \tag{5.24}$$

From figure 5.20, it is clear that in steady state, $v_4 = v_3$ and $v_4 = v_1 + v_2$. If the measured values satisfy $v_3 = v_1 + v_2$, then the system is said to be *consistent*.

However, measurements from a real system will typically be affected by experimental error or neglected side reactions, and so it is unlikely that the measured values will exactly satisfy $v_3 = v_1 + v_2$. The result is an *inconsistent system*:

$$v_4 = v_3 \neq v_1 + v_2 = v_4.$$

What then is the most appropriate estimate for the value of v_4? Because there is no reason to have more confidence in either measured value (i.e., v_3 or $v_1 + v_2$), a reasonable compromise is to split the difference and set v_4 to be the average of the two estimates:

$$v_4 = \frac{v_3 + (v_1 + v_2)}{2}. \tag{5.25}$$

For more complex networks, the appropriate compromise value of $\mathbf{v_u}$ will not be so clear. The best estimate can be calculated by replacing $\mathbf{N_u}^{-1}$ in equation (5.23) with the *pseudoinverse* of $\mathbf{N_u}$, defined as

$$\mathbf{N_u}^{\#} = \left(\mathbf{N_u}^{T}\mathbf{N_u}\right)^{-1}\mathbf{N_u}^{T},$$

where $\mathbf{N_u}^{T}$ is the transpose of $\mathbf{N_u}$. The unknown fluxes $\mathbf{v_u}$ are then estimated as

$$\mathbf{v_u} = -\mathbf{N_u}^{\#}\mathbf{N_k}\mathbf{v_k}.$$

This equation generates a best compromise solution to equation (5.22), in the following sense: it provides the value of $\mathbf{v_u}$ for which the difference $\mathbf{N_u}\mathbf{v_u} - \mathbf{N_k}\mathbf{v_k}$ is as close to zero as possible.

To illustrate this technique, we apply it to the known and unknown vectors in equation (5.24). The stoichiometry matrix is partitioned as

$$\mathbf{N_k} = \begin{bmatrix} 0 & 0 & 1 \\ -1 & -1 & 0 \end{bmatrix} \text{ and } \mathbf{N_u} = \begin{bmatrix} -1 \\ 1 \end{bmatrix}.$$

The pseudo-inverse of $\mathbf{N_u}$ is then

$$\mathbf{N_u}^{\#} = \left(\begin{bmatrix} -1 & 1 \end{bmatrix} \begin{bmatrix} -1 \\ 1 \end{bmatrix} \right)^{-1} \begin{bmatrix} -1 & 1 \end{bmatrix} = (2)^{-1}\begin{bmatrix} -1 & 1 \end{bmatrix} = \begin{bmatrix} -\dfrac{1}{2} & \dfrac{1}{2} \end{bmatrix},$$

and we can solve for the unknown flux as

$$\mathbf{v_u} = [v_4] = -\mathbf{N_u}^{\#}\mathbf{N_k}\mathbf{v_k} = -\begin{bmatrix} -\dfrac{1}{2} & \dfrac{1}{2} \end{bmatrix} \begin{bmatrix} 0 & 0 & 1 \\ -1 & -1 & 0 \end{bmatrix} \begin{bmatrix} v_1 \\ v_2 \\ v_3 \end{bmatrix} = \frac{1}{2}(v_1 + v_2 + v_3),$$

as we had established intuitively in equation (5.25).

Case III   Underdetermined Systems and Flux Balance Analysis   In most applications of metabolic balancing, the constraints provided by measured fluxes are insufficient to determine the values of the remaining fluxes, so equation (5.22) is underdetermined. In this case, pathway analysis can characterize the set of flux profiles that are consistent with the measured flux values, as in the following exercise.

Exercise 5.4.10   Returning to the network in figure 5.18, suppose that the steady-state rates of reactions v_3 and v_4 have been measured. Which other steady-state flux values are determined as a result?   □

A unique prediction for the steady-state flux profile of an underdetermined system can only be reached by imposing additional constraints on the network. One approach for generating additional constraints is to suppose that the flux profile has been optimized for production of some target metabolites. This technique is called flux balance analysis (FBA).

In applying FBA, one typically presumes that the "goal" of a cell is to produce more cells: organisms are optimized (by natural selection) for self-reproduction. By identifying a group of metabolic products that correspond to the building blocks of a new cell (collectively called *biomass*), we may presume that the cell's metabolic network is optimized to produce these target products. (FBA is also used with other optimality criteria. For example, a metabolic engineer interested in improving the yield of a specific metabolite might carry out an FBA optimizing the production rate of the target metabolite. The resulting profile is unlikely to ever be realized in the cell, but it provides an upper bound on achievable production rates.)

To apply FBA, upper bounds on some reaction rates must be provided (otherwise optimal solutions could involve infinite reaction rates). These upper bounds, which correspond to V_{max} values in a kinetic model, are often provided only on exchange fluxes (i.e., uptake and secretion reactions).

FBA involves optimizing an objective of the form $\alpha_1 v_1 + \alpha_2 v_2 + \ldots + \alpha_m v_m$ (corresponding to production of, e.g., biomass) under the following constraints:

- Steady-state balance: $\mathbf{0} = \mathbf{Nv}$.
- Upper or lower bounds on some reaction fluxes: $l_i \leq v_i \leq u_i$.
- Constraints provided by any measured fluxes: $\mathbf{N_u v_u} = -\mathbf{N_k v_k}$.

The resulting optimization problem can be efficiently solved by the technique of *linear programming*, which is commonly featured in computational software packages.

To illustrate FBA, we consider again the network in figure 5.20. Suppose that reaction rate v_1 has been measured and that upper and lower bounds l_3 and u_3 have been provided for reaction rate v_3. Finally, suppose that the network is optimized for yield from reaction v_2. In this case, the analysis is straightforward: the maximal flux through v_2 is achieved when the substrate uptake v_3 is maximal (i.e., $v_3 = u_3$). To satisfy balance, $v_4 = u_3$ as well. This results in an optimal production rate of $v_2 = u_3 - v_1$. Another example is provided in the following exercise.

Exercise 5.4.11

(a) Returning to the network in figure 5.18, suppose all reactions are irreversible ($v_i \geq 0$ for $i = 1 \ldots 10$) and that the uptake rates are constrained by $v_1 \leq 1$ and $v_2 \leq 1$. What is the maximal value of the steady-state rate v_3 under these conditions? What is the maximal value of the steady-state rate v_4?

(b) Repeat part (a) under the additional constraint that $v_7 \leq 1$.

(c) Repeat part (a) under the condition that the enzyme responsible for catalyzing reaction v_8 has been removed (knocked out), so that $v_8 = 0$. $\qquad\square$

5.5 Suggestions for Further Reading

• **Modeling of Metabolic Networks**   Introductions to metabolic modeling and a range of case studies can be found in the books *Kinetic Modeling in Systems Biology* (Demin and Goryanin, 2009) and *Systems Biology: A Textbook* (Klipp et al., 2009).

• **Metabolic Regulation**   A biologically motivated introduction to metabolic control analysis can be found in *Understanding the Control of Metabolism* (Fell, 1997). *The Regulation of Cellular Systems* (Heinrich and Schuster, 1996) contains a thorough description of the mathematical theory of MCA, as well as a comprehensive treatment of stoichiometric network analysis. Regulation of metabolic pathways is also addressed in Michael Savageau's book *Biochemical Systems Analysis: A Study of Function and Design in Molecular Biology* (Savageau, 1976).

• **Stoichiometric Network Analysis**   An introduction to stoichiometric network analysis is provided in "Stoichiometric and Constraint-Based Modeling" (Klamt and Stelling, 2006). The theory and applications of metabolic flux analysis are addressed in the book *Metabolic Engineering: Principles and Methodologies* (Stephanopoulos, 1998). Methods for addressing large-scale metabolic networks, including flux balance analysis, are covered in *Systems Biology: Properties of Reconstructed Networks* (Palsson, 2006).

5.6 Problem Set

5.6.1 Flux Control Coefficients
Consider the metabolic chain $S_0 \overset{v_1}{\leftrightarrow} S_1 \overset{v_2}{\leftrightarrow} S_2 \overset{v_3}{\to}$. Suppose that the concentration of S_0 is fixed, and take the reaction rates as

$$v_1 = e_0 \frac{V_1[S_0] - V_2[S_1]}{1 + [S_0]/K_{M1} + [S_1]/K_{M2}}$$

$$v_2 = e_1 \frac{V_3[S_1] - V_4[S_2]}{1 + [S_1]/K_{M3} + [S_2]/K_{M4}}$$

$$v_3 = e_2 \frac{V_5[S_2]}{1 + [S_2]/K_{M5}}.$$

(a) Use numerical approximation (equation 4.15 of chapter 4) to determine the flux control coefficients of the three reactions at nominal parameter values of (in units of concentration) $e_0 = 1$, $e_1 = 1.5$, $e_2 = 2$, $[S_0] = 1$, $K_{M1} = K_{M2} = K_{M3} = K_{M4} = K_{M5} = 1$; (in concentration$^{-1} \cdot$ time^{-1}) $V_1 = V_2 = 1$, $V_3 = V_4 = 2$, and $V_5 = 0.5$. You can check your calculations by confirming that the flux control coefficients sum to one.

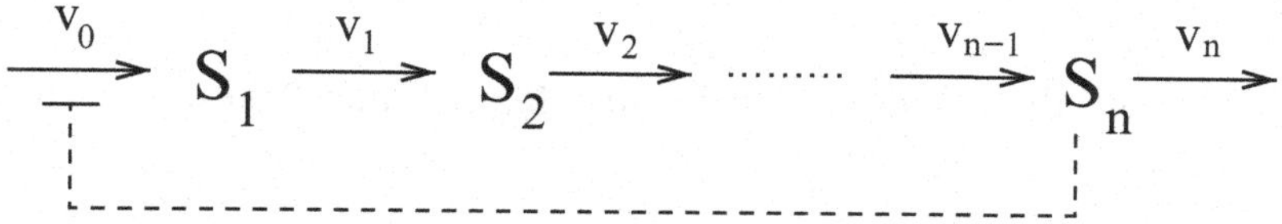

Figure 5.21
Metabolic chain for problem 5.6.2.

(b) Recall that sensitivity coefficients are only valid near a given nominal parameter set. Confirm this fact by repeating part (a) after changing the nominal value of parameter V_1 to 6.

5.6.2 Oscillatory Behavior from End-Product Inhibition

Consider the end-product–inhibited metabolic chain of length n shown in figure 5.21. Take the rate of the first reaction to be

$$v_0 = \frac{v}{1 + [S_n / K]^q}$$

and the others to be described by mass action:

$$v_i = k_i[S_i], \, i = 1, \ldots, n.$$

(a) Take nominal parameter values $v = 10$ (concentration $\cdot$ time^{-1}), $K = 1$ (concentration), $k_i = 1$ (time^{-1}), $i = 1, \ldots, n$. Explore the behavior of the pathway by simulating models for different values of the inhibition strength q and different chain lengths n. From your simulations, determine whether an increase in the length of the chain increases or decreases the range in q-values over which the system exhibits sustained oscillations. (Be sure to run simulations sufficiently long to distinguish sustained oscillations from slowly damped oscillations.) Provide evidence for your conclusion by reporting the range of q-values over which steady oscillations occur.

(b) Next consider an alternative nominal parameter set, for which $k_1 = 3$ (time^{-1}), $k_i = 1$ (time^{-1}), $i = 2, \ldots, n$. Repeat the analysis in part (a). Has this inhomogeneity in reaction rates made the system more or less likely to exhibit oscillations?

(c) Explain your finding in part (b). Hint: Consider the limiting case in which inhomogeneity in reaction rates introduces a significant timescale separation between one reaction and the others. How would this impact the effective length of the chain?

5.6.3 Metabolic Control Analysis: Supply and Demand

Consider the two-step reaction chain $\xrightarrow{v_0} S \xrightarrow{v_1}$, where the reactions are catalyzed by enzymes E_0 and E_1 with concentrations e_0 and e_1. The summation theorem (section 5.2.1) states that

$$C_{e_0}^J + C_{e_1}^J = 1.$$

A complementary result, the *connectivity theorem* (Heinrich and Schuster, 1996) states that

$$C_{e_0}^J \varepsilon_S^0 + C_{e_1}^J \varepsilon_S^1 = 0.$$

(a) Use these two statements to determine the flux control coefficients of the two reactions as

$$C_{e_0}^J = \frac{\varepsilon_S^1}{\varepsilon_S^1 - \varepsilon_S^0} \qquad\qquad C_{e_1}^J = \frac{-\varepsilon_S^0}{\varepsilon_S^1 - \varepsilon_S^0}.$$

(b) In addressing the control of flux through the pathway, we can think of v_0 as the supply rate and v_1 as the demand rate. Given the result in part (a), under what conditions on the elasticities ε_S^0 and ε_S^1 will a perturbation in the rate of supply affect pathway flux more than an equivalent perturbation in the rate of demand?

(c) Suppose the rate laws are given as $v_0 = e_0(k_0 X - k_{-1}[S])$ and $v_1 = e_1 k_1[S]$, where X is the constant concentration of the pathway substrate. Verify that the elasticities are

$$\varepsilon_S^0 = \frac{k_{-1}[S]}{k_0 X - k_{-1}[S]} \qquad \text{and} \qquad \varepsilon_S^1 = 1.$$

Determine conditions on the parameters under which perturbation in the supply reaction v_0 will have a more significant effect than perturbation in the demand reaction v_1. Hint: At steady state, $k_0 X - k_{-1}s = e_1 k_1 s / e_0$.

5.6.4 Metabolic Control Analysis: End-Product Inhibition

The coefficients in equation (5.11) can be derived directly from the steady-state conditions for the network in figure 5.5. However, a simpler derivation makes use of the summation theorem (equation 5.9) and the complementary *connectivity theorem* (Heinrich and Schuster, 1996). In this case, the summation theorem states that

$$C_{e_1}^J + C_{e_2}^J + C_{e_3}^J = 1.$$

The connectivity statements are

$$C_{e_1}^J \varepsilon_{S_1}^1 + C_{e_2}^J \varepsilon_{S_1}^2 = 0 \qquad \text{and} \qquad C_{e_1}^J \varepsilon_{S_2}^1 + C_{e_2}^J \varepsilon_{S_2}^2 + C_{e_3}^J \varepsilon_{S_2}^3 = 0.$$

(Note that $\varepsilon_{S_1}^3 = 0$.) This is a system of three equations in the three unknowns $C_{e_1}^J$, $C_{e_2}^J$, and $C_{e_3}^J$. Solve these equations to arrive at the formulas in equation (5.11).

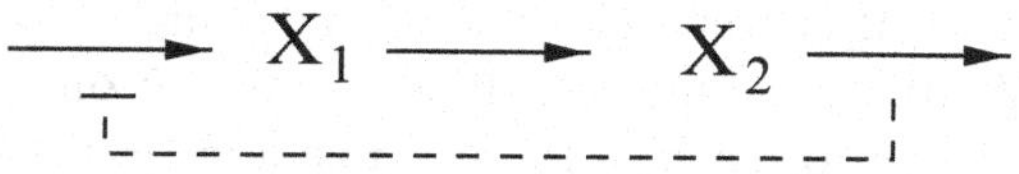

Figure 5.22
Reaction chain for problem 5.6.5.

5.6.5 S-System Analysis of Pathway Regulation

As discussed in section 5.2, Michael Savageau carried out a general analysis of metabolic feedback regulation schemes (Savageau, 1976). He made use of an S-system model formulation (see section 3.5) to derive explicit descriptions of sensitivity coefficients (as in problem 4.8.15 of chapter 4). Here, we consider a simple example to illustrate Savageau's approach. Consider the two-step reaction chain in figure 5.22.

We will compare the behavior of the system in the presence or absence of the negative feedback. An S-system formulation of the model (with $x_i = [X_i]$) is

$$\frac{d}{dt}x_1(t) = \alpha_1 x_2^{g_1} - \alpha_2 x_1^{g_2} \qquad \frac{d}{dt}x_2(t) = \alpha_2 x_1^{g_2} - \alpha_3 x_2^{g_3}.$$

If the negative feedback is absent, the coefficient $g_1 = 0$, otherwise, $g_1 < 0$.

(a) Verify that at steady state,

$$\alpha_1 x_2^{g_1} = \alpha_3 x_2^{g_3}.$$

Take logarithms to verify that at steady state,

$$\log x_2 = \frac{\log \alpha_1 - \log \alpha_3}{g_3 - g_1}.$$

(b) Use the results of problem 4.8.15 to confirm that the relative sensitivities of $[X_2]$ to the rate constants α_1 and α_3 are

$$\frac{\alpha_1}{x_2}\frac{\partial x_2}{\partial \alpha_1} = \frac{1}{g_3 - g_1} \qquad \text{and} \qquad \frac{\alpha_3}{x_2}\frac{\partial x_2}{\partial \alpha_3} = \frac{-1}{g_3 - g_1}.$$

Conclude that the inhibition (which introduces $g_1 < 0$) reduces the sensitivity of $[X_2]$ to perturbations in the rate constants.

5.6.6 Methionine Metabolism

Consider the Martinov model of the methionine cycle in section 5.3.2.

(a) Define the pathway flux as the rate at which methionine is cycled through the pathway (flux $= V_{\text{MATI}} + V_{\text{MATIII}} = V_{\text{GNMT}} + V_{\text{MET}} = V_{\text{D}}$). Simulate the model and determine (numerically; see equation 4.15 of chapter 4) the flux control coefficients for each of the enzymes in the pathway, using parameter values as in figure 5.10 (for which the system is monostable). Verify that control coefficients sum to one. (Be sure to calculate a flux control coefficient for each of the five reactions in the network.) Note: The relative sensitivity to enzyme abundance is equivalent to the relative sensitivity to V_{max} because enzyme abundance is proportional to V_{max}.

(b) The model's bistability is caused by the activity of MATIII and GNMT, both of which show nonlinear dependence on AdoMet levels. When methionine levels rise, flux through GNMT and METIII increases dramatically, whereas flux through the methylation reactions (lumped together in reaction MET) shows only a modest increase. Verify this claim by comparing the reactions fluxes (MATI to MATIII, GNMT to MET) in the two steady states displayed in figure 5.11B.

(c) Bistability depends on a balance between the flux through MATI and MATIII. Verify that bistability is lost when this balance is upset, as follows. Consider the case when [MET] = 51 μM, as in figure 5.11B. With the other parameter values as in figure 5.10, verify that bistability is lost when V_{max}^{MATIII} is perturbed more than 15% up or down from its nominal value.

5.6.7 Branch-Point Control

In a 2003 paper, Gilles Curien and colleagues presented a model of a branch point in the amino acid biosynthesis pathways of the plant *Arabidopsis thaliana* (Curien et al., 2003). They considered the network in figure 5.23, in which phosphohomoserine is converted to either cystathionine or threonine. The two enzymes of interest are cystathionine γ-synthase (CGS), which combines phosphohomoserine (Phser) with cysteine (Cys) to produce cystathionine; and threonine synthase (TS), which produces threonine from phosphohomoserine and is allosterically activated by S-adenosylmethionine (AdoMet). All three reactions in the network are irreversible.

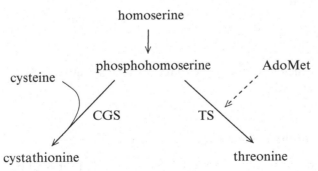

Figure 5.23
Branch-point network for problem 5.6.7. Adapted from figure 1 of Curien et al. (2003).

The rate of phosphohomoserine production is taken as a fixed parameter, J_{Phser}. The other reaction rates are given by

$$v_{\text{CGS}} = \left(\dfrac{k_{\text{catCGS}}}{1 + \dfrac{K_{\text{mCGS}}^{\text{Cys}}}{[\text{Cys}]}} \right) \dfrac{[\text{CGS}] \cdot [\text{Phser}]}{\dfrac{K_{\text{mCGS}}^{\text{Phser}}}{1 + K_{\text{mCGS}}^{\text{Cys}} / [\text{Cys}]} + [\text{Phser}]}$$

$$v_{\text{TS}} = [\text{TS}] \cdot [\text{Phser}] \left(K_1 + \dfrac{K_2 [\text{AdoMet}]^{2.9}}{K_3^{2.9} + [\text{AdoMet}]^{2.9}} \right).$$

Note that CGS follows a ping-pong reaction mechanism (see problem 3.7.6 of chapter 3); TS is described in its first-order (linear) regime. Take parameter values $J_{\text{Phser}} = 0.3$ μM s^{-1}, $k_{\text{catCGS}} = 30$ s^{-1}, $K_{\text{mCGS}}^{\text{Phser}} = 15000$ μM, $K_{\text{mCGS}}^{\text{Cys}} = 460$ μM, $K_1 = 4.9 \times 10^{-6}$ μM^{-1} s^{-1}, $K_2 = 5.6 \times 10^{-4}$, $K_3 = 32$ μM, $[\text{CGS}] = 0.7$ μM, $[\text{TS}] = 5$ μM, $[\text{AdoMet}] = 20$ μM, $[\text{Cys}] = 250$ μM.

(a) Simulate the model. Verify that phosphohomoserine relaxes to its steady state on a timescale of tens of minutes. For the given parameters, which branch carries more flux? What is the steady-state flux ratio?

(b) Verify that for large AdoMet concentrations, the flux through TS is larger than the flux through CGS. Find the AdoMet concentration for which the flux is split equally between the two branches.

(c) AdoMet is produced from cystathionine. If AdoMet production were included in the network, would the allosteric activation of TS by AdoMet act as a positive feedback or a negative feedback on cystathionine levels? Explain your reasoning.

5.6.8* Stoichiometric Network Analysis
Consider the reaction network in figure 5.24.

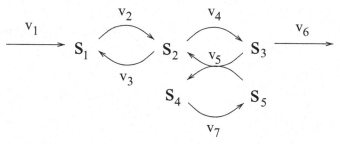

Figure 5.24
Reaction network for problem 5.6.8.

(a) Determine the stoichiometry matrix $\mathbf{N}$. It has rank four.

(b) Suppose all reactions are reversible. Describe all possible steady-state reaction profiles as linear combinations of elements of the kernel of $\mathbf{N}$.

(c) Identify the kernel of the transpose of $\mathbf{N}$ and the corresponding mass conservation(s) in the network.

(d) Suppose now that reactions 1 and 2 are irreversible (i.e., $v_1 \geq 0$ and $v_2 \geq 0$). Identify the four elementary flux modes in the network.

5.6.9*   Stoichiometric Network Analysis: Glycolysis

Consider the reaction network in figure 5.25 (a simplified model of the glycolytic pathway).

Denote the species as s_1 = G6P, s_2 = F6P, s_3 = TP, s_4 = F2,6P2, s_5 = AMP, s_6 = ADP, and s_7 = ATP (where G6P is glucose 6-phosphate, F6P is fructose 6-phosphate, TP is the pool of triose phosphates, F2,6P2 is fructose 2,6-bisphosphate, AMP is adenosine monophosphate, and ADP is adenosine diphosphate). Determine the stoichiometry matrix $\mathbf{N}$ for the system. In this case, the stoichiometry matrix has rank 6.

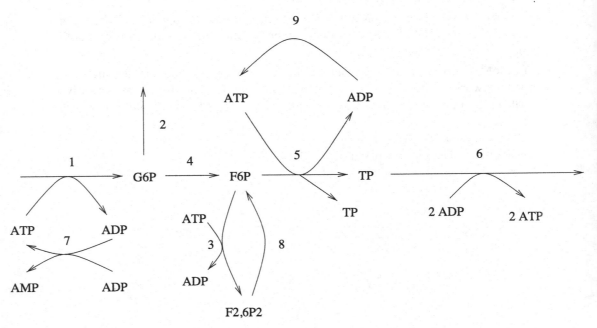

Figure 5.25
Reaction network for problem 5.6.9.

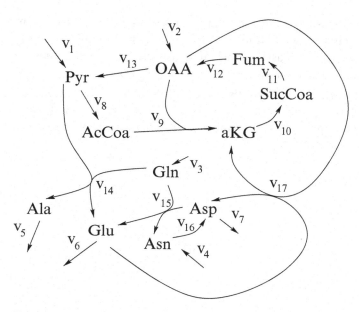

Figure 5.26
Metabolic network for problem 5.6.10. AcCoa, acetyl-CoA; aKG, alpha-ketoglutarate; Ala, alanine; Asn, asparagine; Asp, aspartic acid; Fum, fumarate; Gln, glutamine; Glu, glutamic acid; OAA, oxaloacetate; Pyr, pyruvate; SucCoa, succinyl-CoA. Adapted from figure 3 of Naderi et al. (2011).

(a) Determine the kernel of $\mathbf{N}$. Suppose all reactions are reversible. Describe all possible steady-state reaction profiles as linear combinations of elements of the kernel of $\mathbf{N}$.

(b) Suppose now that reactions 2 and 4 are irreversible. Describe the set of elementary modes for the network.

(c) Determine the kernel of the transpose of $\mathbf{N}$, and use it to identify the mass conservation(s) in the system.

5.6.10* Stoichiometric Network Analysis: Amino Acid Production
Consider the network in figure 5.26, which is a simplified description of amino acid metabolism in Chinese hamster ovary (CHO) cells (Naderi et al., 2011). There are 11 species involved in 17 reactions. Four reactions represent substrate uptake (v_1, v_2, v_3, v_4) and three represent export (v_5, v_6, v_7).

(a) Determine the stoichiometry matrix for the system, and verify that there are no structural conservations.

(b) Suppose all reactions are reversible. Determine the kernel of $\mathbf{N}$. Describe all possible steady-state reaction profiles as linear combinations of elements of the kernel of $\mathbf{N}$.

(c) Suppose the steady-state rates of the exchange reactions (v_1–v_7) have been measured. Identify any consistency conditions that these measurements would be expected to satisfy. Supposing the measurements are consistent, what conditions do they impose on the remaining reaction rates?

(d) Suppose the uptake reactions (v_1–v_4) satisfy upper bounds given by $v_i \leq u_i$, $i =$ 1 ... 4. Suppose further that all reactions are *irreversible*. What is the optimal yield (i.e., rate of export) of glutamic acid (Glu) under these conditions? What is the optimal yield of alanine (Ala)?

6 Signal Transduction Pathways

[T]he ordinary communication system of a mine may consist of a telephone central with the attached wiring and pieces of apparatus. When we want to empty a mine in a hurry, we do not trust to this, but break a tube of a mercaptan [a gas that smells like rotten cabbage] in the air intake. Chemical messengers like this, or like the hormones, are the simplest and most effective for a message not addressed to a specific recipient.
—Norbert Wiener, *Cybernetics: or Control and Communication in the Animal and the Machine*

Cells have evolved to survive in a wide range of conditions. To function in an unpredictable environment, they must be able to sense changes in their surroundings and respond appropriately. Intracellular signal-transduction pathways sense extracellular conditions and trigger cellular responses. Cells sense a myriad of stimuli, ranging from biological signals (e.g., hormones and pheromones) to chemical conditions (e.g., nutrients and toxins) to physical features of the environment (e.g., heat and light). Responses to these signals involve adjustments in cell behavior, often implemented through changes in gene expression (discussed in chapter 7).

Signal transduction pathways are—like metabolic pathways—biochemical reaction networks. However, whereas metabolic pathways shuttle mass and energy through the cell, signal transduction pathways are primarily concerned with propagation of *information*. These two classes of networks thus have distinct functional roles. They also have distinct implementations. Metabolic pathways process material in the form of small molecules (metabolites). In contrast, signal transduction pathways encode information in the configurations of proteins (via conformational shifts or covalent modifications). Changes in protein state are passed through *activation chains*—cascades of enzymes that activate one another in turn. Such reaction chains are made up solely of proteins and are examples of *protein–protein interaction networks*.

Protein–protein interactions are not usually well-described by Michaelis–Menten kinetics. Recall that the Michaelis–Menten rate law was derived under the assumption that reaction substrates are significantly more abundant than the catalyzing enzyme (see section 3.1.1). In signal transduction pathways, the substrates of

enzymatic activity are proteins and are not typically more abundant than the catalyzing enzymes. In this context, when enzyme-catalyzed reactions are treated as
single events, the enzymes are typically presumed to act in the first-order regime.
(So, for example, the reaction $S \to P$ catalyzed by enzyme E has rate $k[E][S]$.)

The term *signal transduction* refers to the transfer of information across spatial
domains (e.g., from the extracellular space to the nucleus). In addition, signaling
pathways perform *information-processing* tasks and are thus analogous to techno-
logical information-processing systems. In this chapter, we will survey pathways that
illustrate a range of information-processing capabilities: amplification and discretization of signals, adaptation to persistent signals, storage of memory, and frequency
encoding.

In this chapter, our focus will be on pathways that are triggered by molecular cues
from the extracellular environment. Some of these molecular signals can diffuse
freely across the cell membrane. The corresponding sensors consist of a cytosolic
protein that binds the signal and consequently elicits a cellular response (most commonly a change in gene expression). These simple sensing systems, called *one-
component* mechanisms, are common in prokaryotes. However, most signaling
molecules cannot diffuse across the cell's bilipid membrane. To sense these molecules in the extracellular environment, cells use *transmembrane receptor proteins*.
As shown in figure 6.1, these proteins span the membrane, exposing an extracellular

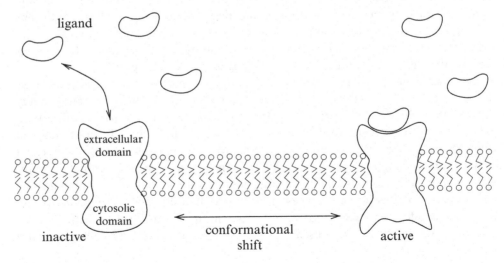

Figure 6.1
Transmembrane receptor protein. The signaling molecule, called a ligand, binds to a site in the protein's
extracellular domain. This binding event causes a conformational shift in the protein, which exposes a
catalytic site in the cytosolic domain. The ligand's presence is thus communicated to the cytosol without
the ligand entering the cell.

protein-domain to the external environment and a cytosolic protein-domain to the cell's interior. Signaling molecules (ligands) bind the extracellular domain of the protein. This binding event causes a conformational change in the receptor, thus activating an enzymatic site at the cytosolic domain. The receptor protein thus transfers information across the membrane while the ligand remains outside the cell.

In the remainder of this chapter, we will address a number of signal transduction pathways, each of which begins with activation of a transmembrane receptor. The receptor typically acts as a *kinase*—an enzyme that catalyzes the covalent addition of a phosphate group (PO_4^{3-}) to its substrate. This event, called *phosphorylation*, typically activates the target protein, which then carries the signal into the cell.

6.1 Signal Amplification

6.1.1 Bacterial Two-Component Signaling Pathways

Transmembrane receptor proteins transmit information across the cell membrane and into the cell. However, because these proteins are lodged in the membrane, they are unable to transmit signals farther than the intracellular membrane surface. Cytosolic "messenger" proteins are required if the information is to be shuttled into the cell's interior. Many bacterial signaling pathways consist of a receptor and a single messenger protein. These are referred to as *two-component signaling pathways* (figure 6.2). The messenger protein, called a *response regulator*, is activated by phosphorylation. Activated response regulators diffuse through the cytosol to perform their function, usually causing a change in gene expression. The response regulator is deactivated by dephosphorylation, thus turning the signal off. This can occur spontaneously (autodephosphorylation) or by the action of a separate enzyme, called a *phosphatase*.

Two-component pathways have been identified in many bacterial species and serve a range of roles, including nitrogen fixation in *Rhizobium*, sporulation in *Bacillus*, uptake of carboxylic acids in *Salmonella*, and porin synthesis in *Escherichia coli*.

A simple reaction scheme for the system in figure 6.2 is the following:

$$R + L \underset{k_{-1}}{\overset{k_1}{\rightleftharpoons}} RL$$

$$P + RL \overset{k_2}{\longrightarrow} P^* + RL$$

$$P^* \overset{k_3}{\longrightarrow} P$$

where R is the receptor, L is the ligand, RL is the active receptor–ligand complex, and P and P^* are the inactive and active response regulator proteins, respectively. The second reaction $P \rightarrow P^*$ is catalyzed by RL. (Writing the reaction as $P + RL \rightarrow P^* + RL$ allows us to describe the first-order catalytic event by a mass-action

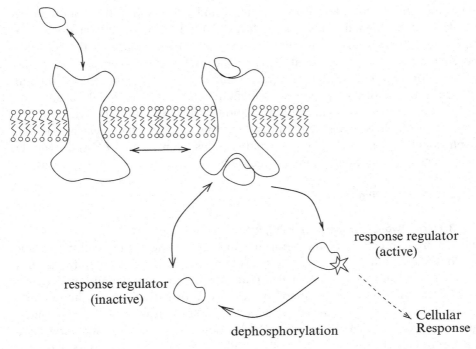

Figure 6.2
Bacterial two-component signaling pathway. Ligand binding activates the kinase activity of the transmembrane receptor's cytosolic domain. The receptor then activates the response regulator protein, which
diffuses into the cytosol and elicits a cellular response. The response regulator protein is deactivated by
dephosphorylation.

rate law.) We have made the simplifying assumptions that (i) the conformational
shift occurs concurrently with ligand binding and (ii) phosphorylation and dephosphorylation each occur as single reaction events. Taking the ligand level as a fixed
input, the model equations are

$$\frac{d}{dt}[R](t) = -k_1[R](t)\cdot[L](t) + k_{-1}[RL](t)$$

$$\frac{d}{dt}[RL](t) = k_1[R](t)\cdot[L](t) - k_{-1}[RL](t)$$

$$\frac{d}{dt}[P](t) = -k_2[P](t)\cdot[RL](t) + k_3[P^*](t)$$

$$\frac{d}{dt}[P^*](t) = k_2[P](t)\cdot[RL](t) - k_3[P^*](t).$$

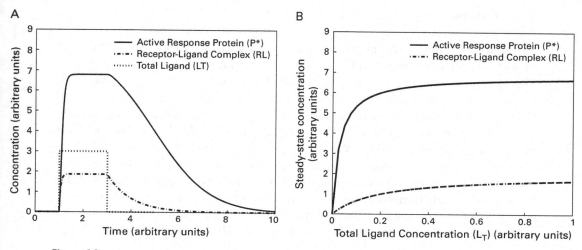

Figure 6.3

Bacterial two-component signaling pathway response. (A) Dynamic response. Ligand is absent until time $t = 1$. When ligand is introduced, the concentration of receptor–ligand complexes quickly comes to equilibrium, as does the population of active response regulator proteins. The ligand is removed at time $t = 3$, resulting in dissociation of receptor–ligand complexes and the decay of the active response regulator pool. (B) Dose response. This continuation diagram shows the steady-state concentrations of active response regulator protein and receptor–ligand complex as functions of ligand abundance. Parameter values are $k_1 = 5$ (concentration^{-1} · time^{-1}), $k_{-1} = 1$ (time^{-1}), $k_2 = 6$ (concentration^{-1} · time^{-1}), $k_3 = 3$ (time^{-1}), total receptor concentration $R_T = 2$ (concentration), and total response regulator protein concentration $P_T = 8$ (concentration). Units are arbitrary.

This system involves two conserved quantities: the total concentration of receptor ($[R] + [RL] = R_T$) and the total concentration of response regulator protein ($[P] + [P^*] = P_T$).

Figure 6.3A illustrates the pathway's behavior. The simulation starts in an inactive steady state with no ligand present. Activity is triggered by an abrupt addition of ligand, and the response reaches steady-state. The system is then inactivated by removal of ligand. Figure 6.3B shows a dose-response curve for the system, indicating the steady-state concentration of active regulator protein and receptor–ligand complex over a range of ligand concentrations.

This two-component cascade achieves amplification of ligand signal through enzyme activation: although the number of active receptor–ligand complexes is restricted by the number of ligand molecules, each complex can activate many response regulator proteins, allowing for amplification of the original molecular signal. (For the parameter values in figure 6.3A, a ligand concentration of 3 units gives rise to a response regulator concentration of 7 units; larger output gains can be achieved by other model parameterizations.)

6.1.2   G-Protein Signaling Pathways

Eukaryotic signal transduction pathways are typically more complex than their bacterial counterparts. A common first step in eukaryotic pathways is the activation of guanosine triphosphate (GTP)-binding regulatory proteins, called *G-proteins*. The G-protein signaling mechanism, sketched in figure 6.4, is similar to the two-component system discussed earlier. The primary distinction is in the character of the response protein — the G-protein itself. It consists of three different polypeptide chains: the α-, β-, and γ-subunits. The α-subunit has a site that binds a molecule of guanosine diphosphate (GDP). When a G-protein–coupled transmembrane receptor is activated by ligand binding, the receptor–ligand complex binds the α-subunit, causing a conformational change that leads to loss of GDP and binding of GTP. This, in turn, leads to dissociation of the α- from the $\beta\gamma$-subunits, exposing a catalytic site on the α-subunit that triggers a downstream response. Eventually, the GTP-bound α-subunit converts the bound GTP to GDP, after which it rebinds the

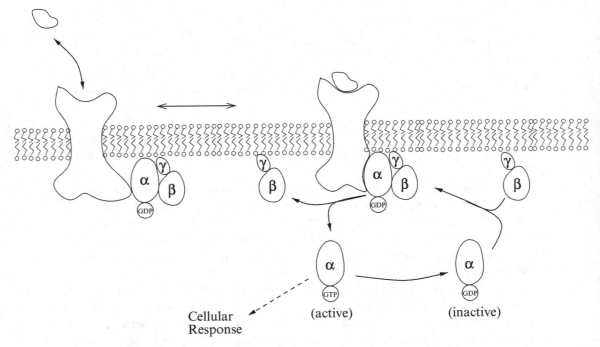

Figure 6.4
G-protein signaling mechanism. The G-protein–coupled transmembrane receptor is activated by ligand binding. It then causes the G-protein to release GDP and bind GTP. The GTP-bound α-subunit dissociates from the $\beta\gamma$-subunits and activates a downstream response. The GTP molecule is subsequently converted to GDP, and the G-protein subunits reassociate, completing the cycle.

$\beta\gamma$-subunits and so is returned to its original state, ready for another activation cycle.

In 2003, Tau-Mu Yi, Hiroaki Kitano, and Mel Simon published a model of G-protein activation in the yeast *Saccharomyces cerevisiae* (Yi et al., 2003). A simplified version of their model describes seven species: ligand (L), receptor (R), bound receptor (RL), inactive G-protein (G), active G_α-GTP (Ga), free $\beta\gamma$-subunit (Gbg), and inactive G_α-GDP (Gd). The reaction scheme is

$$R + L \underset{k_{RLm}}{\overset{k_{RL}}{\rightleftarrows}} RL$$

$$G + RL \xrightarrow{k_{Ga}} Ga + Gbg + RL$$

$$Ga \xrightarrow{k_{Gd0}} Gd$$

$$Gd + Gbg \xrightarrow{k_{G1}} G.$$

The model's behavior is shown in figure 6.5. Overall, the response is similar to that of the two-component signaling mechanism (although for these parameter values there is no amplification from the ligand to the response).

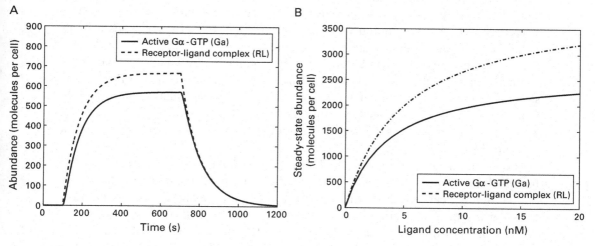

Figure 6.5

G-protein signaling mechanism. (A) Dynamic response. At time $t = 100$ seconds, 1 nM of ligand is introduced. This input signal is removed at time $t = 700$ seconds, causing the response to decay. For these parameter values, there is no amplification of signal. (B) Dose response. The steady-state concentrations of active response regulator protein and receptor–ligand complex are shown as functions of ligand availability. Protein abundance is specified as molecules per cell. Parameter values are $k_{RL} = 2 \times 10^{-3}$ nM^{-1} s^{-1}, $k_{RLm} = 10^{-2}$ s^{-1}, $k_{Ga} = 10^{-5}$ (molecules per cell)$^{-1} \cdot$ s^{-1}, $k_{Gd0} = 0.004$ s^{-1}, $k_{G1} = 1$ (molecules per cell)$^{-1} \cdot$ s^{-1}. The total G-protein population is 10,000 molecules per cell, and the total receptor population is 4000 molecules per cell.

Compared with bacterial two-component signaling pathways, the increased complexity of the G-protein pathway provides additional opportunities for interaction between distinct signaling pathways (called *crosstalk*) and additional avenues for pathway regulation. G-protein pathways can be regulated through a number of means, including genetic control of receptor abundance, covalent modification of receptors, and spatial trafficking of receptors away from G-proteins (sequestration). G-protein pathways can also be regulated by extrinsic sources. As an example, cholera infection produces a toxin that modifies the α-subunit of a G-protein so that it is permanently "on," with dangerous consequences for the affected host. Similar avenues for regulation are exploited for medical purposes: G-protein signaling pathways are the target of about 40% of all prescription drugs.

Exercise 6.1.1 Write out the differential equations for the G-protein pathway model. Treat the ligand concentration as a fixed input. Describe how conservations can be used to reduce the model to three differential equations.   □

6.2 Ultrasensitivity

The dose-response curves in figures 6.3B and 6.5B are hyperbolic. As we saw in chapter 3, this type of response curve is typical for protein-complex formation. Such graded responses are common in signal transduction, but in some cases a more switch-like "all-or-nothing" response is required.

A switch-like response can be achieved by a pathway that triggers activity only when an input signal crosses a threshold. An example from technology is the conversion of continuously-valued (analog) signals into discretely valued (digital) signals, as in the conversion of voltage levels into the binary (0/1) signals used in electronics. This *discretization* process converts the smoothly varying values of a signal into a discrete (ON/OFF) response.

The term *ultrasensitive* is used to describe the behavior of biochemical systems for which the response is steeper than hyperbolic. In section 3.3, we saw that cooperative binding can produce ultrasensitive responses to ligand doses. Cooperative behavior is typically described by Hill functions, with the Hill coefficient reflecting the steepness of the dose-response curve. A more general measure of steepness is the relative difference in input levels between 10% and 90% of full activation. For a hyperbolic curve, an 81-fold increase in input strength is required to span this response range. For ultrasensitive systems, the required increase in input strength is smaller, as verified by the following exercise.

Exercise 6.2.1 Confirm that if response R is given by

$$R = \frac{s}{K+s},$$

then an 81-fold increase in the concentration of s is required to transition from 10% to 90% of full activation. (Note that full activation, $R = 1$, is achieved in the limit as s gets large.) Verify that the corresponding increase in activity of the Hill function

$$R = \frac{s^4}{K + s^4}$$

demands only a threefold increase in activation. (Again, full activation corresponds to $R = 1$.) □

Some signaling pathways use cooperative binding to generate sigmoidal responses. In this section, we will introduce other biochemical mechanisms that can also generate ultrasensitive behavior.

6.2.1 Zero-Order Ultrasensitivity

In a 1981 paper, Albert Goldbeter and Douglas Koshland showed that an ultrasensitive response can be generated in the absence of cooperativity by an activation–inactivation cycle (Goldbeter and Koshland, 1981). Such cycles typically involve activation by covalent modification, most commonly phosphorylation. We saw one such example in the previous section: the phosphorylation–dephosphorylation cycle for the response regulator protein in a two-component signaling pathway. In this section, we consider activation–inactivation cycles as a self-contained signaling systems: the input is the abundance of the activating enzyme (e.g., the receptor-ligand complex); the output is the concentration of the activated protein (e.g., the phosphorylated response regulator).

Goldbeter and Koshland used a simple model to explore the signaling properties of activation–inactivation cycles. They found that if either the activating or the inactivating enzyme (e.g., the kinase or the phosphatase) becomes saturated, then the cycle exhibits an ultrasensitive response to changes in the abundance of the activating enzyme. Because saturation corresponds to the enzyme acting in its zero-order regime, they called this mechanism *zero-order ultrasensitivity*.

Their analysis addresses the reaction network in figure 6.6, where the target protein transitions between its inactive state, W, and its active state, W^*. Activation of W is catalyzed by enzyme E_1; inactivation of W^* is catalyzed by enzyme E_2. The reactions in the cycle are described by

$$W + E_1 \underset{d_1}{\overset{a_1}{\rightleftharpoons}} WE_1 \xrightarrow{k_1} W^* + E_1$$

$$W^* + E_2 \underset{d_2}{\overset{a_2}{\rightleftharpoons}} W^*E_2 \xrightarrow{k_2} W + E_2,$$

(6.1)

where WE_1 and W^*E_2 are the enzyme–substrate complexes. Goldbeter and Koshland made the simplifying assumption that the enzyme concentrations are negligible

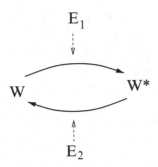

Figure 6.6
Activation–inactivation cycle. The target protein cycles between its inactive (W) and active (W^*) states. Activation is catalyzed by enzyme E_1 and deactivation by enzyme E_2. (The most commonly occurring cycles involve phosphorylation and dephosphorylation; in this case, E_1 is a kinase and E_2 is a phosphatase.) This network can be treated as a signaling pathway: the input is the abundance of E_1, and the output is the concentration of W^*.

compared to the total protein concentration, so the protein conservation need not take the complexes into account. (That is, they assumed that $W_T = [W] + [W^*] + [WE_1] + [W^*E_2] \approx [W] + [W^*]$.) Under this assumption, they derived an equation for the steady-state concentration of W^*. Letting w^* indicate the steady-state fraction of protein in the active state (i.e., $w^* = [W^*]^{ss}/W_T$, so the inactive fraction $w = 1-w^*$), they arrived at the expression

$$\frac{k_1 E_{1T}}{k_2 E_{2T}} = \frac{w^*(w + K_1)}{w(w^* + K_2)} = \frac{w^*(1 - w^* + K_1)}{(1 - w^*)(w^* + K_2)}, \tag{6.2}$$

where E_{1T} and E_{2T} are the total enzyme concentrations, and

$$K_1 = \frac{1}{W_T} \frac{d_1 + k_1}{a_1}, \qquad K_2 = \frac{1}{W_T} \frac{d_2 + k_2}{a_2}.$$

(See problem 6.8.6 for details.) This equation describes w^* as an implicit function of the concentration of activating enzyme, E_{1T}. Figure 6.7 shows the corresponding dose-responses for different values of K_1 and K_2 (with E_{2T} fixed). For small values of K_1 and K_2, the response is quite steep.

To interpret the dose-response, we begin by noting that K_1 and K_2 are scaled versions of the Michaelis constants for the two catalyzed reactions. Recall from section 3.1.1 that the Michaelis constants for the activating and inactivating reactions take the form

$$K_{M1} = \frac{d_1 + k_1}{a_1} \qquad \text{and} \qquad K_{M2} = \frac{d_2 + k_2}{a_2},$$

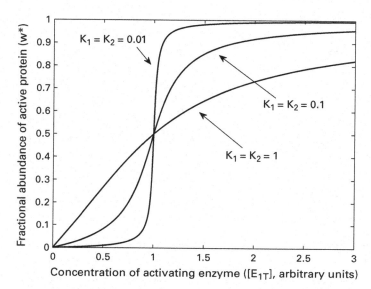

Figure 6.7
Activation–inactivation cycle: dose response. Plots of the implicit equation (6.2) show the input–output behavior of the activation–inactivation cycle in figure 6.6. For small values of K_1 and K_2, the response is ultrasensitive. Parameter values: $k_1 = 1$ (time)$^{-1}$, $k_2 = 1$ (time)$^{-1}$, $E_{2T} = 1$ (concentration). Units are arbitrary. Adapted from figure 1 of Goldbeter and Koshland (1981).

respectively. The values of K_1 and K_2 thus indicate how rapidly these reactions approach saturation as their substrate concentrations increase. This saturating behavior is illustrated in figure 6.8. In each panel of the figure, the solid curve shows the rate of the inactivating reaction as a function of w^*. In panel A, $K_2 = 1$, so the reaction cannot reach saturation. (When all of the protein is in state W^*, the reaction only reaches its half-saturating rate, as in that case $w^* = 1 = K_2$.) The rate curve in panel A is thus rather shallow. In contrast, panel B shows the case $K_2 = 0.1$; the half-saturating concentration is reached when just 10% of the protein is active ($w^* = 0.1$). This rate curve rises quickly to saturation. Next, consider the dashed curves, which show the rate of the activating reaction for different values of E_{1T}. (The fraction of inactive protein is $w = 1 - w^*$, so the rate of the activating reaction is plotted by "flipping" the rate curve.) For each value of E_{1T} (i.e., each dashed curve), the steady state occurs where the two reactions have equal rates—at the intersection point (shown by the dotted vertical line).

As E_{1T} increases, the activating rate law scales vertically. Figure 6.8A shows that when K_1 and K_2 are large, the steady-state fraction of w^* rises gradually as E_{1T} increases. In contrast, when K_1 and K_2 are small (figure 6.8B), there is an abrupt transition in the w^* steady state. This is the zero-order ultrasensitive effect.

A

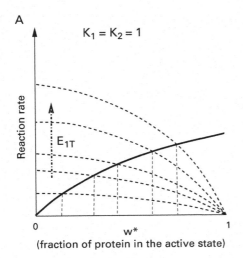

B

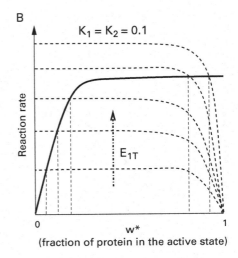

Figure 6.8
Reaction rates in the activation–inactivation cycle. In each panel, the solid curve shows the rate of the inactivating reaction as a function of the fraction w^* of activated protein. The dashed curves show the corresponding rate of the activating reaction. Steady state occurs where the curves intersect (i.e., where the reaction rates are equal—dotted vertical lines). (A) When K_1 and K_2 are large, neither reaction reaches saturation; increases in E_{1T} cause a gradual rise in the steady-state value of w^*. (B) In contrast, when K_1 and K_2 are small, saturation occurs at small concentrations, and so the steady-state value of w^* rises abruptly (from near zero to near one).

Exercise 6.2.2 Verify that for the activation mechanism described by equation (6.2) the ratio of input (E_{1T}) values between 90% and 10% activation is given by

$$R_v = \frac{81(K_1 + 0.1)(K_2 + 0.1)}{(K_1 + 0.9)(K_2 + 0.9)}.$$

Verify that (i) when K_1 and K_2 are both large, R_v tends to 81 (which is the same value observed for a Michaelis–Menten mechanism); and (ii) as K_1 and K_2 shrink to zero, R_v tends to one (infinite sensitivity). □

6.2.2 Ultrasensitive Activation Cascades

Besides the saturation effect that causes zero-order ultrasensitivity, there are other noncooperative mechanisms that can exhibit similarly steep dose-response curves (described in Ferrell (1996)).

One such mechanism is a cascade of activation–inactivation cycles, in which the target protein in each cycle acts as the activating enzyme in the next. The canonical example of this type of pathway is the *mitogen-activated protein kinase* (MAPK) cascade, which is composed of a sequence of phosphorylation–dephosphorylation cycles.

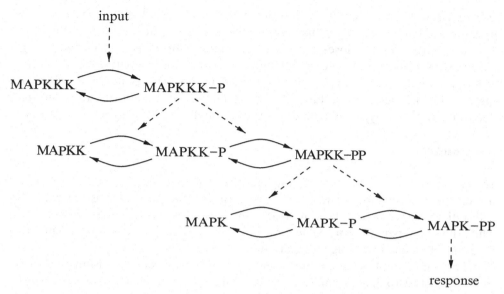

Figure 6.9
MAPK cascade. This three-tier activation cascade is initiated by phosphorylation of MAPKKK (phosphoryl groups are indicated by "-P"). MAPKKK-P phosphorylates MAPKK at two sites, and MAPKK-PP phosphorylates MAPK at two sites. The pathway output, MAPK-PP, triggers a cellular response. At each level, phosphatases lead to deactivation.

MAPK cascades operate in a wide range of eukaryotic species and act in a variety of signaling contexts, including pheromone response in yeast and growth-factor signaling in mammalian cells. The MAPK pathway, shown in figure 6.9, consists of three kinases in series, each activated by phosphorylation: phosphatases catalyze the inactivation reactions. The final kinase in the chain is MAPK. Active MAPK triggers the cellular response. The upstream components are named for their function: MAPK kinase (MAPKK) and MAPK kinase kinase (MAPKKK).

As discussed in section 6.1, a cascade of this form can lead to significant amplification of signal because each activated enzyme can activate many enzymes in the downstream tier. This cascade structure can also result in a sharp sigmoidal response because ultrasensitivity accumulates from one tier to the next (see exercise 6.2.3).

The MAPK cascade exhibits another mechanism for generation of sigmoidal responses: *multistep ultrasensitivity*. This phenomenon can occur in cycles where the activation step involves more than one catalytic event (as in the double-phosphorylation of MAPKK and MAPK). In such cases, the rate of the activation step will depend nonlinearly on the availability of the activating enzyme (details in problem 6.8.7).

Exercise 6.2.3   Consider a three-tiered signaling cascade as in figure 6.9. For simplicity, suppose that the dose-response of tier i is given in functional form as: $output_i = f_i(input_i)$. Verify that the dose-response of the entire pathway is then: $output_3 = f_3(f_2(f_1(input_1)))$. Demonstrate the cumulative effect of ultrasensitivity in this cascade by comparing the steepness (i.e., slope) of this composite function with the slope of the individual dose-response functions $f_i(\cdot)$. Hint: Use the chain rule to relate the derivative of $f_3(f_2(f_1(\cdot)))$ to the derivatives of the individual functions $f_i(\cdot)$.   □

6.3   Adaptation

So far, we have focused on the steady-state response of signaling pathways, summarized by dose-response curves. We next turn to a dynamic response, in which a cell initially responds to an input signal but then shuts that response down, even while the signal persists. This behavior is known as *adaptation* because the signaling mechanism adapts to the signal's continued presence.

A familiar example of adaptation comes from our own sense of vision. Our eyes tune themselves to detect changes in light intensity around the ambient level, whatever that level may be. We can observe this tuning process by leaving a dark building on a bright sunny day—this increases the ambient light level abruptly, and our visual field is briefly saturated. However, the system quickly adapts by tuning itself to detect changes about the new, higher nominal light level.

6.3.1   Bacterial Chemotaxis

Bacterial chemotaxis provides another example of adaptation. Chemotaxis refers to motion induced by the presence of chemical species in the environment. Bacteria swim toward higher concentrations of nutrients—*chemoattractants*—and away from toxins and other noxious substances—*chemorepellents*.

We will consider the *E. coli* chemotaxis network. These cells are attracted to amino acids (e.g., aspartate) and sugars (e.g., maltose, galactose) and are repelled by heavy metals (e.g., Co^{2+} and Ni^{2+}) and bacterial waste products (e.g., indole[1]). *E. coli* cells swim by means of *flagella*—bacterial "tails" (about a half-dozen on each cell) that are rotated by a protein complex called the flagellar motor.

E. coli cells have a rather crude method of steering. The action of the motors is coordinated, so there are only two modes of operation:

1. When the motors all rotate *counterclockwise*, the flagella coil into a bundle, which propels the bacterium in a straight line (figure 6.10A). This is referred to as *running*.

---

1. Indole is a primary odorant in feces and so is a potent chemorepellent for humans as well!

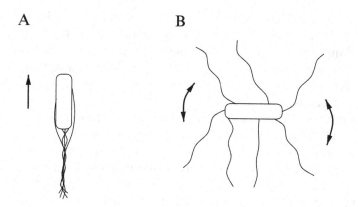

Figure 6.10
The two modes of operation of the *E. coli* flagella. (A) When the flagellar motors turn counterclockwise, the flagella coil together and propel the cell forward in a *run*. (B) When the flagellar motors turn clockwise, the flagella flail apart and randomly reorient the cell in a *tumble*.

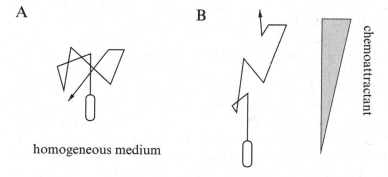

Figure 6.11
Motion of *E. coli* cells. (A) In a homogeneous environment, the cell follows a random walk. (B) In a gradient of attractant, the cell biases the random walk in the direction of increased attractant.

2. When the motors all rotate *clockwise*, the flagella separate and flail about (figure 6.10B). This behavior, called *tumbling*, results in a random reorientation of direction.

When *E. coli* cells are in a homogeneous environment (with no preferred direction of motion), each cell alternates between periods of running and periods of tumbling. (Runs last about 1 second; tumbles about 0.1 second.) The resulting motion is a random walk that samples the local environment (figure 6.11A). In contrast, when exposed to a gradient of chemoattractant (or chemorepellent), these cells bias their random walk by tumbling less frequently when moving in the "good" direction and more frequently when moving in the "bad" direction (figure 6.11B). Should a bacterium find itself once again in a uniform environment, it will return to the original tumbling frequency, even if the homogeneous level of attractant (or repellent) is significantly different from before—the sensory mechanism adapts to the new nominal environment (on a timescale of minutes). The cell is then tuned to respond to changes centered at this new nominal level.

E. coli cells use transmembrane receptor proteins to determine the level of chemoattractants and chemorepellents in their immediate surroundings. These bacterial cells are so small that measurements of chemical gradients across their length are not useful—thermal fluctuations corrupt spatial measurements on that scale. Instead, these cells use a temporal sampling approach: the sensory system compares the current conditions with the past conditions to infer spatial gradients.

Adaptation to changes in the environment can be measured in the laboratory, as follows. Beginning with a colony growing in a uniform environment, attractant is added so that the environment is again uniform, but richer than before. Because the bacteria measure the environment temporally, they are not immediately aware that their new environment is uniform: they wrongly assume they are moving in a "good" direction. The length of time it takes for them to "catch on" and return to the nominal behavior is a called the *adaptation time*.

E. coli cells use a signaling pathway to convert measurements of their environment into an exploration strategy. This pathway, shown in figure 6.12, transduces a signal from the transmembrane receptors (which bind attractant or repellent) to the flagellar motor. The receptors are complexed with a kinase called CheA ("Che" for chemotaxis). CheA phosphorylates the cytosolic protein CheY, which, when activated, binds to the flagellar motor and induces tumbling. Binding of chemoattractant to the receptor inhibits CheA activity: this reduces levels of active CheY and thus inhibits tumbling. Binding of repellent activates CheA and so has the opposite effect.

The pathway's ability to adapt to new environments is conferred by an added layer of complexity. On each receptor there are several sites that can bind methyl groups. Methylation is catalyzed by an enzyme called CheR. Another enzyme, CheB, demethylates the receptors. Receptor methylation enhances CheA activity, so methylation induces tumbling. A feedback loop is provided by an additional catalytic function of CheA—it activates CheB (again by phosphorylation). This is a negative feedback on CheA activity. (When CheA is, for example, inhibited, CheB activity is reduced, which leads to more receptor methylation and hence increased CheA activity.) This feedback acts on a slower timescale than the primary response and causes the system to return to its nominal level of CheA activity (and thus motor behavior) after a change in ligand concentration. This is the mechanism of adaptation.

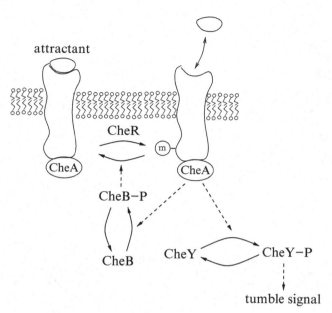

Figure 6.12
The chemotaxis signal-transduction network of *E. coli*. The receptor-bound kinase CheA activates the cytosolic protein CheY by phosphorylation. Active CheY binds the flagellar motor, inducing tumbling. Binding of chemoattractant inhibits CheA activity, thereby reducing the level of active CheY, and thus inhibiting tumbling. Repellent binding has the opposite effect (not shown). Receptors are methylated (m) by CheR and demethylated by active CheB. CheA activates CheB by phosphorylation. Methylation enhances CheA activity, so activation of CheB provides a negative feedback on CheA activity. This feedback causes the system to adapt to its pre-stimulus level after a change in ligand concentration. Adapted from figure 7.6 of Alon (2007).

Exercise 6.3.1 Knockout strains of *E. coli* lack the ability to produce the functional form of specific proteins. Knockouts of chemotaxis proteins often exhibit one of two behaviors (or *phenotypes*): constant running or constant tumbling. Which of these two behaviors would be observed in a CheA knockout? What about a CheB knockout? □

A simple model of the chemotaxis network is shown in figure 6.13 where it is assumed that each receptor has only a single methylation site. In this scheme, A is the CheA–receptor complex, m indicates methylation, L is chemoattractant ligand, B is inactive CheB, and B-P is active (phosphorylated) CheB. For simplicity, it is assumed that CheA is only active when associated with methylated, ligand-free receptors (species Am).

The methylation of receptors by CheR occurs in saturation, so the rate of this reaction is independent of receptor concentration. In contrast, demethylation by CheB follows a Michaelis–Menten rate law. Assuming mass-action kinetics for

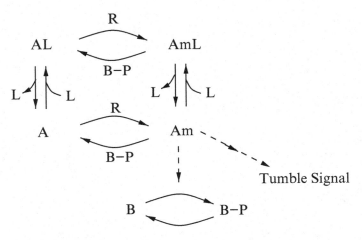

Figure 6.13
A simplified reaction scheme describing the chemotaxis signal-transduction network. Species A is the CheA–receptor complex, m indicates methylation, L is chemoattractant ligand, R is CheR, B is inactive CheB, and B-P is active (phosphorylated) CheB. To simplify the interpretation, only methylated receptor complexes that are not bound to attractant (i.e., species Am) are considered active.

ligand binding/unbinding and first-order kinetics for the activation and deactivation of CheB, the model equations are

$$\frac{d}{dt}[Am](t) = k_{-1}[R] - \frac{k_1[B\text{-}P](t)\cdot[Am](t)}{k_{M1}+[Am](t)} - k_3[Am](t)\cdot[L] + k_{-3}[AmL](t)$$

$$\frac{d}{dt}[AmL](t) = k_{-2}[R] - \frac{k_2[B\text{-}P](t)\cdot[AmL](t)}{k_{M2}+[AmL](t)} + k_3[Am](t)\cdot[L] - k_{-3}[AmL](t)$$

$$\frac{d}{dt}[A](t) = -k_{-1}[R] + \frac{k_1[B\text{-}P](t)\cdot[Am](t)}{k_{M1}+[Am](t)} - k_4[A](t)\cdot[L] + k_{-4}[AL](t)$$

$$\frac{d}{dt}[AL](t) = -k_{-2}[R] + \frac{k_2[B\text{-}P](t)\cdot[AmL](t)}{k_{M2}+[AmL](t)} + k_4[A](t)\cdot[L] - k_{-4}[AL](t)$$

$$\frac{d}{dt}[B](t) = -k_5[Am](t)\cdot[B](t) + k_{-5}[B\text{-}P](t)$$

$$\frac{d}{dt}[B\text{-}P](t) = k_5[Am](t)\cdot[B](t) - k_{-5}[B\text{-}P](t)\,.$$

We take the unbound ligand level $[L]$ as a fixed input.

Figure 6.14 shows the system's response to variations in the input level. Beginning at steady state with a low level of chemoattractant ($[L] = 20$, arbitrary units), the system responds at time $t = 10$ to a doubling of ligand concentration. The initial

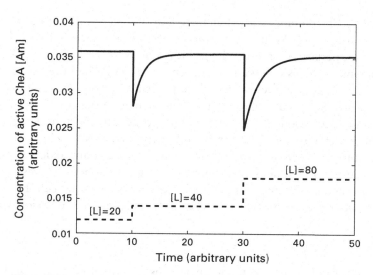

Figure 6.14
Behavior of the chemotaxis signal-transduction pathway. Starting at a low level of chemoattractant ligand ($[L] = 20$), the system responds to a doubling of ligand at time $t = 10$ with an immediate drop in CheA activity (corresponding to a reduction in tumbling) followed by a return to almost the original nominal activity level. A second doubling of ligand concentration at time $t = 30$ produces a similar effect. Parameter values are as follows: (in time^{-1}) $k_1 = 200$, $k_2 = 1$, $k_3 = 1$, $k_4 = 1$, $k_5 = 0.05$, $k_{-1} = 1$, $k_{-2} = 1$, $k_{-3} = 1$, $k_{-4} = 1$, $k_{-5} = 0.005$; (in concentration) $k_{M1} = 1$, $k_{M2} = 1$, $[R] = 1$. Units are arbitrary.

response is a sharp drop in CheA activity (corresponding to less tumbling). Over time, the negative feedback acts to return the system to its pre-stimulus activity level. The next perturbation occurs at time $t = 30$, when the ligand level is doubled again. Again, the activity rate recovers to near the pre-stimulus level. If the pre-stimulus level were exactly recovered, the system would be exhibiting *perfect adaptation*, as explored in the following exercise.

Exercise 6.3.2 As shown by Naama Barkai and Stanislas Leibler (Barkai and Leibler, 1997), this chemotaxis model will exhibit perfect adaptation if we make the assumption that CheB demethylates only active receptors (i.e., $k_2 = 0$). Verify this observation in the case that the level of *B-P* is held constant (i.e., consider only the equations for *Am*, *AmL*, *A*, and *AL*). Hint: Show that the steady-state concentration of *Am* is independent of the ligand level by considering the time derivative of $[Am] + [AmL]$ and solving for the steady state of $[Am]$. □

6.4 Memory and Irreversible Decision-Making

Adaptive systems are able to eventually ignore, or "forget," a persistent signal. In contrast, the opposite behavior can also be useful—some systems *remember* the effect of a transient signal. This memory effect can be achieved by a bistable system (see section 4.2). An input that pushes the state from one basin of attraction to the other causes a change that persists even after the input is removed.

The steady-state response of a bistable system is even more switch-like than the ultrasensitive behaviors addressed in section 6.2—it really is all-or-nothing. Either the state is perturbed a little and then relaxes back to its starting point or it gets pushed into a new basin of attraction and so relaxes to the other steady state. This sort of decision-making mechanism is particularly suited to pathways in which a permanent yes/no decision must be made. For example, when cells commit to developmental pathways, there is no meaningful intermediate response—these are discrete (yes/no) decisions whose consequences persist for the lifetime of the cell. In this section, we will consider a decision that is critical to a cell's fate: the decision to commit suicide.

6.4.1 Apoptosis

The process of programmed cell death—cellular suicide—is called *apoptosis* (from the Greek for "a falling off"). Apoptosis is a necessary part of the development of many multicellular organisms. Some cells play only a transient role during development: when they are no longer needed, they receive signals that induce apoptosis. (A commonly cited example is the tail of a tadpole, which is not needed by the adult frog.) Compared with death caused by stress or injury, apoptosis is a tidy process: rather than spill their contents into the environment, apoptotic cells quietly implode—ensuring there are no detrimental effects on the surrounding tissue.

Apoptosis is invoked by *caspase* proteins, which are always present in the cell but lie dormant until activated. The family of caspase proteins is split into two categories:

Initiator caspases respond to apoptosis-inducing stimuli. They can be triggered externally via transmembrane receptors or internally by stress signals from the mitochondria (e.g., starvation signals).

Executioner caspases are activated by initiator caspases. They carry out the task of cellular destruction by cleaving a number of key proteins and activating DNases that degrade the cell's DNA.

We will consider a model published in 2004 by Thomas Eissing and colleagues (Eissing et al., 2004). The model focuses on *caspase-8*, an initiator, and *caspase-3*, an executioner. Caspase-8 is triggered by external stimuli. When active, it activates caspase-3. Caspase proteins are activated by the removal of a masking domain, revealing a catalytic site. This cleavage is irreversible: the protein can only be inactivated by degradation. Consequently, to describe steady-state behavior, the model includes both production and degradation processes for each protein.

To guarantee that the decision to undergo apoptosis is irreversible, a feedback mechanism is in place: active caspase-3 activates caspase-8. This positive feedback

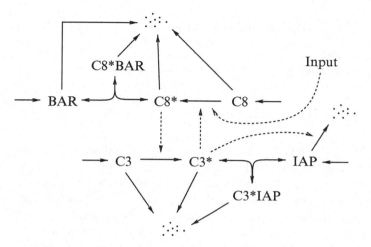

Figure 6.15
Eissing apoptosis model. An extracellular signal triggers activation of caspase-8 (C8*) from its inactive form (C8). Once active, caspase-8 activates caspase-3 (C3 to C3*). Active caspase-3 activates caspase-8, forming a positive feedback loop. Because activation of caspases is irreversible, protein degradation and production are included in the model. Two apoptotic inhibitors are included: BAR and IAP. These proteins bind active caspase-8 and caspase-3, respectively, thus inhibiting the progression to apoptosis. (Dots indicate degraded proteins.) Adapted from figure 1 of Eissing et al. (2004).

ensures that caspase activity is self-perpetuating. The feedback scheme is indicated in figure 6.15, which shows the core reaction network for the Eissing model. In addition to the caspases, the model incorporates two proteins, IAP and BAR, which inhibit apoptosis by binding active caspases, forming inert complexes.

Treating all enzyme-catalyzed reactions as first order, we can write the model as

$$\frac{d}{dt}[C8](t) = k_1 - k_2[C8](t) - k_3([C3^*](t) + [\text{Input}](t)) \cdot [C8](t)$$

$$\frac{d}{dt}[C8^*](t) = k_3([C3^*](t) + [\text{Input}](t)) \cdot [C8](t) - k_4[C8^*](t)$$
$$- k_5[C8^*](t) \cdot [BAR](t) + k_6[C8^*BAR](t)$$

$$\frac{d}{dt}[C3](t) = k_7 - k_8[C3](t) - k_9[C8^*](t) \cdot [C3](t)$$

$$\frac{d}{dt}[C3^*](t) = k_9[C8^*](t) \cdot [C3](t) - k_{10}[C3^*](t) - k_{11}[C3^*](t) \cdot [IAP](t) + k_{12}[C3^*IAP](t)$$

$$\frac{d}{dt}[BAR](t) = k_{13} - k_5[C8^*](t) \cdot [BAR](t) + k_6[C8^*BAR](t) - k_{14}[BAR](t)$$

$$\frac{d}{dt}[\text{IAP}](t) = k_{15} - k_{11}[\text{C3}^*](t) \cdot [\text{IAP}](t) + k_{12}[\text{C3}^*\text{IAP}](t) - (k_{16} + k_{17}[\text{C3}^*](t)) \cdot [\text{IAP}](t)$$

$$\frac{d}{dt}[\text{C8}^*\text{BAR}](t) = k_5[\text{C8}^*](t) \cdot [\text{BAR}](t) - k_6[\text{C8}^*\text{BAR}](t) - k_{18}[\text{C8}^*\text{BAR}](t)$$

$$\frac{d}{dt}[\text{C3}^*\text{IAP}](t) = k_{11}[\text{C3}^*](t) \cdot [\text{IAP}](t) - k_{12}[\text{C3}^*\text{IAP}](t) - k_{19}[\text{C3}^*\text{IAP}](t).$$

This system is bistable. At low levels of activated caspase, the system is at rest in a "life" state. Once caspase activity rises above a threshold, the positive feedback commits the system to reaching a steady state with high levels of active caspase—a "death" state. (Of course, the death state is transient—the cell is being dismantled. We are justified in calling it a steady state on the timescale of the signaling pathway.)

The simulation of the model in figure 6.16A shows the response of the system to an input signal. The system begins at rest with zero input and low caspase activity.

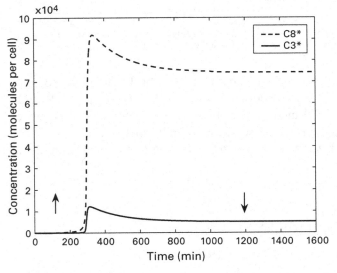

Figure 6.16
Behavior of the Eissing apoptotic pathway model. This simulation begins in the low-caspase-activity "life" state. At time $t = 100$ minutes, an input signal (Input = 200) is introduced, causing an increase in caspase-8 activity. This triggers a positive feedback loop that results in a rapid increase in activated caspase-8 and caspase-3 abundance at about $t = 300$. The system then settles to the high-caspase-activity "death" state. The input stimulus is removed at time $t = 1200$ minutes, but there is no effect: the apoptotic switch is irreversible. Parameter values: (in mpc min^{-1}) $k_1 = 507$, $k_7 = 81.9$, $k_{13} = 40$, $k_{15} = 464$; (in min^{-1}) $k_2 = 3.9 \times 10^{-3}$, $k_4 = 5.8 \times 10^{-3}$, $k_6 = 0.21$, $k_8 = 3.9 \times 10^{-3}$, $k_{10} = 5.9 \times 10^{-3}$, $k_{12} = 0.21$, $k_{14} = 1 \times 10^{-3}$, $k_{16} = 1.16 \times 10^{-2}$, $k_{18} = 1.16 \times 10^{-2}$, $k_{19} = 1.73 \times 10^{-2}$; (in mpc^{-1} min^{-1}) $k_3 = 1 \times 10^{-5}$, $k_5 = 5 \times 10^{-4}$, $k_9 = 5.8 \times 10^{-6}$, $k_{11} = 5 \times 10^{-4}$, $k_{17} = 3 \times 10^{-4}$; mpc = molecules per cell.

At time $t = 100$ minutes, an input is introduced, causing a slow increase in the activity level of caspase-8. This slow activation leads to a rapid rise in caspase activity at about $t = 300$ minutes. The system then settles into the "death" state with high caspase activity. When the input signal is removed (at time $t = 1200$ minutes), this self-perpetuating state persists, confirming that the system is bistable. Because complete removal of the input signal does not cause a return to the initial state, this life-to-death transition is irreversible (recall figure 4.20B).

Exercise 6.4.1 In the model, active caspase-3 promotes degradation of IAP. Does this interaction enhance or inhibit the positive feedback that leads to self-sustained caspase activity? □

6.5 Frequency Encoding

We next consider a signaling system that generates persistent oscillations in response to steady input signals: the strength of the input is encoded in the frequency of the oscillations. Downstream processes then respond to the oscillations in a frequency-dependent manner and elicit an appropriate cellular response. (This principle of frequency encoding is the basis for frequency modulation (FM) broadcast radio.)

6.5.1 Calcium Oscillations

Many types of animal cells use calcium ions, Ca^{2+}, as part of signal transduction cascades. Calcium is used to trigger, for example, the initiation of embryonic development in fertilized egg cells, the contraction of muscle cells, and the secretion of neurotransmitters from neurons.

Calcium signals are sent by rapid spikes in cytosolic Ca^{2+} concentration. Cells that use these signals normally have low levels of cytosolic calcium (about 10–100 nM). These low levels are maintained by ATP-dependent pumps that export cytosolic Ca^{2+} both out of the cell and into the endoplasmic reticulum (ER). The concentration of Ca^{2+} in the ER can reach as high as 1 mM (10^6 nM). Signaling pathways open calcium channels in the ER membrane, leading to rapid (diffusion-driven) surges in cytosolic calcium levels.

However, because calcium is involved in many cellular processes, persistent high concentrations can be detrimental. (For example, failure to remove calcium from muscle cells keeps them in a state of constant tension. This is what causes rigor mortis.) Some cells that use calcium as an intracellular signaling molecule avoid persistently high Ca^{2+} concentrations by generating oscillations in calcium levels. The frequency of the oscillations is dependent on the intensity of the signal, whereas the amplitude is roughly constant. The downstream cellular response is dependent on the oscillation frequency (see problem 6.8.14).

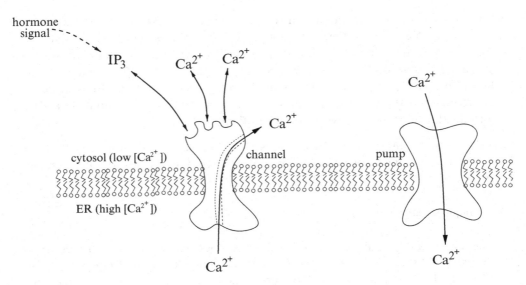

Figure 6.17
Calcium-induced calcium release. A G-protein pathway (not shown) responds to a hormone signal by inducing production of IP$_3$, which activates calcium channels in the ER membrane. These channels bind Ca^{2+} ions at two sites. The first binding event causes the channel to open; the second causes it to close. Calcium pumps continually pump Ca^{2+} ions from the cytosol to the ER.

We will consider an instance of this frequency-encoding mechanism in mammalian liver cells. These cells respond to certain hormones with the activation of G-protein–coupled receptors (section 6.1.2). The G-protein triggers a signaling pathway that results in production of inositol 1,4,5-triphosphate (IP$_3$). These IP$_3$ molecules bind a receptor that is complexed with a calcium channel in the membrane of the ER.

The IP$_3$ binding event exposes two receptor sites at which Ca^{2+} ions can bind (figure 6.17). These two sites have different affinities for Ca^{2+}. At low concentration only one site is occupied, whereas at higher concentrations both sites are bound. The calcium-binding events have opposing effects on the receptor–channel complex. Binding of the first calcium ion causes the channel to open, allowing Ca^{2+} to flow into the cytosol. Binding of the second ion causes the channel to close. This interplay of positive and negative feedback generates oscillations in the cytosolic Ca^{2+} concentration, as follows. When cytosolic calcium levels are low, the channels are primarily in the open state, and so Ca^{2+} rushes into the cytosol from the ER. When high calcium levels are reached, the channels begin to shut. Once most of the channels are closed, the continual action of the Ca^{2+} pumps eventually causes a return to low cytosolic [Ca^{2+}], from which the cycle repeats.

In 1993, Hans Othmer and Yuanhua Tang developed a model of this pathway that focuses on the behavior of the channel (described in Othmer (1997)). Taking $I = [IP_3]$ as the system input, the receptor binding events are described by

$$I + R \underset{k_{-1}}{\overset{k_1}{\rightleftharpoons}} RI$$

$$RI + C \underset{k_{-2}}{\overset{k_2}{\rightleftharpoons}} RIC^+$$

$$RIC^+ + C \underset{k_{-3}}{\overset{k_3}{\rightleftharpoons}} RIC^+C^-,$$

where R is the receptor–channel complex, C is cytosolic calcium, RI is the IP_3-bound receptor–channel complex, RIC^+ is the open (one Ca^{2+}-bound) channel, and RIC^+C^- is the closed (two Ca^{2+}-bound) channel.

The rate of diffusion of calcium into the cytosol depends on the concentration of calcium in the ER (denoted $[C_{ER}]$, and held fixed) and the abundance of open channels. Recall from section 3.4 that the rate of diffusion is proportional to the difference in concentration between the two compartments. This transport rate is modeled as

Rate of Ca^{2+} diffusion into the cytosol $= v_r(\gamma_0 + \gamma_1[RIC^+])([C_{ER}] - [C])$,

where v_r is the ratio of the ER and cytosolic volumes, and γ_0 characterizes a channel-independent "leak."

Calcium is continually pumped from the cytosol to the ER. Presuming strong cooperativity of calcium uptake, the pumping rate is modeled as

Rate of Ca^{2+} pumping out of the cytosol $= \dfrac{p_1[C]^4}{p_2^4 + [C]^4}$,

for parameters p_1 and p_2.

The complete model is then:

$$\frac{d}{dt}[R](t) = -k_1[I](t) \cdot [R](t) + k_{-1}[RI](t)$$

$$\frac{d}{dt}[RI](t) = -(k_{-1} + k_2[C](t)) \cdot [RI](t) + k_1[I](t) \cdot [R](t) + k_{-2}[RIC^+](t)$$

$$\frac{d}{dt}[RIC^+](t) = -(k_{-2} + k_3[C](t)) \cdot [RIC^+](t) + k_2[C](t) \cdot [RI](t) + k_{-3}[RIC^+C^-](t)$$

$$\frac{d}{dt}[RIC^+C^-](t) = k_3[C](t) \cdot [RIC^+](t) - k_{-3}[RIC^+C^-](t)$$

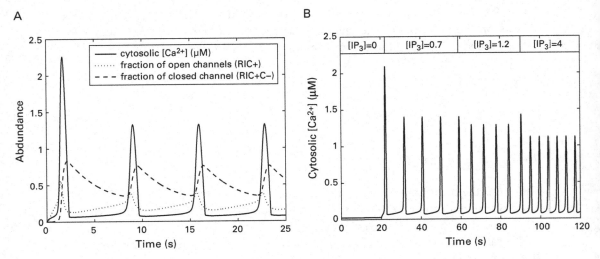

Figure 6.18
Calcium oscillations. (A) This simulation shows the oscillatory behavior of the system. When calcium levels are low, channels open, letting more Ca^{2+} into the cytosol. As calcium levels rise, channels begin to close, leading to a drop in cytosolic Ca^{2+} levels. The IP_3 concentration is fixed at 2 μM. (B) In this simulation, the input level of IP_3 changes, demonstrating the frequency-encoding ability of the system. As the IP_3 level (in μM) increases, the frequency of the oscillations increases considerably, whereas the amplitude is roughly constant. Parameter values: (in μM^{-1} s^{-1}) $k_1 = 12$, $k_2 = 15$, $k_3 = 1.8$; (in s^{-1}) $k_{-1} = 8$, $k_{-2} = 1.65$, $k_{-3} = 0.21$, $\gamma_0 = 0.1$, $\gamma_1 = 20.5$; $[C_{ER}] = 8.37$ μM, $p_1 = 8.5$ μM s^{-1}, $p_2 = 0.065$ μM, $v_r = 0.185$.

$$\frac{d}{dt}[C](t) = v_r(\gamma_0 + \gamma_1[RIC^+](t)) \cdot ([C_{ER}] - [C](t)) - \frac{p_1[C(t)]^4}{p_2^4 + [C(t)]^4}.$$

The simulation in figure 6.18A illustrates the system's oscillatory behavior. When the calcium level is low, the concentration of open channels increases, followed by a rapid increase in $[Ca^{2+}]$. Once the calcium concentration rises, the channels close, and the calcium level falls, setting up a new cycle. Figure 6.18B demonstrates the system's frequency-encoding ability. As the input is increased (in steps), the frequency of oscillations increases while the amplitude changes very little.

6.6*　Frequency Response Analysis

A key feature of any signal transduction pathway is its dynamic response. Although model simulation allows us to predict individual time-varying responses, simulation offers limited insight into the general nature of a system's input–output behavior. In this section, we introduce an analytical tool, the *frequency response*, that provides a succinct characterization of a network's dynamic response to arbitrary inputs and allows direct insights into that behavior. A significant limitation of this analysis is that it applies only to *linear* systems. To use this approach in the investigation of nonlinear biological systems, we will restrict our attention to behavior near a nominal operating point. Recall (as discussed in section 4.2.2) that a nonlinear system will exhibit near-linear behavior when responding to small deviations around a nominal condition. In this case, the response can be described by a linear model.

Frequency response analysis is based on a concise description of the response of linear systems to sine-wave inputs (over a range of frequencies). The character of these responses reveals how the system responds to inputs over a wide range of timescales. In particular, it indicates the system's *bandwidth*—the fastest timescale on which the system is able to respond dynamically.

A system's frequency response can be measured directly from time-series experiments. (This process, called *system identification*, was reviewed in Ang et al. (2011), which also reviews applications of frequency response methods to cellular signaling networks.) In this section, our focus will be on the interpretation of the frequency response and its derivation from differential equation models. We will begin by defining the frequency response and introducing the notion of *frequency filtering*.

6.6.1   Definition of the Frequency Response

The long-term response of a linear system to a persistent sine-wave input can easily be described as follows. Such a response is illustrated in figure 6.19, which shows a simulation of the G-protein pathway model of section 6.1.2. The system input is a steady sine-wave oscillation in ligand abundance, centered about a nominal concentration of 1 nM. The response—abundance of active G_α-GTP—shows a transient rise from the initial state, followed by a steady oscillation at the same frequency as the input. Provided the oscillations in the input do not stray too far from the nominal level, the oscillatory form of the response is guaranteed—the system output will always settle into a steady sine-wave behavior with the *same frequency as the forcing input*. Thus, the only differences between the long-term input and response signals, besides the fact that they are centered at different values, are the amplitude of the oscillations and their relative *phase*—the timing of the peaks and troughs. These two differences are measured by the system *gain* and *phase shift*, respectively.

From the simulation shown in figure 6.19, we see that the gain, defined as the ratio of the amplitude of the input to the amplitude of the response, is 0.4/60 nM/molecule per cell. The time that elapses between peaks is about 50 seconds. The phase shift is defined as ratio of this time difference to the period of the oscillations, expressed in degrees (i.e., multiplied by 360°). In this case, the period is 200 seconds, so the phase shift is 90° (= (50/200) × 360°).

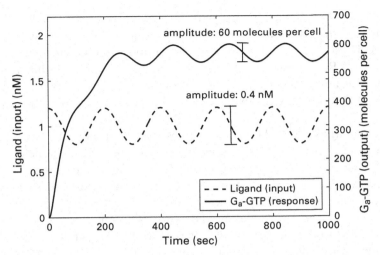

Figure 6.19
G-protein pathway response to an oscillating input. The model from section 6.1.2 is simulated with a sine-wave input of frequency $1/200$ s^{-1}, centered at a nominal level of 1 nM, with amplitude 0.4 nM. The system response settles to a sine wave with the same frequency ($1/200$ s^{-1}) and an amplitude of 60 molecules per cell, as shown. The difference in phase (i.e., the time that elapses between the peak in one curve and the peak in the other) is about 50 seconds. Adapted from figure 10.3 of Ang et al. (2011).

The gain and phase shift depend on the frequency of the input signal, but they are independent of its amplitude, provided the system remains in the linear regime (as explored in problem 6.8.17). The frequency dependence of the gain and phase for the G-protein model are demonstrated in figure 6.20, which shows the long-term response of the model for three different input frequencies. The sine-wave inputs are all centered at the same nominal level and have equal amplitudes, but they oscillate at different frequencies. Each response oscillates at the same frequency as the corresponding input, but the responses exhibit a range of amplitude and phase behavior.

Figure 6.20 demonstrates the system gain and phase shift at three distinct frequencies. In section 6.6.3, we will address a technique for determining—directly from the model equations—the gain and phase shift at all frequencies. The functional dependence of gain and phase shift on frequency is called the *frequency response* of the system.

The frequency response is typically displayed as a pair of plots. The dependence of gain on frequency, in hertz (Hz),[2] is called the *gain Bode plot*; this is normally displayed on a log–log scale. The *phase Bode plot* shows the phase shift as a function of frequency on a semilog scale. Figure 6.21 shows the Bode plots for the G-protein

---

2. Hertz (Hz), equal to 1/second, is a unit of frequency.

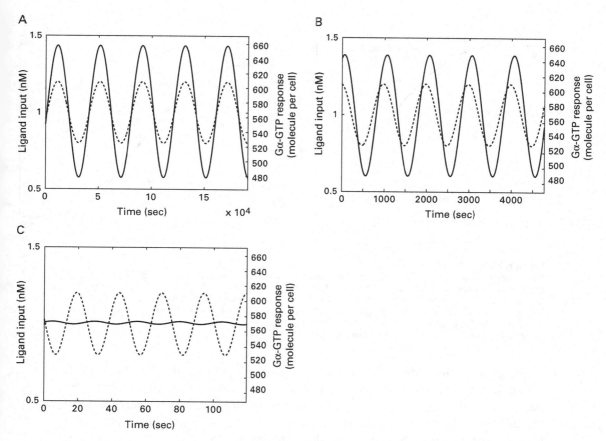

Figure 6.20
Responses of the G-protein model to sine-wave inputs. The inputs (dashed curves) all have amplitude 0.4 nM. The inputs oscillate at frequencies of (A) 2.5×10^{-5} s^{-1}; (B) 10^{-3} s^{-1}; and (C) 4×10^{-2} s^{-1}. (Note the difference in scales on the vertical axis.) The responses (solid curves, all shown on the same scale) exhibit different amplitudes and phase shifts. Adapted from figure 10.4 of Ang et al. (2011).

system (for inputs near a nominal ligand level of 1 nM). The gain and phase behaviors illustrated in figure 6.20 are labeled.

6.6.2 Interpretation of the Frequency Response

Keeping in mind that frequency response analysis applies only to the system's behavior in the neighborhood of a specified operating point, a great deal of information about this local behavior can be gleaned directly from the Bode plots.

The phase plot can be used to predict the results of interconnecting multiple systems (in cascade or in feedback). We will restrict our attention to systems in isolation, and so we will not consider the phase plot further.

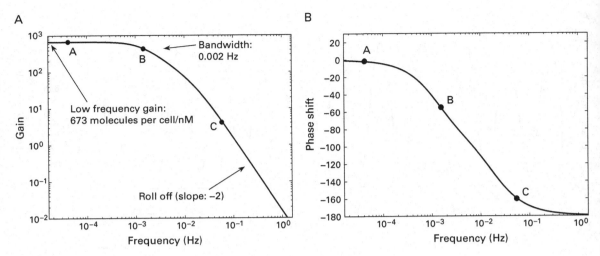

Figure 6.21
Frequency response of the G-protein pathway model. Gain Bode plot (left); phase Bode plot (right). Frequencies corresponding to the simulations shown in panels A, B, and C of figure 6.20 are labeled. The system behaves as a low-pass filter: the low-frequency gain is about 700 molecules per cell/nM, the bandwidth is 0.002 Hz, and the roll-off is 2. (The gain plot does not follow the standard engineering convention of displaying gain on a decibel (dB) scale; this would involve multiplying the logarithm of the gain by a factor of 20.) Adapted from Figure 10.5 of Ang et al. (2011).

The gain plot indicates the strength of the system's response to sine-wave inputs at different frequencies. Because the frequency of oscillations corresponds to the timescale of the signal—low frequency waves oscillate on long timescales; high-frequency oscillations act on short timescales—the gain plot provides a picture of how the system responds to inputs at different timescales.

Crucially, frequency response analysis does not just apply to sine-wave inputs. The behavior described in the Bode plots can be extended to arbitrary inputs via *Fourier decomposition*, which allows us to express any time-varying input as a collection of sine-wave signals.[3] In the remainder of this subsection, we will use the notion of *frequency filtering* to illustrate both the Fourier decomposition and the frequency-domain description of system behavior.

Fourier analysis decomposes an arbitrary input signal into sine-wave components: the signal can then be characterized by the size of the contribution at each frequency—called the frequency *content*. For example, signals that change rapidly are dominated by content at high frequencies, whereas signals that act slowly have

---

3. A periodic signal can be expressed as a sum of sine waves over a discrete set of frequencies in a *Fourier series*. A nonperiodic signal can be written as the integral of sine waves over a continuum of frequencies via the *Fourier transform*.

most of their content at low frequencies. A system's *frequency filtering* properties describe how an input's frequency components are amplified or attenuated in the system response.

As a first example of frequency filtering, consider a resonant system—one that amplifies at a specific frequency and attenuates at all others. The input–output behavior of a resonant system is illustrated in figure 6.22. Panel A shows an input signal that has roughly equal content over all frequencies (a so-called *white noise* signal). Panel B shows the Bode gain plot for a resonant system. The plot has a peak at 0.5 Hz. Input content at that frequency will be amplified by the system. At other frequencies, the gain is less than 1, and so input content at those frequencies will be attenuated—it will be absent from the system response. Panel C shows the response of this resonant system to the input in panel A. The response is essentially a sine wave at frequency 0.5 Hz. The system has "filtered out" all frequency components of the input except at 0.5 Hz. (This sort of resonant behavior is important in technology, e.g., in broadcast radio, and is the basis for some molecular assays, such as fluorescence resonance energy transfer (FRET).)

Figure 6.23 illustrates filtering behaviors more commonly observed in cellular networks. Three different systems are shown, along with their responses to a common input signal. The input (panel A) is a low-frequency sine wave corrupted by high-frequency noise. Panel B shows the Bode gain plot for a *low-pass* filter. This system lets the low-frequency components of the input pass into the output signal. The gain plot has a "corner" at about 1 Hz, above which there is a steady decrease in the gain. The frequency at which this corner occurs is called the system *bandwidth*, defined as the frequency above which the system fails to respond significantly to input signals. In the output from a low-pass filter, the low-frequency components are retained whereas high-frequency components are stripped away. This is illustrated in the response in Panel C: the input's low-frequency oscillation is retained, whereas the high-frequency "chatter" has been considerably smoothed.

Panel D also shows a low-pass filter. Comparing with panel B, the corner frequency in panel D is lower (about 0.01 Hz), meaning that this low-pass filter is more stringent than the system in panel B. The difference is apparent in the response in panel E: the low-frequency oscillations are again retained, but in this case almost all of the higher-frequency noise has been filtered away.

Panel F shows a different behavior. This gain plot is characteristic of a *band-pass filter*. Over a narrow range where the plot peaks (around 100 Hz), frequency components will pass: any content at higher or lower frequencies is attenuated. The response in panel G illustrates these filtering properties: the input's low-frequency oscillations have been eliminated, but the higher-frequency noise has passed through into the response.

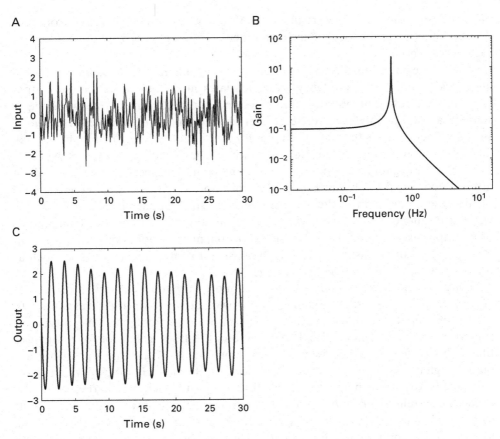

Figure 6.22
Filtering behavior of a resonant system. (A) The input signal is composed of roughly equal content at a wide range of frequencies. (B) The Bode gain plot for a resonant system. The system responds strongly to input oscillations at 0.5 Hz. At all other frequencies, the response is significantly attenuated. (C) The system output. The oscillations at 0.5 Hz have passed through the filter, resulting in a near sine-wave response at this resonant frequency. Adapted from figure 10.7 of Ang et al. (2011).

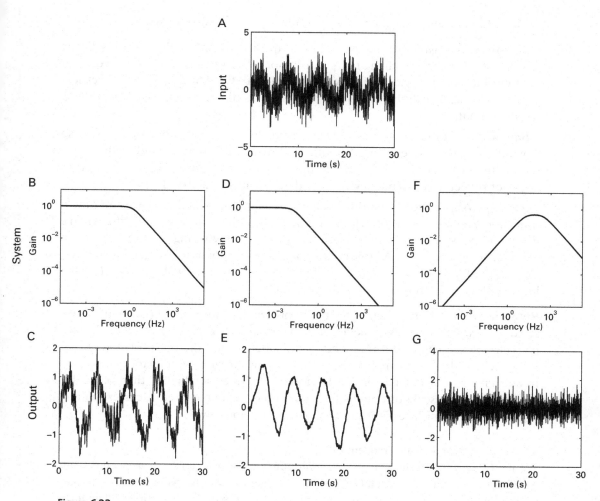

Figure 6.23
Frequency filtering. (A) The input: high-frequency noise added to a low-frequency sine-wave. (B) The Bode gain plot for a low-pass filter. (C) The response of the filter in (B) to the input in (A). The low-frequency oscillation is retained, as are the lower-frequency components of the noise, resulting in a smoother curve. (D) The Bode gain plot for a low-pass filter with a lower bandwidth than in (B). (E) The output response from the system in (D). As in (C), the low-frequency oscillation has been preserved in the response. This more stringent filter has stripped away almost all of the high-frequency noise. (F) The Bode gain plot for a band-pass filter. (G) The response of the band-pass system in (F). This filter has blocked the low-frequency oscillation, leaving only the high-frequency noise in the response. Adapted from figure 10.8 of Ang et al. (2011).

Three features of Bode gain plots are often used to summarize system behavior:

• *Low-frequency gain* The gain at low frequency describes the system's linear response to constant (unchanging) perturbations. The value of this gain is equal to the (absolute) parametric sensitivity coefficient, as defined in section 4.5. (Electrical engineers typically call this value the "DC gain," as direct current (DC) provides a constant input.)

• *Bandwidth* For most systems, the Bode gain plot shows a steady drop in response beyond a certain frequency. That frequency is called the system bandwidth: it represents the fastest timescale on which the system can respond. The system output does not reflect input frequency content that lies above the bandwidth.

• *Roll-off* Above the bandwidth frequency, a gain plot exhibits a consistent downward slope, called the roll-off. The absolute slope of the roll-off is called the *relative degree* of the system and is a measure of how quickly changes in the input are reflected in the response. (A large relative degree corresponds to a sluggish response.) The simple filters shown in figure 6.23 all have roll-off with a slope of −1 (on the log–log plot), indicating a relative degree of 1.

Returning to the Bode gain plot for the G-protein pathway model (figure 6.21), we see that it exhibits low-pass behavior. The low-frequency gain is about 700 molecules per cell/nM, the bandwidth frequency is at 0.002 Hz, and the roll-off has slope −2. We conclude that this signal transduction system provides significant gain at low frequencies and is not responsive to noise at frequencies higher than 0.002 Hz (indicating a timescale of hours). The system has relative degree 2, corresponding to a reasonably quick response to changes in the input level.

6.6.3 Construction of the Frequency Response
We next demonstrate how a system's frequency response can be constructed from a differential equation model. Because the frequency response is used for investigation of linear system behavior, the construction begins with linearization of a general nonlinear model.

Linearization of Input–Output Systems Recall that in section 4.2.2, we addressed the linearization of a system of differential equations about a steady state (as a tool for stability analysis). Here, we extend that analysis to incorporate an input signal and an output response. Using vector notation, suppose the nonlinear system of differential equations

$$\frac{d}{dt}\mathbf{x}(t) = \mathbf{g}(\mathbf{x}(t))$$

has a steady state at $\mathbf{x} = \mathbf{0}$ (i.e., $\mathbf{g}(\mathbf{0}) = \mathbf{0}$). The linearization of this model about the steady state $\mathbf{x} = \mathbf{0}$ is the system

$$\frac{d}{dt}\mathbf{x}(t) = \mathbf{A}\mathbf{x}(t)$$

where $\mathbf{A}$ is the system Jacobian, as introduced in section 4.2.2. In vector notation, $\mathbf{A} = \partial\mathbf{g}/\partial\mathbf{x}$.

To generalize this construction to an input–output system, consider a nonlinear model in which an input signal u affects the dynamics,

$$\frac{d}{dt}\mathbf{x}(t) = \mathbf{f}(\mathbf{x}(t), u(t)), \tag{6.3}$$

and an output response y is identified:

$$y(t) = h(\mathbf{x}(t), u(t)). \tag{6.4}$$

(Most commonly, the response y is simply one of the state variables x_i. In that case, the function h simply returns one of the components of the $\mathbf{x}$ vector, i.e., $h(\mathbf{x}, u) = x_i$.) Suppose that the steady-state nominal operating condition is specified by $\mathbf{x} = \mathbf{0}$, $u = 0$, and $y = 0$. (The input signal can be scaled so that u describes the deviation of the input from the nominal level, likewise for the response y.) In this case, the linearization takes the form of a *linear input–output system*:

$$\frac{d}{dt}\mathbf{x}(t) = \mathbf{A}\mathbf{x}(t) + \mathbf{B}u(t) \tag{6.5}$$

$$y(t) = \mathbf{C}\mathbf{x}(t) + Du(t),$$

where

$$\mathbf{A} = \frac{\partial\mathbf{f}}{\partial\mathbf{x}} \qquad \mathbf{B} = \frac{\partial\mathbf{f}}{\partial u}$$

$$\mathbf{C} = \frac{\partial\mathbf{h}}{\partial\mathbf{x}} \qquad D = \frac{\partial\mathbf{h}}{\partial u}.$$

The derivatives are evaluated at the nominal operating point. The matrix $\mathbf{A}$ is the system Jacobian, $\mathbf{B}$ and $\mathbf{C}$ are linearizations of the input and output maps, respectively, and D is called the *feed-through* (or *feed-forward*) term.

Exercise 6.6.1 Verify that the linearization of the system

$$\frac{d}{dt}x_1(t) = u(t) + x_1(t) - 2(x_1(t))^2$$

$$\frac{d}{dt}x_2(t) = -x_1(t) - 3x_2(t),$$

with output $y = x_2$ at the steady state $(x_1, x_2) = (0, 0)$, with nominal input $u = 0$, is given by

$$\mathbf{A} = \begin{bmatrix} 1 & 0 \\ -1 & -3 \end{bmatrix}, \qquad \mathbf{B} = \begin{bmatrix} 1 \\ 0 \end{bmatrix}, \qquad \mathbf{C} = [0 \ 1], \qquad D = 0. \qquad \square$$

Exercise 6.6.2 Consider the system

$$\frac{d}{dt}s(t) = V(t) - \frac{V_m s(t)}{K + s(t)},$$

which describes the concentration of a species that is produced at rate V and consumed at a Michaelis–Menten rate. Verify that for $V = V_0$, the steady state is $s^{ss} = V_0 K / (V_m - V_0)$, provided $V_0 < V_m$. Take the system input to be $V(t)$, with nominal input value $V = V_0$ (with $V_0 < V_m$). Take the system output to be $y(t) = s(t) - s^{ss}$. Define $x(t) = s(t) - s^{ss}$ and $u(t) = V(t) - V_0$. Then, steady state occurs at $x = 0, u = 0$, and $y = 0$. Verify that the linearization of this input–output system (as in equation 6.5) has

$$A = -\frac{(V_m - V_0)^2}{V_m K}, \qquad B = 1, \qquad C = 1, \qquad D = 0. \qquad \square$$

The Frequency Response Given the linear system description in (6.5), the frequency response, H, is given as a function of frequency ω by

$$H(\omega) = \mathbf{C}(i\omega\mathbf{I} - \mathbf{A})^{-1}\mathbf{B} + D, \tag{6.6}$$

where $i = \sqrt{-1}$, and $\mathbf{I}$ is the identity matrix of the same dimension as $\mathbf{A}$. (This formula can be derived using the *Laplace transform*, which is used to construct the system *transfer function*—a valuable tool for system analysis and design.)

The frequency response $H(\omega)$ is a complex-valued function. When written as

$$H(\omega) = a(\omega) + b(\omega)i,$$

with a and b real-valued, the system gain takes the form

$$\text{Gain}(\omega) = \sqrt{a(\omega)^2 + b(\omega)^2},$$

which is the *modulus* (or *magnitude*) of the complex number $H(\omega)$. The phase shift, given by $\tan^{-1}(a(\omega)/b(\omega))$, is the *argument* (or *phase*) of the complex number $H(\omega)$.

Example: Derivation of the Frequency Response of the G-Protein Signaling Model As an example, we will derive the G-protein signaling model's frequency response (shown in figure 6.21). From section 6.1.2, the nonlinear model can be written as (see exercise 6.1.1):

$$\frac{d}{dt}[RL](t) = k_{RL}[L](t) \cdot (R_T - [RL](t)) - k_{RLm}[RL](t)$$

$$\frac{d}{dt}[G](t) = -k_{Ga}[RL](t) \cdot [G](t) + k_{G1}(G_T - [G](t) - [Ga](t)) \cdot (G_T - [G](t))$$

$$\frac{d}{dt}[Ga](t) = k_{Ga}[RL](t) \cdot [G](t) - k_{Gd0}[Ga](t).$$

Here, the input signal is the ligand level $[L]$, whereas the output is the concentration of active G_α-GTP: $[Ga]$. Comparing with equation (6.3), with $\mathbf{x} = ([RL], [G], [Ga])$ and $u = [L]$, this system defines $\mathbf{f}$ as a vector function with three coefficients:

$$f_1([RL], [G], [Ga], [L]) = k_{RL}[L](R_T - [RL]) - k_{RLm}[RL]$$

$$f_2([RL],[G],[Ga],[L]) = -k_{Ga}[RL][G] + k_{G1}(G_T^2 - 2[G]G_T + [G]^2 - [Ga]G_T + [Ga][G])$$

$$f_3([RL], [G], [Ga], [L]) = k_{Ga}[RL][G] - k_{Gd0}[Ga].$$

The output function (equation 6.4) takes the simple form

$$h([RL], [G], [Ga], [L]) = [Ga].$$

The Jacobian

$$\mathbf{A} = \begin{bmatrix} \dfrac{\partial f_1}{\partial [RL]} & \dfrac{\partial f_1}{\partial [G]} & \dfrac{\partial f_1}{\partial [Ga]} \\[2mm] \dfrac{\partial f_2}{\partial [RL]} & \dfrac{\partial f_2}{\partial [G]} & \dfrac{\partial f_2}{\partial [Ga]} \\[2mm] \dfrac{\partial f_3}{\partial [RL]} & \dfrac{\partial f_3}{\partial [G]} & \dfrac{\partial f_3}{\partial [Ga]} \end{bmatrix}$$

is given by

$$\mathbf{A} = \begin{bmatrix} -k_{RL}[L]-k_{RLm} & 0 & 0 \\ -k_{Ga}[G] & -k_{Ga}[RL]+k_{G1}(2[G]-2G_T+[Ga]) & k_{G1}([G]-G_T) \\ k_{Ga}[G] & k_{Ga}[RL] & -k_{Gd0} \end{bmatrix}$$

The linearized input map takes the form

$$\mathbf{B} = \begin{bmatrix} \dfrac{\partial f_1}{\partial [L]} \\[2mm] \dfrac{\partial f_2}{\partial [L]} \\[2mm] \dfrac{\partial f_3}{\partial [L]} \end{bmatrix} = \begin{bmatrix} k_{RL}(R_T - [RL]) \\ 0 \\ 0 \end{bmatrix}.$$

The linearized output map is

$$\mathbf{C} = \begin{bmatrix} \dfrac{\partial h}{\partial [RL]} & \dfrac{\partial h}{\partial [G]} & \dfrac{\partial h}{\partial [Ga]} \end{bmatrix} = [0 \quad 0 \quad 1].$$

There is no feed-through:

$$D = \frac{\partial h}{\partial [L]} = 0.$$

Substituting the parameter values indicated in figure 6.5 into these matrices, and taking the nominal input value to be $[L] = 1$ nM, the system's frequency response can be calculated from equation (6.6):

$$H(\omega) = \frac{359 + 0.626i\omega}{-i\omega^3 - 571\omega^2 + 71.3i\omega + 0.533} \quad \text{molecules per cell/nM} .$$

Key features of the Bode plots can be gleaned directly from this formula: the low-frequency gain corresponds to the magnitude at frequency $\omega = 0$, which is 673 (= 359/0.533) molecules per cell/nM; the roll-off is the difference in degree between the denominator and numerator, which is 2. (The bandwidth is more difficult to discern: it is determined by the location of the roots of the denominator.) MATLAB commands for computing the frequency response and generating the Bode plots are described in appendix C.

Exercise 6.6.3 Consider the linear one-dimensional input–output model:

$$\frac{d}{dt} x(t) = -ax(t) + u(t)$$

$$y(t) = x(t).$$

(a) Verify that the frequency response of this system is (from equation 6.6)

$$H(\omega) = \frac{1}{i\omega - a}.$$

(b) Verify that the gain of the system is

$$\frac{1}{\sqrt{a^2 + \omega^2}}.$$

The low-frequency gain (attained at $\omega = 0$) is then $1/a$. The bandwidth is $\omega = a$ (because for $\omega > a$, the gain shrinks rapidly). $\qquad \square$

6.7 Suggestions for Further Reading

• **Signal Transduction Pathways** Most introductory cell biology texts address signal transduction pathways. A comprehensive treatment is provided by *Cellular Signal Processing: An Introduction to the Molecular Mechanisms of Signal Transduction* (Marks et al., 2009). A range of prokaryotic signal transduction systems is presented in *The Physiology and Biochemistry of Prokaryotes* (White, 2000). Several models of signal transduction pathways are surveyed in *Systems Biology: A Textbook* (Klipp et al., 2009).

• **Bacterial Chemotaxis** The book *E. coli in Motion* (Berg, 2004) provides an accessible account of bacterial chemotaxis. A model-based analysis of the robustness of chemotaxis signaling can be found in *An Introduction to Systems Biology: Design Principles of Biological Circuits* (Alon, 2007).

• **Calcium Oscillations** Models of calcium oscillations are covered in the book *Biochemical Oscillations and Cellular Rhythms: The Molecular Bases of Periodic and Chaotic Behavior* (Goldbeter, 1996).

• **The Frequency Response** Frequency response analysis is a component of systems control theory. An accessible introduction to this area can be found in *Feedback Systems: An Introduction for Scientists and Engineers* (Aström and Murray, 2008). An introduction to the use of these tools in systems biology can be found in *Feedback Control in Systems Biology* (Cosentino and Bates, 2011).

6.8 Problem Set

6.8.1 Cooperativity in the Bacterial Two-Component Signaling Pathway

Modify the two-component signaling pathway model in section 6.1.1 so that ligand-bound receptors dimerize before becoming active. A strongly cooperative binding mechanism can be approximated by replacing the ligand-binding reaction with

$$2L + R \underset{k_{-1}}{\overset{k_1}{\rightleftharpoons}} RL_2$$

where R now represents a receptor dimer. With parameter values as in figure 6.3, prepare a dose-response curve for this modified model. Compare with figure 6.3B and verify that the cooperative model exhibits an ultrasensitive response. Next, confirm that the ultrasensitivity is more pronounced if the dimers tetramerize with strong cooperativity upon ligand binding (i.e., $4L + R \leftrightarrow RL_4$).

6.8.2 The Two-Component KdpD–KdpE Signaling Pathway

When cells of the bacterium *E. coli* need to increase the rate at which they take up K$^+$ ions from the environment, they increase production of a high-affinity K$^+$ uptake system. Production of this system is under the control of the protein KdpE, which is the response regulator in a two-component signaling pathway. KdpE is activated by a sensor protein called KdpD. Activation of KdpE is a two-step process: first, activated KdpD undergoes autophosphorylation; next, the phosphate group is transferred to KdpE. Inactivation is also mediated by KdpD: it acts as a phosphatase, removing the phosphate group from activated KdpE. In a 2004 paper, Andreas Kremling and his colleagues published a model of the KdpD–KdpE pathway (Kremling et al., 2004). A simplified version of their model network is

$$ATP + KdpD \xrightarrow{k_1} ADP + KdpD^p$$

$$KdpD^p + KdpE \underset{k_{-2}}{\overset{k_2}{\rightleftharpoons}} KdpD + KdpE^p$$

$$KdpE^p + KdpD \xrightarrow{k_3} KdpE + KdpD + Pi,$$

where the superscript "p" indicates phosphorylation, and "Pi" is a free phosphate group. The parameter k_1 can be used as an input to the system.

(a) Treating the concentration of ATP as constant, write a set of differential equations describing the system behavior. Suppose the total concentrations of the proteins (in the unphosphorylated and phosphorylated states) are fixed at KdpE^T and KdpD^T.

(b) The parameter values reported by Kremling and colleagues are as follows: (in μM) [ATP] = 1500, KdpE^T = 4, KdpD^T = 1; (in μM^{-1} hr^{-1}) $k_1 = 0.0029$, $k_2 = 108$, k_{-2} = 1080, $k_3 = 90$. This value of k_1 corresponds to an activated sensor. Run a simulation from initial condition of inactivity (no phosphorylation). How long does it take for the system to reach its steady-state response? Run another simulation to mimic inactivation of the activated system (i.e., by decreasing k_1 to zero from the active steady state). Does the system inactivate on the same timescale?

(c) Kremling and colleagues conducted in vitro experiments on this system at a low ATP level of [ATP] = 100 μM. How does that change affect the system behavior? Does it impact the activation and inactivation timescales?

(d) How would the system behavior be different if the inactivating phosphatase were not KdpD? Does this dual role of KdpD enhance or diminish the system's response? (Note that only the unphosphorylated form of KdpD has phosphatase activity.)

(e) In addition to activating production of the high-affinity K$^+$ uptake system, KdpE^p also causes increased production of the KdpD and KdpE proteins. How would the model's behavior change if this feedback were included?

6.8.3   G-Protein Signaling Pathway: Gain

Recall that for the parameter values in figure 6.5, the G-protein signaling pathway model in section 6.1.2 does not exhibit amplification from activated receptors (RL) to active G-protein output (Ga). Find an alternative set of parameter values for which the pathway output amplifies the activated receptor signal (i.e., for which the steady-state concentration of Ga is larger than that of RL).

6.8.4   G-Protein Signaling Pathway: Inhibition

Consider the G-protein signaling pathway model in section 6.1.2. Suppose you would like to design a drug to inhibit the G-protein signaling pathway response and that your putative drug would have the effect of changing the value of one of the kinetic constants by 50%. Which parameter provides the best target (based on the steady-state response at $L = 1$ nM)? How might the drug work? (Describe a potential biochemical mechanism of action.)

6.8.5   G-Protein Signaling Pathway: Receptor Recycling

The G-protein signaling pathway model in section 6.1.2 is a simplification of the original model of Yi and colleagues (Yi et al., 2003). The original model includes an additional feature: production and degradation of receptor molecules. The rate of production of receptors is constant: k_{Rs}. Degradation of receptors depends linearly on concentration, with rate constants k_{Rd0} for unbound receptors and k_{Rd1} for ligand-bound receptors. Extend the model to include these effects. (Note that the total receptor abundance will no longer be conserved.)

Run a simulation with $k_{Rs} = 4$ molecules per cell s^{-1}, $k_{Rd0} = 4 \times 10^{-4}$ s^{-1}, and $k_{Rd1} = 2 \times 10^{-2}$ s^{-1}. Compare your simulation to the simplified model illustrated in section 6.1.2. Does the expansion of the model have a significant impact on the response? If not, can you think of an experiment that can be simulated by the complete model but not by the simplified version?

6.8.6   Ultrasensitivity

Derive equation (6.2) in section 6.2.1, as follows. Begin by writing the steady-state conditions for each of the four species in network (6.1) . Use the steady-state conditions for the complexes WE_1 and W^*E_2 to write the steady-state concentration $[WE_1]$ in terms of $[W]^{ss}$, E_{1T}, and the rate constants. Likewise, write $[W^*E_2]^{ss}$ in terms of $[W^*]^{ss}$, E_{2T}, and the rate constants. Finally, use the steady-state condition $k_1[WE_1]$ = $k_2[W^*E_2]$ and the approximation $W_T = [W] + [W^*]$ to derive equation (6.2).

6.8.7 Multistep Ultrasensitivity

In a 2005 paper, Jeremy Gunawardena presented a straightforward analysis of multistep ultrasensitivity and revealed that it is not well-described by Hill functions (Gunawardena, 2005).

Consider a protein S that undergoes a sequential chain of n phosphorylations, all of which are catalyzed by the same kinase E, but each of which requires a separate collision event. Let S_k denote the protein with k phosphate groups attached. Suppose the phosphatase F acts in a similar multistep manner. The reaction scheme is then

$$S_0 \rightleftharpoons S_1 \rightleftharpoons S_2 \rightleftharpoons \cdots \rightleftharpoons S_n.$$

In steady state, the net reaction rate at each step is zero.

(a) Consider the first phosphorylation–dephosphorylation cycle. Expanding the steps in the catalytic mechanism, we have

$$S_0 + E \underset{k_{-1E}}{\overset{k_{1E}}{\rightleftharpoons}} ES_0 \overset{k_{catE}}{\longrightarrow} E + S_1 \quad \text{and} \quad S_1 + F \underset{k_{-1F}}{\overset{k_{1F}}{\rightleftharpoons}} FS_1 \overset{k_{catF}}{\longrightarrow} F + S_0.$$

Apply a rapid equilibrium assumption to the association reactions ($S_0 + E \leftrightarrow ES_0$ and $S_1 + F \leftrightarrow FS_1$), to describe the reaction rates as

$$k_{catE}[ES_0] = \frac{k_{catE}}{K_{ME}}[S_0][E] \quad \text{and} \quad k_{catF}[FS_1] = \frac{k_{catF}}{K_{MF}}[S_1][F],$$

where $K_{ME} = k_{-1E}/k_{1E}$, $K_{MF} = k_{-1F}/k_{1F}$, and $[E]$ and $[F]$ are the concentrations of free kinase and phosphatase.

(b) Use the fact that at steady state the net phosphorylation–dephosphorylation rate is zero to arrive at the equation

$$\frac{[S_1]}{[S_0]} = \lambda \frac{[E]}{[F]},$$

where

$$\lambda = \frac{k_{catE}}{k_{catF}} \frac{K_{MF}}{K_{ME}}.$$

(c) Suppose that the kinetic constants are identical for each phosphorylation–dephosphorylation step. In that case, verify that

$$\frac{[S_2]}{[S_1]} = \lambda \frac{[E]}{[F]}, \quad \text{so} \quad \frac{[S_2]}{[S_0]} = \lambda^2 \left(\frac{[E]}{[F]}\right)^2, \quad \text{and more generally} \quad \frac{[S_j]}{[S_0]} = \lambda^j \left(\frac{[E]}{[F]}\right)^j.$$

(d) Use the result in part (c) to write the fraction of protein S that is in the fully phosphorylated form as

$$\frac{[S_n]}{[S_{\text{total}}]} = \frac{(\lambda u)^n}{1 + \lambda u + (\lambda u)^2 + \cdots (\lambda u)^n}, \tag{6.7}$$

where $u = [E]/[F]$ is the ratio of kinase to phosphatase concentrations. Hint: $[S_{\text{total}}] = [S_0] + [S_1] + [S_2] + \ldots + [S_n]$.

(e) Use the ratio $[E]/[F]$ to approximate the ratio of total concentrations $[E_{\text{total}}]/[F_{\text{total}}]$, so that equation (6.7) describes the system's dose-response. Plot this function for various values of n. Take $\lambda = 1$ for simplicity. Use your plots to verify Gunawardena's conclusion that for high values of n, these dose-response curves exhibit a threshold (in this case, at $u = 1$) but their behavior cannot be described as switch-like, as they show near-hyperbolic growth beyond the threshold, regardless of the values of n.

6.8.8 Feedback in MAPK Cascades

As discussed in section 6.2, MAPK cascades are common features of eukaryotic signaling pathways. In the year 2000, Boris Kholodenko published a simple model of the mammalian Ras–Raf–ERK activation cascade to explore dynamic behaviors of MAPK pathways (Kholodenko, 2000). His model follows the reaction scheme in figure 6.24, which includes a feedback from the pathway output (MAPK-PP) to the first activation reaction.

For simplicity, Michaelis–Menten kinetics are used to describe the reaction rates:

$$v_1 = \frac{V_1[\text{MAPKKK}](1 + K_a[\text{MAPK-PP}])}{(K_{M1} + [\text{MAPKKK}])(1 + K_i[\text{MAPK-PP}])} \qquad v_2 = \frac{V_2[\text{MAPKKK-P}]}{K_{M2} + [\text{MAPKK-P}]}$$

$$v_3 = \frac{k_{\text{cat3}}[\text{MAPKKK-P}][\text{MAPKK}]}{K_{M3} + [\text{MAPKK}]} \qquad v_4 = \frac{k_{\text{cat4}}[\text{MAPKKK-P}][\text{MAPKK-P}]}{K_{M4} + [\text{MAPKK-P}]}$$

$$v_5 = \frac{V_5[\text{MAPKK-PP}]}{K_{M5} + [\text{MAPKK-PP}]} \qquad v_6 = \frac{V_6[\text{MAPKK-P}]}{K_{M6} + [\text{MAPKK-P}]}$$

$$v_7 = \frac{k_{\text{cat7}}[\text{MAPKK-PP}][\text{MAPK}]}{K_{M7} + [\text{MAPK}]} \qquad v_8 = \frac{k_{\text{cat8}}[\text{MAPKK-PP}][\text{MAPK-P}]}{K_{M8} + [\text{MAPK-P}]}$$

$$v_9 = \frac{V_9[\text{MAPK-PP}]}{K_{M9} + [\text{MAPK-PP}]} \qquad v_{10} = \frac{V_{10}[\text{MAPK-P}]}{K_{M10} + [\text{MAPK-P}]}.$$

The parameters K_a and K_i characterize activation or inhibition of the initial reaction by the output MAPK-PP. Kholodenko took parameter values of (in nM s^{-1})

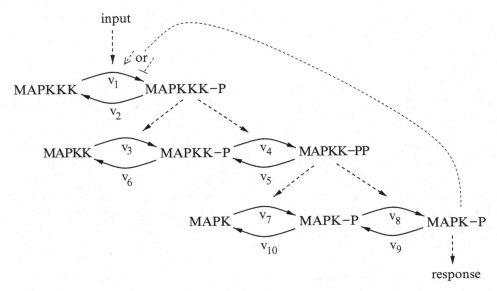

Figure 6.24
MAPK cascade for problem 6.8.8. Adapted from figure 1 of Kholodenko (2000).

$V_1 = 2.5$, $V_2 = 0.25$, $V_5 = V_6 = 0.75$, $V_9 = V_{10} = 0.5$; (in s^{-1}) $k_{cat3} = k_{cat4} = k_{cat7} = k_{cat8} = 0.025$; (in nM) $K_{M1} = 10$, $K_{M2} = 8$, $K_{M3} = K_{M4} = K_{M5} = K_{M7} = K_{M8} = K_{M9} = K_{M10} = 15$.

(a) Set $K_a = K_i = 0$ so that there is no feedback in the system. Run simulations with initial concentrations of (in nM) [MAPKKK] = 100, [MAPKK] = 300, [MAPK] = 300, and all other concentrations zero. Take V_1 as the system input and confirm the system's switch-like dose-response by determining the steady-state output ([MAPK-PP]) for values of V_1 in the range from 0 to 0.4 nM s^{-1}.

(b) To explore the effect of feedback on the dose-response curve, simulate the model for the case $K_a = 0.02$, $K_i = 0$ and the case $K_a = 0$, $K_i = 0.0002$. Interpret your results.

(c) Kholodenko used the model to demonstrate that negative feedback could lead to an oscillatory response. Verify his conclusion by simulating the system for $K_i = 0.1$, and input $V_1 = 2.5$ (with $K_a = 0$). What is the period? Do the oscillations persist at lower input levels?

6.8.9 Chemotaxis: Saturation of Adaptation

The bacterial chemotaxis signaling model in section 6.3.1 was constructed under the assumption that enzyme CheR is fully saturated (which is why the rate of methylation does not depend on the concentration of unmethylated receptors). This assumption ensures that the system adapts to ligand inputs over a wide range.

(a) Use the model to generate a dose-response curve showing the system's steady-state response ($[Am]^{ss}$) as a function of ligand level for $[L]$ between 20 and 100. Use parameter values as in figure 6.3. Verify that the system shows near-perfect adaptation over this input range.

(b) When the assumption of CheR saturation is removed, the system does not adapt well at high ligand levels. Confirm this claim by replacing the zero-order methylation rates ($k_{-1}[R], k_{-2}[R]$) with Michaelis–Menten rate laws. Use parameter values for the rest of the model as in figure 6.3, and select parameter values for the methylation kinetics so that CheR is not fully saturated. Generate dose-response curves for your modified model and compare with figure 6.3.

(c) The modified model in part (b) fails to adapt when the receptors become saturated by methyl groups. This effect can be alleviated by incorporating multiple methylation sites on each receptor. Modify your model from part (b) so that each receptor has two methylation sites (each receptor can then be unmethylated, once-methylated, or twice-methylated). Treat the kinetics of the second methylation–demethylation reactions as identical to the first. To simplify your analysis, assume that only the twice-methylated, non-ligand-bound receptor complexes are active. Verify that in comparison with the model in part (b), this extended model shows improved adaptation at high ligand levels.

6.8.10 Apoptosis: Duration of Triggering Signal

Consider the model of apoptosis presented in section 6.4.1. In the simulation shown in figure 6.16, the input signal was maintained for 900 minutes (until after the system had settled to the caspase-active "death" state.) Re-run this simulation to determine how short the input pulse can be while still triggering the irreversible life-to-death transition. Use the same input size as in the figure (Input = 200).

6.8.11 Apoptosis: Model Reduction

In a 2007 paper, Steffen Waldherr and colleagues presented a reduced version of the apoptosis model presented in section 6.4.1 (Waldherr et al., 2007). They determined that bistability is retained when a quasi–steady state assumption is applied to four of the state variables. Verify their finding, as follows. Apply a quasi–steady state assumption to the species [C8], [C3], [IAP] and [BAR] and demonstrate bistability in the reduced model. (This model reduction was not motivated by a separation of timescales. Instead, Waldherr and colleagues determined the significance of each state variable for bistability and eliminated those that were not needed.)

6.8.12 Calcium-Induced Calcium Release: Frequency Encoding

Consider the model of calcium oscillations presented in section 6.5.1. Run simulations to explore the dependence of frequency and amplitude on the strength of the input. Prepare plots showing frequency and amplitude as a function of model input I over the range 1–10 μM. Does the amplitude vary significantly over this range? What about for higher input levels?

6.8.13 Calcium-Induced Calcium Release: Parameter Balance
Consider the model of calcium oscillations presented in section 6.5.1. The model's oscillatory behavior depends on a balance between the positive and negative effects of calcium binding. Oscillations are easily lost if parameter values vary. Explore this sensitivity by choosing one of the model parameters and determining the range of values over which the system exhibits oscillations (with $I = 1$ μM). Provide an intuitive explanation for why oscillations are lost outside the range you have identified.

6.8.14 Calcium-Induced Calcium Release: Frequency Decoding
One mechanism by which cells can "decode" the information encoded in calcium oscillations is by the activity of Ca^{2+}/calmodulin-dependent protein kinases (CaM-kinases). Calmodulin is a protein that mediates a number of calcium-dependent cellular processes. It has four high-affinity Ca^{2+} binding sites and is activated when saturated by Ca^{2+}. CaM-kinases are activated (by autophosphorylation) upon binding to active calmodulin. The CaM-kinase activity is "turned off" by phosphatases.

Extend the model in section 6.5.1 to include CaM-kinase activity, and verify that for persistent oscillations, the higher the frequency of Ca^{2+} oscillations, the higher the average CaM-kinase activity level. To keep the model simple, suppose that four calcium ions bind calmodulin simultaneously (i.e., with high cooperativity) and that CaM-kinase autophosphorylation occurs immediately upon calmodulin binding. (Hint: The frequency-dependent effect is strongest when the timescales of deactivation of calmodulin and CaM-kinase are slow, so that high-frequency inputs cause near-constant activity levels.)

6.8.15 cAMP Oscillations in *Dictyostelium*
The slime mold *Dictyostelium discoideum* is used as a model organism in the study of development. *Dictyostelium* is eukaryotic: it feeds on bacteria. When bacterial prey are plentiful, the cells forage independently. When food is scarce, the population comes together to form first a motile "slug," and then a fruiting body, which scatters *Dictyostelium* spores into the environment. Aggregation into a slug involves cell-to-cell communication mediated by secretion of cyclic AMP (cAMP). The concentration of cAMP oscillates in waves across the colony, thus directing the cells to an aggregation point.

In 1998, Michael Laub and William Loomis proposed a model of the intracellular signaling pathway that generates oscillations in cAMP levels (reviewed in Maeda

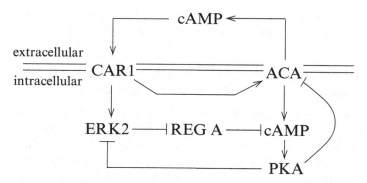

Figure 6.25
cAMP signaling in *Dictyostelium* (problem 6.8.15). Adapted from figure 2 of Maeda et al. (2004).

et al., 2004). A version of their model involves the network in figure 6.25. In this model, extracellular cAMP activates the transmembrane receptor CAR1, which activates the signaling molecule ERK2 and the membrane-associated enzyme adenylyl cyclase (ACA). ACA catalyzes production of cAMP, some of which is directly secreted. ERK2 inhibits the enzyme REG A, which degrades intracellular cAMP. Feedback is provided by protein kinase A (PKA), which is activated by cAMP and inhibits both ERK2 and ACA. The model accounts for activation/production and deactivation/degradation of each species. The equations are

$$\frac{d}{dt}[\mathrm{ACA}](t) = k_1[\mathrm{CAR1}](t) - k_2[\mathrm{ACA}](t)\cdot[\mathrm{PKA}](t)$$

$$\frac{d}{dt}[\mathrm{PKA}](t) = k_3[\mathrm{cAMP}]_{\mathrm{int}}(t) - k_4[\mathrm{PKA}](t)$$

$$\frac{d}{dt}[\mathrm{ERK2}](t) = k_5[\mathrm{CAR1}](t) - k_6[\mathrm{ERK2}](t)\cdot[\mathrm{PKA}](t)$$

$$\frac{d}{dt}[\mathrm{REG\ A}](t) = k_7 - k_8[\mathrm{REG\ A}](t)\cdot[\mathrm{ERK2}](t)$$

$$\frac{d}{dt}[\mathrm{cAMP}]_{\mathrm{int}}(t) = k_9[\mathrm{ACA}](t) - k_{10}[\mathrm{REG\ A}](t)\cdot[\mathrm{cAMP}]_{\mathrm{int}}(t)$$

$$\frac{d}{dt}[\mathrm{cAMP}]_{\mathrm{ext}}(t) = k_{11}[\mathrm{ACA}](t) - k_{12}[\mathrm{cAMP}]_{\mathrm{ext}}(t)$$

$$\frac{d}{dt}[\mathrm{CAR1}](t) = k_{13}[\mathrm{cAMP}]_{\mathrm{ext}}(t) - k_{14}[\mathrm{CAR1}](t).$$

Parameter values are as follows: (in min^{-1}) $k_1 = 2.0$, $k_3 = 2.5$, $k_4 = 1.5$, $k_5 = 0.6$, $k_7 = 1.0$, $k_9 = 0.3$, $k_{11} = 0.7$, $k_{12} = 4.9$, $k_{13} = 23.0$, $k_{14} = 4.5$; (in min^{-1} μM^{-1}) $k_2 = 0.9$, $k_6 = 0.8$, $k_8 = 1.3$, $k_{10} = 0.8$.

(a) Simulate this model and determine the period and amplitude of the oscillations of intracellular cAMP. (You can take all initial conditions to be 1 μM.)

(b) These oscillations are brought about by coupled positive and negative feedback. Describe the positive feedback loop that causes external cAMP to amplify its own concentration. Describe the negative feedback loop that keeps internal cAMP levels low. Run simulations that cut each of these feedback loops and verify that persistent oscillations are not maintained in either case. (For example, to cut the positive feedback in extracellular cAMP, you could replace the production term $k_{11}[\text{ACA}](t)$ with a constant: k_{11}.)

(c) Extend the model to address two neighboring cells that share the same pool of extracellular cAMP. Your extended model will need to include separate compartments for each cell, with separate descriptions of each internal species. The shared extracellular pool of cAMP can be described by a single equation:

$$\frac{d}{dt}[\text{cAMP}]_{\text{ext}}(t) = k_{11}[\text{ACA}]_1(t) + k_{11}[\text{ACA}]_2(t) - k_{12}[\text{cAMP}]_{\text{ext}}(t),$$

where $[\text{ACA}]_1$ and $[\text{ACA}]_2$ are the ACA concentrations in each of the two cells. Choose initial conditions for the two cells so that they begin their oscillations out of phase (i.e., so that the peaks and troughs are not aligned). Confirm that the shared cAMP pool synchronizes the oscillations in the two cells. How long does it take for synchrony to occur?

6.8.16   The Cell Division Cycle: Bistability

The cell division cycle is the process by which cells grow and replicate. The eukaryotic cell cycle is driven by a relaxation oscillator that consists of a negative feedback wrapped around a bistable system. The negative feedback causes periodic transitions between the quasi-stable states. In *Xenopus* cells, the bistable system has been identified as the Cdc2 activation switch. (*Xenopus laevis* is an African clawed frog.)

Cdc2 is activated by the protein cyclin B and promotes its own activation via a positive feedback. A simple model of the Cdc2 activation switch was published by Ibrahima Ndiaye, Madalena Chaves, and Jean-Luc Gouzé in 2010 (Ndiaye et al., 2010). The model network is shown in figure 6.26.

They treated the concentration of the activator cyclin B as a fixed input parameter (u). A Hill function was used to describe the positive feedback that enhances Cdc2 activation. Presuming that the total abundance of Cdc2 remains fixed, the model

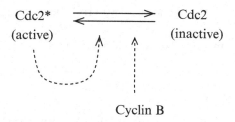

Figure 6.26
Bistable Cdc2 activation network (problem 6.8.16).

describes the evolution of the fraction of Cdc2 in the active state, denoted x. The inactive fraction is then $1 - x$. The activation–inactivation dynamics are modeled as

$$\frac{d}{dt}x(t) = \left(k(w_0 + u) + V\frac{(x(t))^n}{\theta^n + (x(t))^n} \right)(1 - x(t)) - \gamma x(t). \tag{6.8}$$

Here, w_0 is a basal level of cyclin B. Inactivation is presumed to occur at a steady rate γ.

(a) Take parameter values $k = 1$ nM^{-1} time^{-1}, $w_0 = 61$ nM, $V = 430$ time^{-1}, $\gamma = 843$ nM^{-1} time^{-1}, $\theta = 0.27$, and $n = 4$. Explore the bistable nature of the system by constructing phase-lines (as in problem 4.8.1 of chapter 4) for $u = 30, 45,$ and 60 nM. (These can be generated by plotting the rate of change (dx/dt) against x. Points at which the rate of change is zero are steady states.)

(b) Produce a bifurcation diagram showing the steady-state fraction of active Cdc2 (i.e., x) as a function of the cyclin B concentration (u). Determine the two bifurcation points.

(c) Ndiaye and colleagues also considered an expanded version of model (6.8) that describes periodic changes in cyclin B levels. Cyclin levels are linked to Cdc2 by another feedback — active Cdc2 activates the anaphase promoting complex (APC), which causes degradation of cyclin B. With y denoting the level of APC activity, the model takes the form

$$\frac{d}{dt}x(t) = \left(k\left(w_0 + \frac{a_1}{y(t)} \right) + V_1\frac{(x(t))^n}{\theta^n + (x(t))^n} \right)(1 - x(t)) - \gamma_1 y(t)x(t)$$

$$\frac{d}{dt}y(t) = V_2\frac{x(t)}{x(t) + \theta_2} - \gamma_2 y(t).$$

Simulate this model with parameter values $k = 3.77 \times 10^{-4}$ nM^{-1} time^{-1}, $w_0 = 2$ nM, $a_1 = 0.015$ nM^2, $V_1 = 162$ time^{-1}, $V_2 = 0.25$ time^{-1}, $\gamma_1 = 0.36$ nM^{-1} time^{-1}, $\gamma_2 = 0.026$ time^{-1}, $\theta_1 = 0.27$, $\theta_2 = 0.27$, and $n = 4$. Verify that the system exhibits limit-cycle oscillations. Provide an intuitive explanation for the cyclic rise-and-crash behavior of Cdc2.

More detailed models of the cell cycle are introduced in Goldbeter (1996) and Fall et al. (2002).

6.8.17* Frequency Response Analysis: Linear Regime

(a) Consider the G-protein signaling model presented in section 6.1.2. As shown in figure 6.19, this model displays near-linear behavior for small-amplitude input oscillations. Verify that this near-linear (sinusoidal) response is maintained even when the input oscillations have an amplitude of 2 nM (again centered at $L = 1$ nM).

(b) To illustrate the breakdown of the frequency response for nonlinear systems, consider the input/output system:

$$\frac{d}{dt}x(t) = -x(t) + u(t) + x(t)u(t), \qquad y(t) = x(t).$$

For $u = 0$, the steady state is $x = 0$. Simulate the system's response to input signals of the form $u(t) = u_0 \sin(t)$. Verify that (i) for $u_0 < 0.1$, the response is approximately a sine wave centered at $x = 0$; (ii) for $0.1 < u_0 < 1$, the response is roughly sinusoidal but is no longer centered at $x = 0$; and (iii) for $u > 1$, the response is periodic but is not a smooth sine-wave oscillation.

6.8.18* Frequency Response Analysis of a Two-Component Signaling Pathway

(a) Following the procedure in section 6.6.3, determine the linearization of the two-component signaling pathway model of section 6.1.1 at an arbitrary nominal input value. Use species conservations to reduce the model before linearizing.

(b) Simulate the model to determine the steady state corresponding to a nominal input of $L_T = 0.04$. Use MATLAB to generate the magnitude Bode plot of the corresponding frequency response (details in appendix C).

(c) Repeat part (b) for a nominal input of of $L_T = 0.4$. Use figure 6.3 to explain the difference in the frequency response at these two nominal input values.

7 Gene Regulatory Networks

[In the ancestral Indo-European language, the word] *gene* signified beginning, giving birth.... [It later] emerged as genus, genius, genital, and generous; then, still holding on to its inner significance, it became "nature" ([from the Latin] *gnasci*).
—Lewis Thomas, *The Lives of a Cell*

As demonstrated in the previous chapters, all cellular functions are driven by proteins. Protein production occurs through *gene expression*—a process that involves reading information encoded in the DNA. The cellular abundance of each protein is controlled primarily by its production rate: these production rates are, in turn, controlled by specialized proteins called *transcription factors*. A set of genes whose protein products regulate one another's expression rates is referred to as a *gene regulatory network*. In this chapter, we will address gene regulatory networks that implement switch-like responses, store memory, generate oscillations, and carry out logical computations and cell-to-cell communication.

Gene expression is a two-step process. The first step, **transcription**, occurs when the coding region of a gene is "rewritten" in the form of a complementary RNA strand called a messenger RNA (mRNA). Transcription is carried out by a protein complex called *RNA polymerase,* which binds the *promoter region* of the gene and then "walks" along the DNA, catalyzing formation of the mRNA strand from nucleotide precursors.

The second step of gene expression is **translation**, in which the mRNA molecule binds a protein–RNA complex called a ribosome, which reads the nucleotide sequence and produces a corresponding polypeptide chain. Translation, like transcription, involves information transfer: the ribosome "reads along" the mRNA and catalyzes the formation of a protein from amino acid building blocks.

Although the organization and behavior of gene regulatory networks share a number of similarities with metabolic networks and signal transduction pathways, the underlying processes are significantly different. The biochemical interactions in metabolic or signal transduction systems are each decomposable into a handful of elementary chemical events. In contrast, transcription and translation are complex processes that each involve a very large number of biochemical reactions (many of which have not been fully characterized).

We will use a mass action–based formalism to develop models of gene regulatory networks, but we will be applying this rate law in a more abstracted sense than in the previous chapters: our descriptions of gene expression processes will be far more coarse-grained than our earlier models of biochemical networks.

A further complication in modeling genetic systems is that the molecules involved in the regulation of gene expression are often present in very small numbers. Recall from section 2.1.2 that the continuum hypothesis (which justifies our use of smoothly varying concentration values) should only be applied when there are large numbers of molecules present (so that individual reaction events produce near-infinitesimal changes in abundance). Proteins that impact gene expression are often present in small quantities: the copy number of each protein species is regularly in the hundreds or less. Moreover, genes themselves are almost always present at very low copy number—rarely more than dozens, and often as few as one or two. Cells typically have a small number of copies of their inherent (chromosomal) genes—typically one or two, in some cases four or more. Genes that are introduced to a bacterial cell from its environment (e.g., in the laboratory) are typically carried on small circular DNA molecules called *plasmids*, which can be maintained at much higher copy numbers—as many as a few hundred.

In addressing the behavior of systems with low molecule counts, we can justify the mass-action formalism by interpreting differential-equation models as descriptions of the average behavior over a large population of cells. This interpretation is useful when addressing cultures or tissues composed of comparable cells that exhibit similar behaviors.

An alternative modeling framework—one that describes individual reaction events—can be adopted in cases where we seek truly to capture the behavior of individual cells (e.g., where single-cell measurements are available). This stochastic framework, which incorporates the probabilistic (i.e., noisy) effects that play a significant role at these small scales, will be taken up in section 7.6.

7.1 Modeling of Gene Expression

7.1.1 Unregulated Gene Expression

The fundamental processes that constitute gene expression are sketched in figure 7.1. Transcription and translation involve information transfer from DNA to RNA to protein, while degradation results in turnover of the RNA and protein pools. Each of these processes relies on a significant amount of background cellular "machinery"—including nucleic acids, RNA polymerases, amino acids, and ribosomes. In develop-

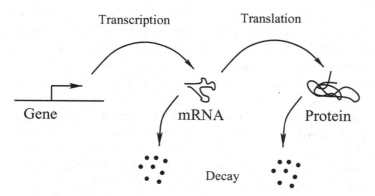

Figure 7.1
Gene expression. Transcription of the gene results in the formation of mRNA molecules, which can then be translated by ribosomes to produce proteins. These production processes are balanced by degradation of mRNA and protein molecules.

ment of models of gene expression, we will assume that the activity of these "housekeeping" elements is fixed.

To simplify our discussion, we will focus on prokaryotic gene expression. Eukaryotic gene regulatory networks are also modeled using the procedure we will develop here, but eukaryotic gene expression involves a number of additional processes that we will not address explicitly (such as splicing of mRNA and transport of mRNA across the nuclear membrane).

Unregulated, or *constitutive*, gene expression involves the four processes shown in figure 7.1. Application of mass action to arrive at simple rate laws gives

$$\frac{d}{dt}m(t) = k_0 - \delta_m m(t)$$

$$\frac{d}{dt}p(t) = k_1 m(t) - \delta_p p(t),$$
(7.1)

where m is the concentration of mRNA molecules and p is the concentration of the gene's protein product. The population-averaged transcription rate k_0 depends on a number of factors, including the gene copy number, the abundance of RNA polymerase, the strength of the gene's promoter (e.g., its affinity for RNA polymerase), and the availability of nucleotide building blocks. Parameter k_1, the per-mRNA translation rate, likewise depends on a range of factors, including the availability of ribosomes, the strength of the mRNA's ribosome binding site (i.e., the mRNA's affinity for ribosomes), and the availability of transfer RNAs and free amino acids.

Transcription and translation are balanced by decay of the mRNA and protein pools, characterized by the degradation rates δ_m and δ_p. Several factors contribute to the decay process. Generally, mRNA and protein molecules may be unstable and so decay spontaneously (with characteristic half-lives). Additionally, the cell contains ribonucleases and proteases that specifically degrade mRNA and protein molecules.

The parameters δ_m and δ_p can also be used to describe the dilution of mRNA and protein pools caused by cell growth. (Constant growth causes the overall cell volume to increase exponentially, resulting in an exponential decrease in concentrations: model (7.1) applies if the cell maintains constant concentrations of the background expression machinery.) In rapidly growing bacterial cells, dilution is often more significant than degradation. In model (7.1), we will use the parameters δ_m and δ_p as combined degradation/dilution rates.

The steady-state concentrations in model (7.1) are easily determined:

$$m^{ss} = \frac{k_0}{\delta_m} \qquad p^{ss} = \frac{k_1 k_0}{\delta_p \delta_m}.$$

Models of gene expression are often simplified by taking advantage of the fact that mRNA decay is typically much faster than protein decay. (mRNA half-lives are typically measured in minutes, whereas proteins often have half-lives of hours. In rapidly growing bacterial cells, protein degradation is often negligible, and so the protein decay rate is dictated solely by the cell's growth rate.) This separation of timescales justifies a quasi-steady-state approximation for the mRNA levels. The reduced model is

$$\frac{d}{dt} p(t) = \frac{k_1 k_0}{\delta_m} - \delta_p p(t). \tag{7.2}$$

The parameter $\alpha = k_1 k_0 / \delta_m$ is the *expression rate*.

7.1.2   Regulated Gene Expression

Gene expression can be regulated at many stages, including RNA polymerase binding, elongation of the mRNA strand, translational initiation (i.e., mRNA–ribosome binding), and polypeptide elongation. In addition, mRNA and protein molecules can be specifically targeted for — or protected from — degradation.

Despite this range of control points, the majority of gene regulation occurs through control of the initiation of transcription. In prokaryotes, this is achieved primarily through regulating the association of RNA polymerase with gene promoter regions. Proteins that bind DNA and affect polymerase association are called *transcription factors*. In prokaryotes, transcription factor–binding sites, called *operator regions*,

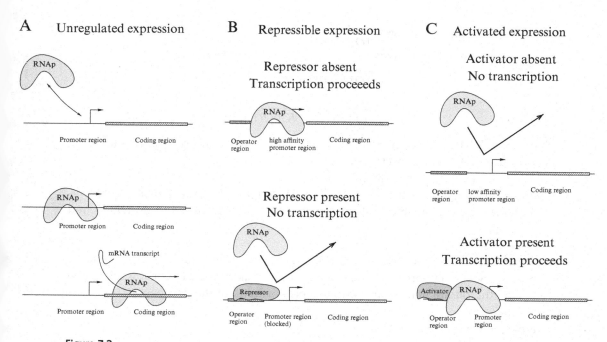

Figure 7.2
Transcriptional regulation. (A) Unregulated gene expression. RNA polymerase (RNAp) binds the gene's promoter region, then slides along the DNA to the coding region, where it produces the mRNA transcript. (B) A repressor binds to its operator region and blocks access to the promoter. When the repressor is bound, transcription cannot occur. (C) An activator enhances transcription. In this case, the promoter region has a low affinity for RNA polymerase, and so transcription does not occur from the unregulated gene. Once bound at the operator, the activator binds RNA polymerase and recruits it to the promoter site.

are typically situated close to the promoter of the affected gene. If a transcription factor increases the rate of RNA polymerase binding, it is called an *activator* of gene expression; if it inhibits binding, it is called a *repressor* (figure 7.2).

Transcription Factor Binding The binding of a transcription factor to an operator site can be described by

$$O + P \underset{d}{\overset{a}{\rightleftharpoons}} OP \, ,$$

where O is the unbound operator, P is the transcription factor protein, and the complex OP is the bound operator. This association–dissociation event occurs on a much faster timescale than gene expression, so it can be treated in equilibrium when modeling gene expression. Setting $K = d/a$, we find (compare with equation 3.17 of chapter 3):

$$\text{Fraction of bound operators} = \frac{[OP]}{[O]+[OP]}$$

$$= \frac{[O][P]/K}{[O]+[O][P]/K} \tag{7.3}$$

$$= \frac{[P]/K}{1+[P]/K} = \frac{[P]}{K+[P]}.$$

Note that K, the dissociation constant of the binding event, is the half-saturating concentration for the transcription factor P.

Equation (7.3) describes the *promoter occupancy*: it represents the fraction of a population of operators that are bound to transcription factor proteins (or, equivalently, the fraction of time that any given operator spends in the protein-bound state).

Rates of Transcription from Regulated Genes   The rate of transcription from a regulated gene depends on the promoter occupancy. If the transcription factor P is an activator, then the rate of gene transcription is proportional to the occupancy:

$$\text{Rate of activated transcription} = \alpha\frac{[P]/K}{1+[P]/K}. \tag{7.4}$$

The constant of proportionality, α, is the *maximal transcription rate*. Formula (7.4) suggests that the rate of transcription will be zero when the activator P is absent. Most activated genes are transcribed at a low (so-called *basal*) rate even when the activator is unbound. Incorporating a basal expression rate of α_0 gives

$$\text{Rate of activated transcription} = \alpha_0 + \alpha\frac{[P]/K}{1+[P]/K}. \tag{7.5}$$

In this case, the maximal transcription rate is $\alpha_0 + \alpha$.

When the transcription factor P acts as a repressor, the regulated transcription rate is proportional to the fraction of unbound operators $1/(1+[P]/K)$. If we allow for a small transcription rate α_0 from the repressed promoter (a "leak"), we have

$$\text{Rate of repressible transcription} = \alpha_0 + \alpha\frac{1}{1+[P]/K}.$$

In many cases, this leak is small (and is often neglected), but complete repression can never be achieved: thermal fluctuations cause continual unbinding and rebinding at the operator, so there is always some chance that RNA polymerase will find its way to the operator.

Regulation by Multiple Transcription Factors   Genes are commonly regulated by multiple transcription factors and by multiple copies of each factor. To introduce the general technique for analyzing these multiple-regulator schemes, we next provide an analysis of the promoter occupancy for a gene regulated by two transcription factors.

Consider a promoter with two non-overlapping operator sites: O_A binds transcription factor A; O_B binds transcription factor B. The promoter can then be found in four states:

O: A and B unbound.

OA: A bound at O_A, B unbound.

OB: B bound at O_B, A unbound.

OAB: A at bound O_A, B bound at O_B.

If the binding events at O_A and O_B are independent of one another, then the reaction scheme is

$$O + A \underset{d_1}{\overset{a_1}{\rightleftharpoons}} OA \qquad O + B \underset{d_2}{\overset{a_2}{\rightleftharpoons}} OB$$

$$OB + A \underset{d_1}{\overset{a_1}{\rightleftharpoons}} OAB \qquad OA + B \underset{d_2}{\overset{a_2}{\rightleftharpoons}} OAB$$

which results in the following steady-state distribution of promoters:

Fraction in state O: $\quad\dfrac{1}{1 + \dfrac{[A]}{K_A} + \dfrac{[B]}{K_B} + \dfrac{[A][B]}{K_A K_B}}$

Fraction in state OA: $\quad\dfrac{\dfrac{[A]}{K_A}}{1 + \dfrac{[A]}{K_A} + \dfrac{[B]}{K_B} + \dfrac{[A][B]}{K_A K_B}}$

Fraction in state OB: $\quad\dfrac{\dfrac{[B]}{K_B}}{1 + \dfrac{[A]}{K_A} + \dfrac{[B]}{K_B} + \dfrac{[A][B]}{K_A K_B}}$ \hfill (7.6)

Fraction in state OAB: $\quad\dfrac{\dfrac{[A][B]}{K_A K_B}}{1 + \dfrac{[A]}{K_A} + \dfrac{[B]}{K_B} + \dfrac{[A][B]}{K_A K_B}}$,

where $K_A = d_1/a_1$ and $K_B = d_2/a_2$ are the dissociation constants for the two binding events.

Exercise 7.1.1 Derive equations (7.6) by treating the binding events in steady state. $\square$

The rate of transcription from this regulated promoter depends on the nature of the transcription factors A and B. For instance, if both are repressors and the binding of *either* factor blocks polymerase binding, then transcription can only occur from state O. The corresponding transcription rate can be written as

$$\frac{\alpha}{1+\dfrac{[A]}{K_A}+\dfrac{[B]}{K_B}+\dfrac{[A][B]}{K_A K_B}}.$$

Alternatively, the two repressors might inhibit transcription only when *both* are bound, in which case transcription occurs from all states except OAB. The resulting transcription rate takes the form

$$\alpha\frac{1+\dfrac{[A]}{K_A}+\dfrac{[B]}{K_B}}{1+\dfrac{[A]}{K_A}+\dfrac{[B]}{K_B}+\dfrac{[A][B]}{K_A K_B}}.$$

Exercise 7.1.2 Which promoter state(s) would allow transcription if A is an activator of the gene while B is a repressor that completely blocks the RNA polymerase binding site? Suppose that no expression occurs if A is unbound. Formulate a description of the rate of transcription in this case. $\square$

Cooperativity in Transcription Factor Binding Recall from section 3.3 that cooperativity occurs when multiple ligands bind a single protein and the association of ligand at a binding site affects the affinity of the other binding sites. Cooperativity also occurs among transcription factors binding multiple operator sites along a length of DNA.

Consider the case in which two transcription factors bind at non-overlapping operator sites with positive cooperativity. Suppose the dissociation rate of the second DNA-binding event is reduced by a factor K_Q. In this case, the dissociation constant for A binding to OB is $K_A K_Q$ ($<K_A$), indicating enhanced affinity. (Likewise the dissociation constant for B binding to OA is $K_B K_Q$, which is less than K_B.) The steady-state distribution of promoters is then:

Fraction in state O: $\dfrac{1}{1+\dfrac{[A]}{K_A}+\dfrac{[B]}{K_B}+\dfrac{[A][B]}{K_A K_B K_Q}}$

Fraction in state OA: $\dfrac{\dfrac{[A]}{K_A}}{1+\dfrac{[A]}{K_A}+\dfrac{[B]}{K_B}+\dfrac{[A][B]}{K_A K_B K_Q}}$

Fraction in state OB:
$$\dfrac{\dfrac{[B]}{K_B}}{1+\dfrac{[A]}{K_A}+\dfrac{[B]}{K_B}+\dfrac{[A][B]}{K_A K_B K_Q}}$$

Fraction in state OAB:
$$\dfrac{\dfrac{[A][B]}{K_A K_B K_Q}}{1+\dfrac{[A]}{K_A}+\dfrac{[B]}{K_B}+\dfrac{[A][B]}{K_A K_B K_Q}}\ .$$

If the cooperativity is strong, the second operator site will almost always be occupied when the first binding event has occurred. Consequently, the states OA and OB will be negligible. This case corresponds to $K_Q \ll 1$, which allows the approximation

$$1+\dfrac{[A]}{K_A}+\dfrac{[B]}{K_B}+\dfrac{[A][B]}{K_A K_B K_Q}\approx 1+\dfrac{[A][B]}{K_A K_B K_Q}\ .$$

The distribution of promoter states is then:

Fraction in state O:
$$\dfrac{1}{1+\dfrac{[A][B]}{K_A K_B K_Q}}$$

Fraction in state OAB:
$$\dfrac{\dfrac{[A][B]}{K_A K_B K_Q}}{1+\dfrac{[A][B]}{K_A K_B K_Q}}\ .$$

In the particular case that the two transcription factors are identical ($A = B = P$), the distribution becomes

Fraction in state O:
$$\dfrac{1}{1+\dfrac{[P]^2}{K_P^2 K_Q}}$$

Fraction in state OAB:
$$\dfrac{\dfrac{[P]^2}{K_P^2 K_Q}}{1+\dfrac{[P]^2}{K_P^2 K_Q}}\ .$$

The promoter occupancy thus takes the familiar form of a Hill function (as introduced in section 3.3).

When N transcription factors bind with strong cooperativity, promoter occupancy can be written as

$$\frac{\left(\dfrac{[P]}{K}\right)^{N}}{1+\left(\dfrac{[P]}{K}\right)^{N}} = \frac{[P]^{N}}{K^{N}+[P]^{N}}, \tag{7.7}$$

where K is the half-saturating concentration. This functional form is often used as an empirical fit when the details of transcription factor binding are unknown.

Transcription factors commonly bind DNA as multimers (e.g., dimers or tetramers). Formula (7.7) can sometimes be used to describe occupancy by a multimer. However, because the multimerization process occurs in the cytosol—rather than at the operator site—the analysis in this section does not apply directly. See problem 7.8.4 for details.

7.1.3   Gene Regulatory Networks

A gene regulatory network, also called a *genetic circuit*, is a group of genes whose protein products regulate one another's expression. The simplest genetic circuit consists of a single gene that regulates its own activity (figure 7.3). If the gene's protein product enhances expression, the gene is called as *autoactivator*; if the product inhibits expression, the gene is an *autoinhibitor*.

Autoinhibition To construct a simple model of an autoinhibitor, we treat mRNA in quasi–steady state and presume that the transcription factor P binds to a single operator site. The resulting model is

$$\frac{d}{dt}p(t) = \alpha \frac{1}{1+p(t)/K} - \delta_{p}p(t), \tag{7.8}$$

where $p = [P]$.

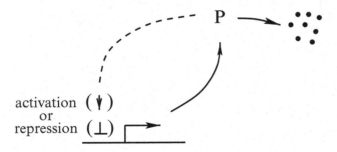

Figure 7.3
Autoregulatory gene circuit. (The dashed arrow indicates regulation.) The protein product of the gene regulates its own expression by acting as an activator or as an inhibitor.

Many genes are autoinhibitory. One advantage of this regulatory scheme is reduction of sensitivity to certain perturbations: autoinhibition decreases the sensitivity to variation in the maximal expression rate α (see problem 7.8.2). Because α depends on a host of background processes, this increased robustness can provide a significant advantage over unregulated expression.

Another advantage of autoinhibition is a fast response to changes in demand for protein product. Consider an unregulated gene whose product attains a particular concentration. If an autoinhibitory gene is to generate an equivalent abundance of protein, it must have a higher maximal expression rate α and will consequently respond more quickly when changes in protein level are required (see problem 7.8.1).

Autoactivation The behavior of an autoactivating gene can be modeled as

$$\frac{d}{dt}p(t) = \alpha \frac{p(t)/K}{1+p(t)/K} - \delta_p p(t). \tag{7.9}$$

Positive feedback of this sort can lead to runaway behavior. However, because the expression rate cannot rise above α, autoactivation typically results in quick convergence to a state in which the gene is expressing at a high rate—an "on" state. (This simple model also exhibits a steady "off" state, at $p = 0$.)

Exercise 7.1.3 Derive a formula for the nonzero steady-state solution of model (7.9). Confirm that this steady state does not occur when $\alpha < K\delta_p$. Verify that when the nonzero steady state exists, it is stable and the zero steady state is unstable. □

An autoactivator P that binds cooperatively at multiple operator sites can be described by

$$\frac{d}{dt}p(t) = \alpha \frac{(p(t)/K)^N}{1+(p(t)/K)^N} - \delta_p p(t). \tag{7.10}$$

This nonlinear model can exhibit bistability, in which the OFF state ($p = 0$) and the high-expressing ON state are both stable. The basin of attraction of the OFF state tends to be small, meaning that once turned ON, it takes a significant effect to transition to the OFF state. In the next section, we will consider genetic switches that exhibit more balanced behaviors.

Exercise 7.1.4 Verify that when $N = 2$, the system in equation (7.10) exhibits three nonnegative steady states provided that $\alpha > 2K\delta_p$. □

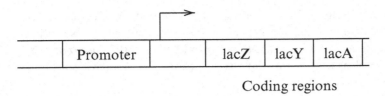

Coding regions

Figure 7.4
The *lac* operon. The coding regions for the three *lac* genes—*lacZ*, *lacY*, and *lacA*—follow one another on the DNA. Transcription of these coding regions begins at a shared promoter, so the genes are transcribed together.

7.2   Genetic Switches

7.2.1   The *lac* Operon

One of the best understood gene circuits involves a set of genes found in *Escherichia coli* whose products allow the bacterium to metabolize lactose (milk sugar). This set of genes is contained in an *operon*—a single promoter region followed by a set of coding regions, one for each protein product (figure 7.4). The genes in an operon are expressed simultaneously and are coordinately regulated via the shared promoter.

The *lac* operon contains coding regions for three proteins:

- β-galactosidase—coded by the gene *lacZ*
- β-galactoside permease—coded by the gene *lacY*
- β-galactoside transacetylase—coded by the gene *lacA*.

Together, these proteins allow *E. coli* to metabolize lactose. In this bacteria's natural environment (the mammalian gut), lactose is typically far less abundant than other sugars, and so the cell represses expression from the *lac* operon to conserve resources. This repression is caused by a transcription factor called LacI (or simply *lac* repressor), which binds to an operator region near the operon's promoter and blocks expression (figure 7.5A).

The *lac* repressor is constitutively present: when lactose is scarce, the repression is almost complete—leaked expression maintains only a few copies of each of the *lac* protein products in the cell. When lactose is abundant, it is converted to *allolactose*, which binds *lac* repressor, reducing its affinity for the operator site (by about 1000-fold). Thus, expression from the operon is triggered by the presence of lactose, with allolactose as the *inducer* (figure 7.5B).

In addition to the direct induction by allolactose, the presence of lactose sets off a positive feedback that leads to rapid expression of the *lac* genes. This feedback is implemented by the *lac* proteins themselves: β-galactoside permease is a transmembrane protein that transports lactose into the cell; β-galactosidase is an enzyme that

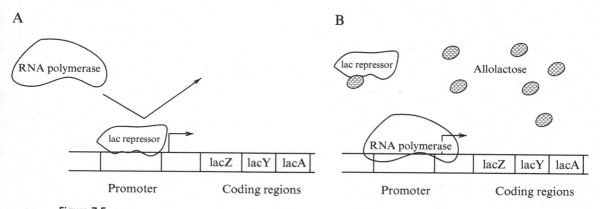

Figure 7.5
The *lac* repressor, LacI. (A) The *lac* repressor binds at the *lac* promoter, interfering with RNA polymerase binding. (B) When bound to allolactose, the *lac* repressor is unable to bind DNA. RNA polymerase is then free to bind the *lac* promoter, and transcription proceeds.

catalyzes the conversion of lactose to allolactose. β-Galactosidase also catalyzes the metabolism of lactose. (The third protein in the operon, β-galactoside transacetylase, chemically modifies β-galactosidase, but its role in lactose metabolism is currently unclear.) These *lac* proteins, as mentioned, are present at very low levels when lactose is absent. When lactose is introduced to the environment, the action of this handful of *lac* protein molecules leads to a small amount of allolactose being present in the cell, and the repression of the *lac* operon weakens. This weakened repression leads to increased expression from the operon, resulting in increased levels of allolactose, and further increases in expression, as in figure 7.6. (This is not the whole story: transcription from the *lac* operon requires the presence of an activator called catabolic gene activating protein (CAP), which is only active when glucose is absent from the environment. Glucose is the preferred energy and carbon source for these cells. The CAP mechanism ensures that when glucose is present, the *lac* genes are not expressed, regardless of the availability of lactose.)

A Model of the *lac* Operon   In 2007, Moisés Santillán, Michael Mackey, and Eduardo Zeron published a model of *lac* operon activity (Santillán et al., 2007). We will address a simplified version of their model.

To begin, we describe binding of the *lac* repressor (R) to the operon's operator site by

$$O + R \underset{k_{-1}}{\overset{k_1}{\rightleftharpoons}} OR ,$$

where O is the unbound operator and OR is the repressor-bound operator. (In fact, the repressor binds at three distinct operator sites, but we will ignore that

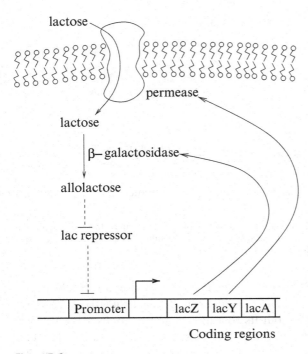

Figure 7.6
Feedback in the *lac* operon. (Dashed blunted arrows indicate repression.) The protein products of the operon—permease and β-galactosidase—bring lactose into the cell and convert it to allolactose. Allolactose activates gene expression by de-repression (inactivation of the *lac* repressor).

complication in this simple model.) Treating the repressor binding event in steady state, the fraction of unrepressed genes can be written as

$$\frac{1}{1+r(t)/K_1},$$

where r is the concentration of *lac* repressor, and $K_1 = k_{-1}/k_1$ is the dissociation constant. Letting m denote the concentration of operon mRNA, we can write

$$\frac{d}{dt}m(t) = a_1 \frac{1}{1+r(t)/K_1} - \delta_M m(t), \tag{7.11}$$

where a_1 is the maximal rate of transcription, and δ_M is the mRNA degradation/dilution rate.

We next consider the permease (coded by *lacY*). Letting y denote its concentration, we have

$$\frac{d}{dt} y(t) = c_1 m(t) - \delta_Y y(t),$$ (7.12)

where c_1 is the rate of translation, and δ_Y is the protein's degradation/dilution rate.

Because the operon's protein products are translated from the same mRNA transcript, these proteins are translated at comparable rates. Supposing that β-galactosidase and permease also share the same degradation/dilution rate, the model for β-galactosidase is identical to equation (7.12). However, the β-galactosidase protein is a tetramer, whereas permease is a monomer. Ignoring the dynamics of tetramer formation, we can write the β-galactosidase tetramer concentration, denoted b, as one fourth that of the permease:

$$b(t) = \frac{y(t)}{4}.$$ (7.13)

Lactose uptake is mediated by the permease. Assuming Michaelis–Menten kinetics for the transport event gives

$$\text{Lactose uptake} = \frac{k_L y(t) L_e}{K_{\text{ML}} + L_e},$$

where k_L is the maximal (per permease) transport rate, L_e is the external lactose concentration, and K_{ML} is the Michaelis constant. Once it has been transported across the membrane, lactose is either converted to allolactose or is metabolized (into the simpler sugars glucose and galactose); both reactions are catalyzed by β-galactosidase. Denoting the intracellular concentration of lactose by L, we write

$$\frac{d}{dt} L(t) = \frac{k_L y(t) L_e}{K_{\text{ML}} + L_e} - \frac{k_g b(t) L(t)}{K_{\text{Mg}} + L(t)} - \frac{k_a b(t) L(t)}{K_{\text{Ma}} + L(t)} - \delta_L L(t),$$

where δ_L is the dilution rate. The parameters k_g and k_a are the maximal (per β-galactosidase) rates at which lactose can be metabolized or converted to allolactose, respectively, and K_{Mg} and K_{Ma} are the corresponding Michaelis constants.

Making the assumption that the two reactions catalyzed by β-galactosidase have identical kinetics (i.e., $k_g = k_a$, $K_{\text{Mg}} = K_{\text{Ma}}$), we arrive at a simplified description of the lactose dynamics:

$$\frac{d}{dt} L(t) = \frac{k_L y(t) L_e}{K_{\text{ML}} + L_e} - 2 \frac{k_g b(t) L(t)}{K_{\text{Mg}} + L(t)} - \delta_L L(t).$$ (7.14)

Santillán and his colleagues made the further simplifying assumption that the concentration of allolactose, denoted A, is equivalent to the concentration of lactose (justified in exercise 7.2.1), so

$$A(t) = L(t). \tag{7.15}$$

Finally, we address inactivation of *lac* repressor by allolactose. The repressor is a homotetramer (i.e., a complex of four identical monomers). It is inactivated by the binding of an allolactose molecule to any of its four monomers. Taking these binding events as independent, the fraction of unbound monomers is

$$\text{Fraction of unbound monomers} = \frac{K_2}{K_2 + A(t)}, \tag{7.16}$$

where K_2 is the dissociation constant. The concentration of active repressor tetramers (in which all four monomers are unbound) is then

$$r(t) = R_T \left(\frac{K_2}{K_2 + A(t)} \right)^4, \tag{7.17}$$

where R_T is the total concentration of repressor protein (presumed constant).

Equations (7.11)–(7.17) comprise a model with three independent state variables: m, y, and L. Figure 7.7 illustrates the model's response to changes in the external lactose level, L_e. Panel A shows the time-varying response to a series of increases in the external lactose level. When L_e rises from 0 to 50 μM, there is a negligible response in system activity, as evidenced by the minor increase in β-galactosidase abundance. When the lactose level doubles to 100 μM, a dramatic increase in enzyme level is triggered. A further step in the lactose level to ($L_e = 150$ μM) elicits no response—the system has already switched to its fully "on" state. When the external lactose is removed, the system abruptly returns to its "off" state of low activity.

The dose-response curve in figure 7.7B shows the switch-like nature of the system's response. The solid curve shows that the model's transition from low activity to high activity occurs at a threshold lactose concentration. This all-or-nothing response results in a discrete (yes/no) response to the lactose input. Figure 7.7B also shows the behavior of a modified model in which the positive feedback loop has been cut (details in problem 7.8.5). The dashed dose-response curve for this hypothetical model shows a graded response to lactose availability, in contrast to the switch-like behavior of the true *lac* system.

The *lac* operon is a sensory gene regulatory network. Its function is to provide the cell with an appropriate response to the current environmental conditions. When conditions change, the response changes as well; in this case, the switch turns "on" when lactose is present and "off" when lactose is absent. In the next section, we consider a system in which a yes/no decision must persist after the activating stimulus has been removed.

Exercise 7.2.1 The rate of allolactose production is given in equation (7.14) as $k_g b(t) L(t)/(K_{Mg} + L(t))$. Allolactose is almost identical to lactose in structure, and

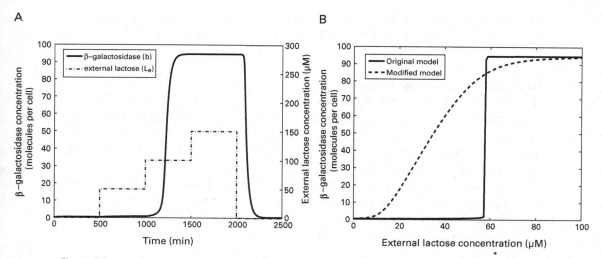

Figure 7.7
Switching behavior in the *lac* operon. (A) Dynamic response. A series of steps in the external lactose concentration (L_e) lead to changes in the activity of the *lac* operon, reflected in the abundance of β-galactosidase. The initial increase from 0 to 50 μM (at time 500 minutes) causes a negligible rise in activity. A later increase to 100 μM (at 1000 min) has a significant impact: the system shifts abruptly to its "on" state. A further increase to 150 μM (at 1500 min) elicits no response—the system is already fully active. Finally, at time 2000 min, the external lactose is removed. Once the internal lactose level drops, the system rapidly switches to the "off" state. (B) Dose response. The solid line shows the behavior of the *lac* operon model. The steady-state level of operon activity (indicated by β-galactosidase abundance) switches abruptly at a threshold lactose concentration. The dashed dose-response curve corresponds to a hypothetical model in which the positive feedback loop has been cut (details in problem 7.8.5). The resulting graded response contrasts sharply with the switch-like behavior of the true *lac* model. Parameter values: $\delta_M = 0.48$ min^{-1}, $\delta_Y = 0.03$ min^{-1}, $\delta_L = 0.02$ min^{-1}, $a_1 = 0.29$ molecules min^{-1}, $K_2 = 2.92 \times 10^6$ molecules, $K_1 R_T = 213.2$, $c_1 = 18.8$ min^{-1}, $k_L = 6.0 \times 10^4$ min^{-1}, $k_{ML} = 680$ μM, $k_g = 3.6 \times 10^3$ min^{-1}, $K_{Mg} = 7.0 \times 10^5$ molecules.

so is metabolized by β-galactosidase in the same manner as lactose (and consequently with the same kinetics). Denoting the concentration of allolactose by A and allowing for dilution, this gives

$$\frac{d}{dt} A(t) = \frac{k_g b(t) L(t)}{K_{Mg} + L(t)} - \frac{k_a b(t) A(t)}{K_{Ma} + A(t)} - \delta_L A(t).$$

Taking $k_g = k_a$ and $K_{Mg} = K_{Ma}$ (as above), Santillán and colleagues made the assumptions that (i) dilution of allolactose is negligible and (ii) allolactose is in quasi–steady state. Verify that these assumption lead to the equivalence

$$A^{qss}(t) = L(t).$$

         □

Exercise 7.2.2 Experimental studies of the *lac* operon often make use of isopropyl β-D-1-thiogalactopyranoside (IPTG), which is a molecular mimic of allolactose but is not metabolized in the cell. Extend the model to include the effect of IPTG. For simplicity, suppose that IPTG is present at a fixed intracellular concentration.   □

7.2.2 The Phage Lambda Decision Switch

Developmental gene regulatory networks are responsible for guiding the differentiation processes that occur as a single fertilized egg cell develops into a multicellular organism. Many of the decisions made during development require discrete (yes/no) responses to environmental conditions. Moreover, because the triggering signals do not continue indefinitely, these responses need to be persistent.

In this section, we will address the lysis–lysogeny decision-switch in phage lambda. This genetic circuit that has a discrete (on/off) character and retains a memory of past stimuli. This is a viral response process that occurs in host bacterial cells. Nevertheless, it serves as a biological model of more complex differentiation processes in multicellular organisms.

Phage lambda is a bacteriophage—a virus that infects bacterial cells. Phage particles, like all viruses, consist of a small genome encased in a protein shell. Upon penetrating the membrane of an *E. coli* host cell, the phage follows one of two infection processes (figure 7.8):

• **Lytic growth** The host's genetic machinery is used to produce about a hundred new phages, which then lyse (burst) the host cell.

• **Lysogenic growth** The phage's genetic material is integrated into the host cell's genome. (The viral genome is then called a *prophage*.) When the lysogenized host divides, it makes a copy of the prophage along with its own DNA. The phage thus dormantly infects all progeny of the host cell.

The phage senses the host's condition and chooses the appropriate infection mechanism: if the host is growing well, the phage integrates and multiplies lysogenically along with the host and its progeny; if the host cell is starving or damaged, the phage grows lytically—an "abandon ship" response. This decision is based on a genetic switch.

The Decision Switch We will not address the initial infection process, which involves several phage genes. Instead, we will model the simpler situation in which a prophage "chooses" whether to continue to grow lysogenically or to begin the lytic process. This decision switch can be described in terms of two genes and their protein products:

• gene *cI* codes for protein cI, also called repressor;

• gene *cro* codes for protein cro (an acronym for control of repressor and others).

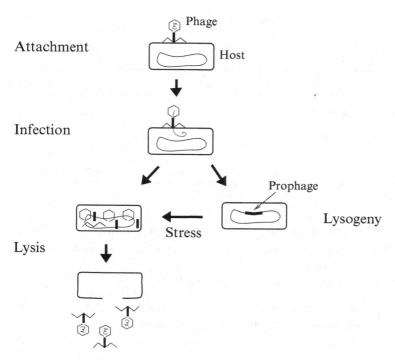

Figure 7.8
Phage lambda infection. Upon attaching to a host cell, the phage injects its genome and then follows one of two infection processes. If the host is healthy, the phage genome is incorporated into the host's DNA as a so-called prophage. The prophage is copied when the cell divides, so all the host's progeny are dormantly infected. Alternatively, if the host cell is under stress, expression of phage genes leads to the production of new phage particles. The cell wall is then ruptured (lysed), releasing the phage particles to infect new hosts. Adapted from figure 1.2 of Ptashne (2004).

These two genes are adjacent to one another on the phage DNA. They lie on opposite strands of the double helix and are consequently transcribed in opposite directions. Their promoters lie back-to-back, as shown in figure 7.9.

Both the cI and cro proteins regulate their own and each other's expression. This regulation occurs through the binding of these proteins to an operator region that overlaps both the *cI* and *cro* promoters. This operator region is called O_R. The O_R region contains three binding sites called O_R1, O_R2, and O_R3. Both cI and cro bind to all three of these sites, but in different manners and with opposing effects, as we next describe.

Regulation by cI The cI protein is a homodimer. These dimers bind strongly to O_R1 and weakly (with about 10 times less affinity) to O_R2 and O_R3. However, a cI dimer bound to O_R1 interacts with another at O_R2. This cooperativity greatly increases the

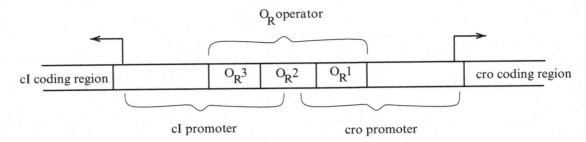

Figure 7.9
Phage lambda decision genes. The *cI* and *cro* genes lie on opposite strands of the DNA double helix and so are transcribed in opposite directions. Their promoter regions lie back-to-back. The shared operator, called O_R, overlaps both promoters: it contains three sites at which both cI and cro bind.

affinity of cI dimers for O_R2. Consequently, at low concentrations, cI dimers are found bound to O_R1 and O_R2, whereas at high concentrations, cI dimers will be bound to all three operator sites.

The effects of cI binding are as follows (figure 7.10A):

• When no proteins are bound at the operator site, there is strong expression of cro. The cI promoter has only weak affinity for RNA polymerase, so there is minimal expression of cI in this case.

• When cI dimer is bound to O_R1, it inhibits cro expression by blocking access to the cro promoter.

• When cI dimer is bound to O_R2, it upregulates its own expression (about 10-fold) by binding to RNA polymerase at the cI promoter, effectively increasing the affinity of the docking site. Thus, cI is an autoactivator.

• When cI dimer is bound to O_R3, it blocks access to the cI promoter and thus represses its own expression. Thus, at high concentrations, cI is an autoinhibitor.

The interplay of autoactivation at low levels and autoinhibition at high levels results in tight regulation of the cI concentration.

Regulation by cro   The cro protein is also a homodimer. These dimers bind to all three operator sites, with affinity opposite to that of cI, and with no cooperative effects. The cro dimer has a high affinity for O_R3 and so binds there at low concentrations; it has lower affinity for O_R2 and O_R1 (roughly equal) and so is found at these sites only at higher concentrations.

The effects of cro binding are as follows (figure 7.10B):

• As described earlier, when no proteins are bound at the operator site, cro is expressed strongly whereas only weak expression of cI occurs.

A

B

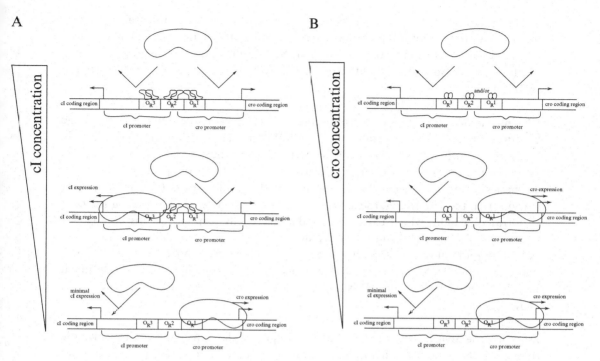

Figure 7.10
Regulation by cI and cro. (A) Dimers of cI bind to O_R1 and O_R2 at low concentration and at O_R3 at higher concentration. Basal expression of cI is weak, but polymerase binding is enhanced by a cI dimer bound at O_R2. When bound to O_R1, cI blocks cro expression. When cI binds O_R3, it blocks expression from the *cI* gene. (B) Dimers of cro bind strongly to O_R3 and weakly to O_R2 and O_R1. When bound to O_R3, cro blocks expression of cI. When bound to O_R2 or O_R1, cro blocks its own expression.

• When cro dimer is bound to O_R3, it blocks the cI promoter. Hence, cro inhibits expression of cI.

• When cro dimer is bound to O_R2 or O_R1, it inhibits its own expression by blocking access to the cro promoter. Thus, at high concentrations, cro is an autoinhibitor.

Bistability The *cI* and *cro* genes are antagonists—each represses the other. Consequently, we expect the system to exhibit two steady states: either cI will be abundant, repressing cro; or cro will be abundant, repressing cI. These two states characterize the two pathways of infection.

• In the lysogenic state, the cI concentration is high, and the cro concentration is low.

• Lysis begins when the cI concentration is low and the cro concentration is high.

The system is bistable. The lysogenic state is stable on a long timescale: it can be maintained for generations of hosts. In contrast, the lytic state is necessarily transient—it leads to the host cell's death. Nevertheless, we are justified in calling the lytic condition a steady state on the relatively short timescale of the decision switch itself.

Flipping the Switch Once it has integrated into the host's genome, the prophage continuously monitors the state of the cell. When it senses that the host cell is in jeopardy, it "flips the switch" to begin lytic growth. There is no mechanism for a switch in the opposite direction—lytic growth is an irreversible process.

The switch to lysis occurs when the host cell is under stress (e.g., is injured or starving). In the laboratory, the simplest way to induce lysis is by exposing the cells to ultraviolet light, which causes DNA damage. Cells respond to this damage by invoking expression of a number of repair proteins—this is called the *SOS response*. A key component of the SOS response is the bacterial protein RecA, which triggers expression of DNA repair genes. Once activated, RecA cleaves cI, rendering it unable to bind the O_R sites. This frees *cro* from repression, leading to lytic phage growth.

Modeling the Switch We present here a simple model that captures the bistable nature of the system with minimal detail. (A detailed model that incorporates descriptions of DNA-binding and expression processes can be found in Reinitz and Vaisnys (1990).)

We will neglect mRNA dynamics and incorporate only two state variables: the concentrations of cI and cro protein. Because cI is also called repressor, we will write $r = [\text{cI}]$ and $c = [\text{cro}]$. The model takes the form

$$\frac{d}{dt}r(t) = f_r(r(t), c(t)) - \delta_r r(t)$$

$$\frac{d}{dt}c(t) = f_c(r(t), c(t)) - \delta_c c(t),$$

(7.18)

where δ_r and δ_c account for dilution and degradation, and f_r and f_c describe the rates of expression from the cI and cro promoters, respectively.

To characterize the operator occupancy function, we begin by making the following simplifying assumptions: (i) cro and cI will never bind the operator simultaneously; (ii) strong cooperativity causes the binding of cI at $O_R 2$ to happen concurrently with cI binding to $O_R 1$; and (iii) states in which cro dimer is bound to $O_R 1$ or $O_R 2$ can be lumped together. These assumptions result in five distinct DNA-binding states (figure 7.10), as summarized in table 7.1, which indicates the rates of expression

Table 7.1
Rates of Expression of cI and cro

State	Notation	Rate of cI Expression	Rate of cro Expression
Unbound operator	O	a	b
cI at O_R1 and O_R2	$O(cI_2)_2$	$10a$	0
cI at O_R1, O_R2, and O_R3	$O(cI_2)_3$	0	0
cro at O_R3	$O(cro_2)$	0	b
cro at O_R3 and O_R1 and/or O_R2	$O(cro_2)_{2+}$	0	0

of cI and cro from each state. (The parameters a and b are the expression rates from the unregulated genes.)

We have assumed that there is no expression from the repressed states (i.e., leakage is negligible). Next, we consider the occupancy functions for these binding states. For simplicity, we assume that all cI and cro protein is in the dimer form, so that

$$[cI_2] = \frac{r(t)}{2} \quad \text{and} \quad [cro_2] = \frac{c(t)}{2}.$$

Treating the corresponding DNA-binding events in quasi–steady state, we arrive at expression rates for cI and cro (see exercise 7.2.3):

$$\text{cI expression rate: } f_r(r,c) = \frac{a + 10aK_1(r/2)^2}{1 + K_1(r/2)^2 + K_2K_1(r/2)^3 + K_3(c/2) + K_4K_3(c/2)^2}$$

$$\text{cro expression rate: } f_c(r,c) = \frac{b + bK_3(c/2)}{1 + K_1(r/2)^2 + K_2K_1(r/2)^3 + K_3(c/2) + K_4K_3(c/2)^2}.$$

(7.19)

The model behavior is illustrated in figure 7.11. The phase portrait in panel A reveals the bistable nature of the system. Both stable states exhibit near-zero levels of the repressed protein. The lytic state has a small basin of attraction: only initial conditions with overwhelmingly large concentrations of cro will end up in this high-cro, low-cI condition. Panel B shows the system behavior when RecA is active (simulated by a 10-fold increase in δ_r). This parameter change shifts the r-nullcline so that the system is monostable—only the lytic steady state (high-cro, low-cI) remains.

The lambda decision network fulfills the two requirements of a developmental switch: a threshold transition from one condition to another, and a persistent memory. This response is consistent with Lewis Wolpert's "French Flag" developmental model, in which a nascent tissue is exposed to a gradient of a chemical signal—called a morphogen—that induces differentiation into specific cell types.

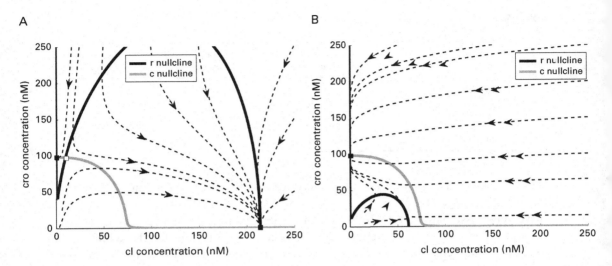

Figure 7.11
Behavior of the decision switch model. (A) This phase portrait shows the bistable nature of the system. The nullclines intersect three times (boxes). The two stable steady states are close to the axes: in each case, the repressed protein is virtually absent. (B) When RecA activity is included (by increasing δ_r 10-fold), the system becomes monostable—all trajectories are attracted to the lytic (high-cro, low-cI) state. Parameter values: $a = 5$ min^{-1}, $b = 50$ min^{-1}, $K_1 = 1$ nM^{-2}, $K_2 = 0.1$ nM^{-1}, $K_3 = 5$ nM^{-1}, $K_4 = 0.5$ nM^{-1}, $\delta_r = 0.02$ min^{-1} (0.2 in panel (B)), $\delta_c = 0.02$ min^{-1}.

The signal strength varies continuously over the tissue domain, and does not persist indefinitely. In response, each cell makes a discrete decision (as to how to differentiate) and internalizes that decision so that the effect persists after the signal is removed.

The phage lambda decision circuit is a valuable model of developmental gene circuits. However, because it is irreversible, it cannot serve as an example of a generic on/off switch that could be used as part of a larger decision-making circuit. In the next section, we consider a genetic switch that was designed to be reversible.

Exercise 7.2.3 Derive the expression rates in equation (7.19) as follows. Note that the DNA-binding events are

$$O + cI_2 + cI_2 \underset{k_{-1}}{\overset{k_1}{\rightleftharpoons}} O(cI_2)_2 \qquad O(cI_2)_2 + cI_2 \underset{k_{-2}}{\overset{k_2}{\rightleftharpoons}} O(cI_2)_3$$

$$O + cro_2 \underset{k_{-3}}{\overset{k_3}{\rightleftharpoons}} O(cro_2) \qquad O(cro_2) + cro_2 \underset{k_{-4}}{\overset{k_4}{\rightleftharpoons}} O(cro_2)_{2+}.$$

For $i = 1, 2, 3, 4$, define the association constants $K_i = k_i/k_{-i}$ and determine the equilibrium conditions for each binding event. Next, use these equilibrium equations, along with conservation of operator sites,

$$[O] = [O(cI_2)_2] + [O(cI_2)_3] + [O(cro_2)] + [O(cro_2)_{2+}] = O_T,$$

to determine the occupancy function for each of the five states. Finally, use the expression rates for each state (in table 7.1) to derive the expression rates in equation (7.19).   □

7.2.3   The Collins Toggle Switch

As discussed in chapter 1, in the year 2000 Timothy Gardner, Charles Cantor, and Jim Collins designed and constructed a genetic toggle switch by rewiring the components of existing gene regulatory networks (Gardner et al., 2000).

Their engineered circuit (figure 7.12) uses the same mutual repression scheme that we saw in the phage lambda decision switch. However, the toggle switch can be flipped in both directions: the transcription factors were chosen so that each could be inhibited by an appropriate intervention.

The toggle switch design includes a *reporter* gene, which allows for direct observation of the system's activity. The reporter is *green fluorescent protein* (GFP), which fluoresces green when exposed to blue light. The coding region for the GFP reporter was attached downstream of the coding region for one of the two repressors, creating an operon. The concentration of GFP—and intensity of fluorescence—is then correlated with the concentration of that repressor.

Gardner and his colleagues constructed multiple instances of the toggle switch network. They used only genes and promoters whose behavior had been well

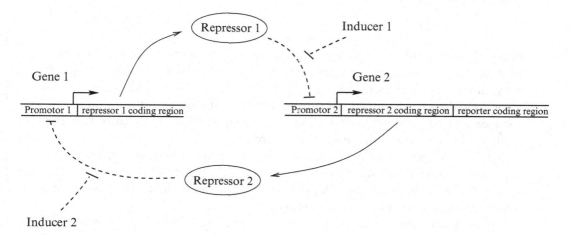

Figure 7.12
Collins toggle switch. (Dashed blunted arrows indicate repression.) Two genes repress each other's expression, leading to a bistable system. Each transcription factor was chosen so that it could be deactivated by an experimental intervention.

characterized: the *lac* repressor (LacI) and the *tet* repressor (TetR) from *E. coli*, and cI from phage lambda. (TetR inhibits expression from the *tet* genes, which are responsible for protection from tetracycline, an antibiotic.) The target of cI repression was not the cro promoter studied in the previous section, but another phage promoter, P_L, whose repression mechanism is simpler. For each of these transcription factors, expression from the target gene could be induced by inactivating the repression. (LacI is inactivated by IPTG—a nonmetabolizable analogue of allolactose. Likewise, TetR can be inactivated by anhydrotetracycline (aTc)—a nontoxic analogue of its natural inactivator tetracycline. The phage protein cI does not have a native inactivation mechanism. Gardner and colleagues made use of a mutated form of cI that is temperature-sensitive: it functions normally at 30°C but is nonfunctional when the temperature is raised to 42°C.)

Gardner and his colleagues developed a simple model to explore the behavior of the switch circuit. The model was not meant to accurately reflect the specifics of their proposed construction but was used to investigate the nature of bistability in such a device.

Neglecting mRNA dynamics, the model can be written as

$$\frac{d}{dt} p_1(t) = \frac{\alpha_1}{1 + \left(\frac{p_2(t)}{1 + i_2} \right)^\beta} - p_1(t) \tag{7.20}$$

$$\frac{d}{dt} p_2(t) = \frac{\alpha_2}{1 + \left(\frac{p_1(t)}{1 + i_1} \right)^\gamma} - p_2(t), \tag{7.21}$$

where p_1 and p_2 are the concentrations of the two proteins, α_1 and α_2 are their maximal expression rates, β and γ indicate the degree of nonlinearity (i.e., cooperativity) in the repression mechanisms, and i_1, i_2 characterize the two inducers. Dilution is considered to be dominant over degradation, so the decay rates are identical. Gardner and colleagues scaled the time and concentration units to reduce the number of parameters in the model (see exercise 7.2.4).

Figure 7.13 shows the model behavior. The two inducers have the desired effect of causing transitions between the stable steady states. Figure 7.14A shows a phase portrait for the uninduced (bistable) system. The portrait in figure 7.14B shows the monostable system that occurs when inducer 2 is present.

Gardner and his colleagues used their model to predict features of the circuit that would result in bistability. The two-dimensional bifurcation plot in figure 7.15 shows the results of such an analysis. This plot subdivides the α_1–α_2 parameter space into regions in which the system is monostable or bistable for various values of β and γ. As expected, when $\alpha_1 = \alpha_2$, the system is perfectly balanced and so is bistable, pro-

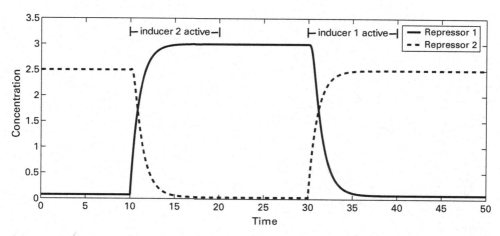

Figure 7.13
Behavior of the toggle switch model. Repressor 2 is abundant in the initial condition. At time $t = 10$, inducer 2 is introduced, rendering repressor 2 inactive, and so inducing expression of repressor 1. Repressor 1 is then able to repress expression of repressor 2. The high level of repressor 1 is maintained after the inducer is removed at $t = 20$. The opposite effect occurs on introduction of inducer 1 (at $t = 30$, removal at $t = 40$). Parameter values: $\alpha_1 = 3$ (concentration/time), $\alpha_2 = 2.5$ (concentration/time), $\beta = 4$, and $\gamma = 4$. Inducer activity is simulated by increasing i_1 or i_2 from 0 to 10 in each case. Units are arbitrary.

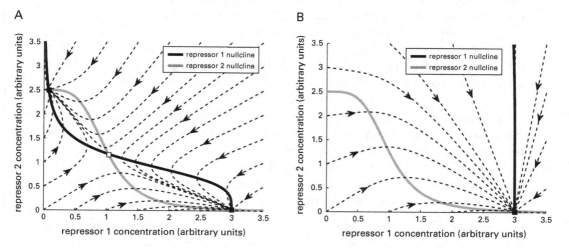

Figure 7.14
Phase portraits for the toggle switch. (A) The uninduced switch ($i_1 = i_2 = 0$). The nullclines intersect three times: at two stable steady states and one intermediate unstable steady state. (B) Under the influence of inducer 2 ($i_2 = 10$), the nullcline for repressor 1 has shifted so there is a single steady state, to which all trajectories converge. Parameter values as in figure 7.13. Adapted from figure 2 of Gardner et al. (2000).

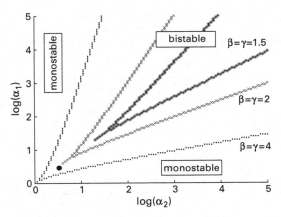

Figure 7.15
Dependence of bistability on parameter values. For each value of β and γ, the α_1–α_2 parameter space is divided into regions in which the system is monostable or bistable ($i_1 = i_2 = 0$). As the degree of nonlinearity (i.e., β and γ) increases, the bistable region (wedge) grows, indicating that bistability is preserved despite asymmetry between the two components of the switch. The parameter set for the simulation in figure 7.13 is indicated by the dot near the bottom left-hand corner. Adapted from figure 2 of Gardner et al. (2000).

vided that the maximal expression rates are sufficiently large (otherwise a single balanced steady state occurs). When the two expression rates are not balanced, bistability may be lost as one gene dominates the other. The degree of imbalance that is allowed within the bistability domain depends strongly on γ and β, which reflect the degree of nonlinearity in repressor–DNA binding. The greater the nonlinearity, the more allowance the switch has for unequal expression rates. These observations suggest that (i) bistability is favored by strong expression (i.e., strong promoters and strong ribosome binding sites); and (ii) the more cooperativity involved in repressor–DNA binding, the more robust the switch's bistability will be to asymmetry between the two component genes. Gardiner and colleagues used these principles to construct multiple functioning instances of the genetic switch (implemented in *E. coli* cells). They successfully confirmed the bistable behavior of the device by monitoring the GFP readout in the lab.

Exercise 7.2.4 When constructing a generic model, rescaling of units can absorb parameters into the definition of time or concentration scales, thus reducing the number of free parameters. For instance, the concentration profile $s(t) = e^{-t/60}$ nM, where t is measured in seconds, can be written as $s(\tau) = e^{-\tau}$ nM, where τ is measured in minutes (i.e., in time units of 60 seconds).

Consider a model of gene expression:

$$\frac{d}{dt} p(t) = \frac{\alpha}{K + p(t)} - \delta p(t).$$

Describe the rescaling of time and concentration units in which the model can be written as

$$\frac{d}{d\tau}\tilde{p}(\tau) = \frac{\tilde{\alpha}}{1+\tilde{p}(\tau)} - \tilde{p}(\tau).$$ □

Exercise 7.2.5 An analysis of bistability as in figure 7.15 can be carried out analytically in the special case of $\beta = \gamma = 1$. Verify that in this case the system is monostable (i.e., it exhibits a single steady state) when $\alpha_1 = \alpha_2$. (Take $i_1 = i_2 = 0$.) □

7.3 Oscillatory Gene Networks

We next consider examples of gene regulatory networks that generate persistent oscillations. These networks allow cells to maintain internal clocks that can be used to predict periodic changes in conditions (such as the night–day cycle).

7.3.1 The Goodwin Oscillator

In 1965, Brian Goodwin proposed a generic model of an oscillatory genetic circuit (Goodwin, 1965). The model, sketched in figure 7.16, involves a single gene. The mRNA, X, is translated into enzyme Y, which catalyzes production of metabolite Z, which causes inhibition of expression (by activating an unmodeled repressor). Neglecting the specifics of catalysis and inhibition, Goodwin formulated the model in terms of concentrations x, y, and z as

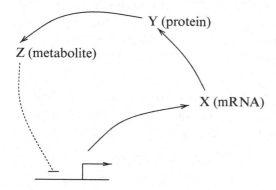

Figure 7.16
The Goodwin oscillator. (The dashed blunted arrow indicates repression.) The mRNA (X) is translated into an enzyme (Y), which catalyzes production of a metabolite (Z), which (indirectly) represses gene expression. This negative feedback, coupled with the delay inherent in the three-step loop, can result in oscillatory behavior.

$$\frac{d}{dt}x(t) = \frac{a}{k^n + (z(t))^n} - bx(t)$$

$$\frac{d}{dt}y(t) = \alpha x(t) - \beta y(t) \tag{7.22}$$

$$\frac{d}{dt}z(t) = \gamma y(t) - \delta z(t).$$

The model was not meant to describe a particular system: it was constructed to demonstrate how persistent oscillations could be generated by an autoinhibitory gene circuit.

Goodwin included three states in the model to impose sufficient delay in the negative feedback loop. As discussed in section 4.3, oscillations can arise from negative feedback if the effect of the feedback is delayed and if there is sufficient nonlinearity in the loop. Indeed, a two-state model that arises from applying the quasi-steady-state assumption to the Goodwin model cannot exhibit sustained oscillations, as verified by J. S. Griffith (Griffith, 1968).

Even with three steps providing a lag in the feedback, a high degree of nonlinearity is required to generate limit-cycle oscillations in this model. In his paper, Griffith showed that the system cannot exhibit sustained oscillations unless the Hill coefficient n is higher than eight, and even then, oscillations only occur for certain values of the other parameters (see problem 7.8.8).

The system's oscillatory behavior is shown in figure 7.17. The mechanism of oscillations is apparent in panel A. In each cycle, the mRNA concentration rises, followed by a rise in enzyme concentration, and then a rise in metabolite concentration. The rise in z causes a crash in x, which causes y and z to drop, allowing x to rise again. Panel B shows a three-dimensional phase portrait, confirming that the system trajectories all settle to a periodic (limit cycle) behavior.

Exercise 7.3.1 Goodwin offered multiple interpretations of his model. In addition to the description given here (X is mRNA, Y is enzyme, Z is metabolite), he also suggested that the model could be used to describe the following feedback loops:

(a) X is nuclear mRNA, Y is cytoplasmic mRNA, Z is protein product.

(b) X is mRNA, Y is inactive protein product, Z is active protein product.

Under what assumptions can the model apply to each of these cases?   □

7.3.2   Circadian Rhythms

The Goodwin model demonstrates that an autoinhibitory gene can generate persistent oscillations. A specific instance of this behavior is provided by the circadian rhythm generator in the fruit fly *Drosophila melanogaster*.

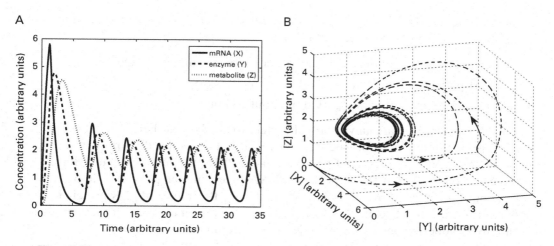

Figure 7.17
The Goodwin oscillator. (A) This simulation shows relaxation to sustained (limit-cycle) oscillations. (B) A phase portrait showing convergence to a limit cycle in the three-dimensional phase space. Parameter values are $a = 360$ (concentration $\cdot$ time^{-1}), $k = 1.368$ (concentration), $b = 1$ (time^{-1}), $\alpha = 1$ (time^{-1}), $\beta = 0.6$ (time^{-1}), $\gamma = 1$ (time^{-1}), $\delta = 0.8$ (time^{-1}), $n = 12$. Units are arbitrary.

We are familiar with the circadian rhythms of our own bodies: they regulate our sleep–wake cycles and are disrupted by jet-lag when we travel across time zones. Because they allow prediction of periodic changes in temperature and light, these internal rhythms are an important aspect of the biology of many organisms.

Behavioral studies of these internal clocks have shown them to have a free-running period of roughly 24 hours (i.e., in the absence of external cues). Moreover, these rhythms are readily entrained to light and temperature cues and are remarkably robust to changes in ambient temperature.

In mammals, the primary circadian pacemaker has been identified as a group of about 8000 neurons in the suprachiasmatic nucleus (located in the hypothalamus), which have a direct connection to the retina (in the eye). A model of the gene network responsible for generation of circadian rhythms in mammals is provided in LeLoup and Goldbeter (2003).

Here, we consider the first dynamic mathematical model that was proposed for a circadian oscillator: Albert Goldbeter's model of circadian rhythms in *Drosophila* (reviewed in Goldbeter (1996)).

Studies of *Drosophila* have yielded many advances in genetics. In 1971, Ronald Konopka and Seymour Benzer published a study in which they identified flies with mutations that caused changes in the period of the free-running circadian rhythm (Konopka and Benzer, 1971). These mutations occurred in a gene named *per* (for period): the protein product is called PER. In contrast to wild-type (i.e., nonmutant) flies, whose rest/activity patterns demonstrated a roughly 24-hour free-running period, they reported on three mutations:

- an arrhythmic mutant that exhibits no discernible rhythm in its activity;
- a short-period mutant with a period of about 19 hours;
- a long-period mutant with a period of about 28 hours.

Additional molecular analysis provided clues to the dynamic behavior of *per* gene expression. Observations of wild-type flies revealed that total PER protein levels, *per* mRNA levels, and levels of phosphorylated PER protein all oscillate with the same 24-hour period, with the peak in mRNA preceding the peak in total protein by about 4 hours. Moreover, it was shown that when the import of PER protein into the nucleus was blocked, the oscillations did not occur. On the basis of these observations, Goldbeter constructed a model of an autoinhibitory *per* circuit.

Goldbeter's model, sketched in figure 7.18, has the same basic structure as the Goodwin model: a gene codes for a product that, after a delay, represses its own expression. In this case, the delay is caused by transport across the nuclear membrane and a two-step activation process (by phosphorylation).

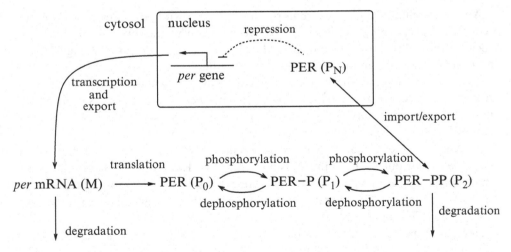

Figure 7.18
Goldbeter's circadian oscillator model. (The dashed blunted arrow indicates repression.) The *per* gene is transcribed in the nucleus; *per* mRNA (*M*) is exported to the cytosol, where is it translated and is subject to degradation. PER protein (P_0) is activated by two reversible rounds of phosphorylation. Active PER (P_2) is subject to degradation and can cross the nuclear membrane. Once in the nucleus, PER (P_N) represses transcription of the *per* gene. Delay oscillations arise from the combination of autoinhibitory feedback, nonlinear repression kinetics, and delay. Adapted from figure 11.6 of Goldbeter (1996).

The feedback loop begins with the production of *per* mRNA (M), which is exported from the nucleus to the cytosol. In the cytosol, the mRNA is translated into protein and is subject to degradation. Newly translated PER protein (P_0) is inactive. It undergoes two rounds of phosphorylation to become active PER (P_2), which is reversibly transported across the nuclear membrane. Once in the nucleus (P_N), PER represses transcription of the *per* gene. Degradation of PER is assumed to occur only in the cytosol, and degradation of inactive PER is assumed negligible.

Using lowercase letters to denote concentrations, Goldbeter's model takes the form

$$\frac{d}{dt}m(t) = \frac{v_s}{1+(p_N(t)/K_I)^n} - \frac{v_m m(t)}{K_{m1}+m(t)}$$

$$\frac{d}{dt}p_0(t) = k_s m(t) - \frac{V_1 p_0(t)}{K_1+p_0(t)} + \frac{V_2 p_1(t)}{K_2+p_1(t)}$$

$$\frac{d}{dt}p_1(t) = \frac{V_1 p_0(t)}{K_1+p_0(t)} - \frac{V_2 p_1(t)}{K_2+p_1(t)} - \frac{V_3 p_1(t)}{K_3+p_1(t)} + \frac{V_4 p_2(t)}{K_4+p_2(t)}$$

$$\frac{d}{dt}p_2(t) = \frac{V_3 p_1(t)}{K_3+p_1(t)} - \frac{V_4 p_2(t)}{K_4+p_2(t)} - k_1 p_2(t) + k_2 p_N(t) - \frac{v_d p_2(t)}{K_d+p_2(t)}$$

$$\frac{d}{dt}p_N(t) = k_1 p_2(t) - k_2 p_N(t).$$

The model is based on first-order kinetics for transport across the nuclear membrane and Michaelis–Menten kinetics for the degradation and phosphorylation/dephosphorylation processes. Transcription and export of mRNA are lumped into a single process, which is cooperatively repressed by P_N with Hill coefficient n. As with the Goodwin model, this model only exhibits oscillatory behavior if the repression kinetics is sufficiently nonlinear. Goldbeter carried out his analysis with $n = 4$: he found that the model can exhibit oscillations with $n = 2$ or even $n = 1$, but only under restrictive conditions on the other parameter values.

The oscillatory behavior of the model is illustrated in figure 7.19. Panel A shows the periodic behavior of *per* mRNA, total PER protein, and nuclear PER protein. The period is roughly 24 hours, and the mRNA peak precedes the total PER peak by about 4 hours. This behavior is consistent with experimental observation but does not provide direct validation of the model because Goldbeter chose parameter values to arrive at this behavior. Nevertheless, the model represented a valuable hypothesis as to how circadian rhythms could be generated by the activity of the *per* gene.

Goldbeter used the model to explore possible mechanisms for the effects of the short- and long-period *per* mutations. To explore the hypothesis that these mutations

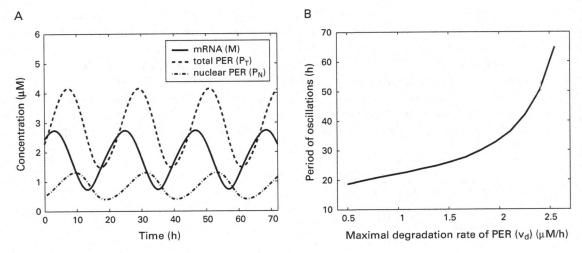

Figure 7.19
Behavior of the Goldbeter circadian oscillator model. (A) The simulated concentrations of mRNA (m), total PER protein ($p_T = p_0 + p_1 + p_2 + p_N$), and nuclear PER protein (p_N). The period of the oscillation is about 24 hours, with a lag of about 4 hours between the peak in mRNA and protein levels. (B) This continuation diagram shows the effect of changes in the maximal PER degradation rate (v_d) on the oscillation period. Within the range over which oscillations occur, the period ranges from about 20 to more than 60 hours. Parameter values are $v_s = 0.76$ μM/hr, $v_m = 0.65$ μM/hr, $v_d = 0.95$ μM/hr (panel (A), $k_s = 0.38$ hr^{-1}, $k_1 = 1.9$ hr^{-1}, $k_2 = 1.3$ hr^{-1}, $V_1 = 3.2$ μM/hr, $V_2 = 1.58$ μM/hr, $V_3 = 5$ μM/hr, $V_4 = 2.5$ μM/hr, $K_1 = K_2 = K_3 = K_4 = 1$ μM, $K_I = 1$ μM, $K_{m1} = 0.5$ μM, $K_d = 0.2$ μM, $n = 4$. Adapted from figures 11.7 and 11.9 of Goldbeter (1996).

affect the rate of PER degradation, he determined the effect of changes in the maximal PER degradation rate (v_d) on the oscillation period. His findings, reproduced in figure 7.19B, show that as v_d varies (between 0.45 and 2.6 μM/hr), the period ranges between 20 and 62 hours (beyond this range the oscillations are lost). The mutant periods fall roughly into this range, indicating that alterations in the protein degradation rate could be the cause of the observed changes.

In the years since Goldbeter's model was published, additional experiments have led to a better understanding of the circadian clock in *Drosophila*. A model that incorporates more recent findings is explored in problem 7.8.10.

7.3.3 Synthetic Oscillatory Gene Networks
In this section, we address two gene circuits that were engineered to display oscillatory behavior.

A Synthetic Delay Oscillator: The Repressilator   In the year 2000, Michael Elowitz and Stanislas Leibler announced the construction of an oscillatory synthetic circuit (Elowitz and Leibler, 2000). Elowitz and Leibler called their device the *repressilator*,

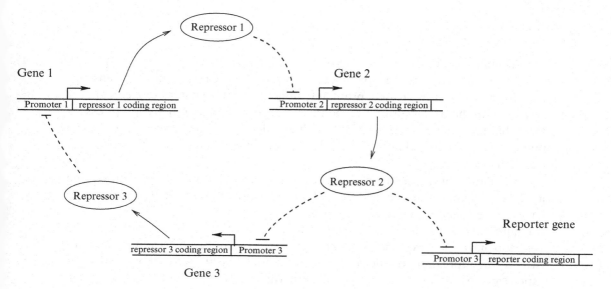

Figure 7.20
Repressilator gene network. (Dashed blunted arrows indicate repression.) Three genes each repress expression of the next around a loop. This network architecture can give rise to sustained oscillations—the protein levels rise and fall in succession. The reporter gene, GFP, is under the control of a separate copy of promoter 3.

in homage to the theoretical oscillating chemical system known as the *Brusselator* (see exercise 4.3.1 in chapter 4).

The repressilator design, like the Collins toggle switch, involves stringing together promoter–repressor pairs; in this case, there are three genes in the loop (figure 7.20). This three-repressor loop does not lend itself to steady-state behavior. When any one protein dominates over the others, it leads to its own repression—the dominant protein de-represses its own repressor, which then becomes dominant. When this process continues around the loop, the result is sustained oscillations in the protein concentrations. This is a delay oscillator—each protein inhibits its own expression through the chain of three inhibitions.

Elowitz and Leibler constructed a simple model of the network as part of the design process. Because they needed to capture the network's time-varying behavior, they included mRNA dynamics explicitly. Assuming that all three genes have identical characteristics, they arrived at the following model:

$$\frac{d}{dt}m_1(t) = \alpha_0 + \frac{\alpha}{1+(p_3(t))^n} - m_1(t) \qquad \frac{d}{dt}p_1(t) = \beta m_1(t) - \beta p_1(t)$$

$$\frac{d}{dt}m_2(t) = \alpha_0 + \frac{\alpha}{1+(p_1(t))^n} - m_2(t) \qquad \frac{d}{dt}p_2(t) = \beta m_2(t) - \beta p_2(t)$$

$$\frac{d}{dt}m_3(t) = \alpha_0 + \frac{\alpha}{1+(p_2(t))^n} - m_3(t) \qquad \frac{d}{dt}p_3(t) = \beta m_3(t) - \beta p_3(t).$$

The six state-variables are the mRNA concentrations (m_1, m_2, m_3) and the protein concentrations (p_1, p_2, p_3). The parameter α_0 represents the rate of "leaky" transcription from the fully repressed promoter; $\alpha_0 + \alpha$ is the maximal expression rate (achieved in the absence of repression). The degree of cooperativity in repressor–DNA binding is characterized by the Hill coefficient n. Parameter β is the decay rate for the proteins. Additional parameters were eliminated by scaling of the time and concentration units.

A simulation of the model is shown in figure 7.21A. The symmetric nature of the model is apparent: all three protein profiles follow identical cycles. The parameters have been chosen so that the period is about 150 minutes.

Figure 7.21B shows a set of two-dimensional bifurcation plots demonstrating the system behavior. As expected, oscillatory behavior is favored by stronger cooperativity in the repression kinetics (i.e., increased nonlinearity n). Moreover, stronger expression (α) results in a more robust oscillator. The plot also shows that the value

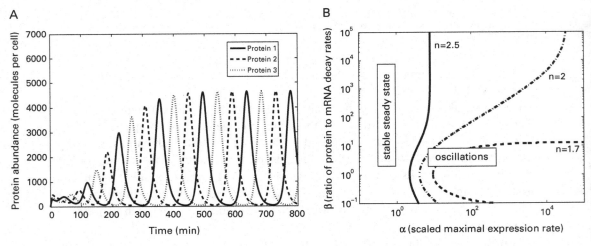

Figure 7.21
Behavior of the repressilator model. (A) Simulation of the model. The peaks in protein concentration are about 50 minutes apart, giving an overall period of about 150 minutes. Parameter values: $\alpha_0 = 0.03$ (molecules per cell $\cdot$ min^{-1}), $\alpha = 298.2$ (molecules per cell $\cdot$ min^{-1}), $\beta = 0.2$ (min^{-1}), $n = 2$. The model outputs are scaled as follows: protein concentration $= 40\,p_i(t)$ (corresponding to a half-saturating constant of 40 molecules per cell); time $= t/0.3485$ (corresponding to a mRNA half-life of 2 minutes). (B) A set of two-dimensional bifurcation plots showing the range of α and β values for which the model exhibits sustained oscillations. Oscillations are favored by β near one, and α and n large. Adapted from figure 1 of Elowitz and Leibler (2000).

of β (the ratio of protein decay rate to mRNA decay rate) has a significant impact on the behavior of the system. Oscillatory behavior is easier to attain when this ratio is close to one. This finding is consistent with the need for a significant delay in the loop: if mRNA dynamics are very fast, they will not contribute to the overall delay. A similar analysis shows that low leakiness (α_0) favors oscillations (see problem 7.8.12).

Elowitz and Leibler used model-based observations in their design process. First, they chose promoters that were known to be cooperatively repressed (high n values) and selected strong versions of those promoters (high α) with tight repression (low α_0). Second, to bring the protein decay rate closer to the (typically much faster) mRNA decay rate, they added a "degradation tag" to the proteins in the network, reducing their half-lives by as much as 15-fold.

Elowitz and Leibler constructed their circuit from the same promoter–repressor pairs that were used in the Collin's toggle switch: LacI and TetR from *E. coli*, and cI from phage lambda. They were able to synchronize a population of *E. coli* cells hosting the network by exposing them to a pulse of IPTG (which inhibits LacI) and successfully demonstrated oscillations. The period of the oscillations (about 150 minutes) was considerably longer than the doubling time of the cells (about 30 minutes): the state of the oscillations was passed from mother to daughter cells after division.

Although the repressilator design resulted in oscillatory behavior, the oscillations themselves were irregular: the cells exhibited significant variation in amplitude, period, and phase. We next consider an engineered gene network that acts as a relaxation oscillator and consequently exhibits less variability in its periodic behavior.

A Synthetic Relaxation Oscillator   Relaxation oscillators typically exhibit more robust behavior than delay oscillators. To implement an oscillator with robust periodic behavior, Jesse Stricker and colleagues designed and constructed a relaxation oscillator involving two genes: an activator and a repressor (Stricker et al., 2008). They used a promoter that is regulated by both of these transcription factors. The network, sketched in figure 7.22, incorporates two identical copies of this promoter, separately driving expression of the repressor and activator.

Stricker and co-workers used both deterministic (differential equation–based) and stochastic models in designing the system. A preliminary model, published earlier by Jeff Hasty and colleagues (Hasty et al., 2002), takes the form (details in exercise 7.3.2)

$$\frac{d}{dt}x(t) = \frac{1+x(t)^2+\alpha\sigma x(t)^4}{\left(1+x(t)^2+\sigma x(t)^4\right)\left(1+y(t)^4\right)} - \gamma_x x(t) \tag{7.23}$$

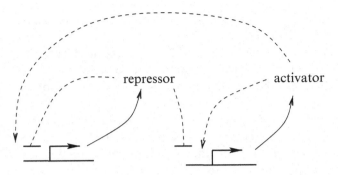

Figure 7.22
Genetic relaxation oscillator network. (Dashed arrows indicate regulation.) Identical promoters drive expression of the activator and the repressor. The interplay between positive and negative feedback can lead to sustained oscillations (characterized by bursts of expression followed by periods of repression).

$$\frac{d}{dt} y(t) = a_y \frac{1 + x(t)^2 + \alpha \sigma x(t)^4}{\left(1 + x(t)^2 + \sigma x(t)^4\right)\left(1 + y(t)^4\right)} - \gamma_y y(t),$$

where x and y are the concentrations of the activator, X, and repressor, Y, respectively.

The model's behavior is shown in figure 7.23. Panel A shows the persistent oscillations exhibited by the system: the activator X and repressor Y are expressed together. Their concentrations grow until the repressor cuts off expression. Concentrations then fall until repression is relieved and the next burst of expression begins. The oscillations exhibit sharp peaks—particularly for X. This relaxation behavior is displayed in panel B, which shows the limit-cycle trajectory in the phase plane, along with the nullclines. The horizontal motions are rapid—the trajectory spends most of its time following the x nullcline at low X concentration (in the repressed state). Because this behavior is dependent on positive feedback (autoactivation), it is relatively robust to parameter variation.

Stricker and colleagues successfully implemented their relaxation oscillator design using LacI and an activator called AraC. They used a microfluidic platform to observe individual cells and saw steady, persistent oscillations over several periods. They also found that they could tune the period of the oscillator through partial inactivation of LacI.

Both the repressilator and the Stricker oscillator successfully generate single-cell oscillations. However, when implemented in a population, deviations in phase between individual cells tend to cancel out the population-averaged oscillatory signal. In the next section, we will address gene networks that involve cell-to-cell communication, providing a mechanism to synchronize populations of cellular oscillators.

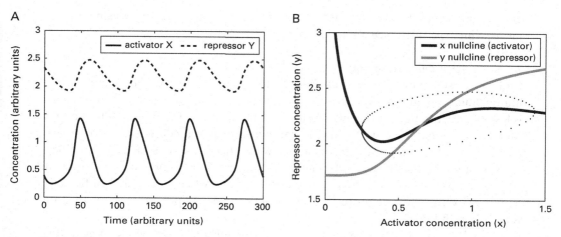

Figure 7.23
The Hasty relaxation oscillator model. (A) This simulation of the system's periodic behavior shows concentrations of the activator X and the repressor Y rising and falling in near unison. Sharp peaks in the concentration of X indicate that this is a relaxation oscillator. (B) This phase portrait shows the nullclines along with the limit-cycle trajectory. The points are plotted along a single period and are spaced equally in time. The trajectory transits rapidly around the cycle but moves slowly in the repressed (low $[X]$) condition. Parameter values: $\alpha = 11$, $\sigma = 2$, $\gamma_x = 0.2$ (time^{-1}), $\gamma_y = 0.06$ (time^{-1}), $a_y = 0.2$ (concentration · time^{-1}). Units are arbitrary.

Exercise 7.3.2 Derive the expression rates in the model (7.23). The repressor is assumed to bind with strong cooperativity at four sites. The activator X binds with strong cooperativity at each of two distinct pairs of sites. Assume that the activator and repressor binding events are independent of one another. Expression, which is completely blocked by Y binding, occurs at a basal rate unless all four X sites are occupied. Rescale the concentration units so that the half-saturating constants for DNA-binding of the first two X molecules and of the four Y molecules are both one. Rescale the time units so that the basal expression rate for X is one. The parameter α is the degree to which the expression rate increases when the second activator pair is bound, and σ is the ratio of the binding affinities at the two pairs of activator sites. □

7.4 Cell-to-Cell Communication

Gene networks operating in individual cells can communicate their states to one another by producing a signaling molecule that can pass from one cell to another—providing an *intercellular* connection. In this section, we will consider two examples of cell-to-cell communication in engineered gene circuits: the passing of signals between two distinct populations of cells, and the synchronization of a population of cellular oscillators. These circuits are based on bacterial quorum sensing mechanisms, which we introduce next.

7.4.1 Bacterial Quorum Sensing

Cell-to-cell signaling is crucial to the development and proper functioning of all multicellular organisms. For bacterial cells, the need for intercellular communication is less critical, but these cells nevertheless use a multitude of such signals to monitor their environment. One well-studied example of bacterial cell-to-cell communication is *quorum sensing*—a mechanism by which bacterial cells measure the local density of their population. Bacteria use this information to enhance their survival. (One example is the formation of bacterial *biofilms* when cells reach sufficiently high density. Biofilms are protective layers of proteins and polysaccharides that are secreted from the cells.)

To implement quorum sensing, each cell communicates its presence by secreting a signaling molecule, called an *autoinducer*, into the local environment. These molecules are taken up by neighboring cells and activate gene expression—including genes that lead to production of the autoinducer itself. This positive feedback results in a switch-like response (as in section 7.2) to changes in the local population density.

Quorum sensing was first identified in the bioluminescent marine bacterium *Vibrio fischeri*. These cells live freely in seawater but can also take up residence in specialized light organs of some squid and fish. In seawater, *V. fischeri* are usually found at low densities (less than 100 cells per milliliter) and produce only a small amount of light (less than 0.8 photons/cell/second). In light organs, the cells reach densities of more than 10^{10} cells per milliliter and increase their per-cell light output more than 1000-fold.

The quorum sensing mechanism that controls light output makes use of a signaling molecule called acyl-homoserine lactone (AHL). This autoinducer is a small nonpolar molecule that diffuses freely across the cell membrane. As shown in figure 7.24, production of AHL is catalyzed by an enzyme called LuxI. Intracellular AHL (whether self-generated or imported from the environment) binds to the constitutively expressed protein LuxR. When complexed with AHL, LuxR binds to an operator called the lux box, enhancing production of LuxI. The light-producing protein—called *luciferase*—is coded in an operon with LuxI (along with enzymes needed to fuel its activity). Thus increased LuxI expression leads to increased light production.

LuxI and AHL form a positive feedback loop: expression of LuxI enhances AHL production and so enhances LuxI expression. This system responds to external AHL with a steep switch-like response in LuxI expression—and in light production.

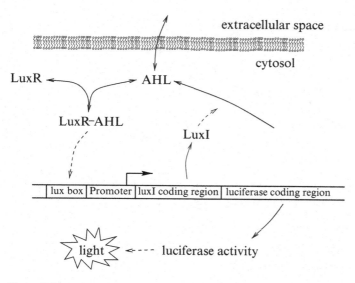

Figure 7.24
Quorum sensing in *Vibrio fischeri*. (Dashed arrows indicate activation.) The autoinducer AHL diffuses freely across the cell membrane. Its production is catalyzed by LuxI. AHL binds to the regulator LuxR, causing it to enhance transcription of LuxI and of the luciferase genes, whose protein products generate light. Adapted from figure 18 of Weiss et al. (2003).

Sally James and her colleagues published a model of the *V. fischeri* quorum sensing mechanism in the year 2000 (James et al., 2000). A simplified version of their model is the following:

$$\frac{d}{dt}A(t) = k_0 I(t) - r(A(t) - A_{ext}(t)) - 2k_1 (A(t))^2 (R_T - 2R^*(t))^2 + 2k_2 R^*(t)$$

$$\frac{d}{dt}R^*(t) = k_1 (A(t))^2 (R_T - 2R^*(t))^2 - k_2 R^*(t)$$

$$\frac{d}{dt}I(t) = a_0 + \frac{aR^*}{K_M + R^*(t)} - bI(t) \tag{7.24}$$

$$\frac{d}{dt}A_{ext}(t) = pr(A(t) - A_{ext}(t)) - dA_{ext}(t).$$

The state variable A is the (averaged) intracellular concentration of free autoinducer (AHL), R^* is the concentration of active LuxR–AHL complexes (each composed of a LuxR homodimer bound to two molecules of AHL), I is the concentration of LuxI, and A_{ext} is the extracellular concentration of autoinducer. The rate of diffusion of AHL across each cell membrane is given by $r(A(t) - A_{ext}(t))$. This results

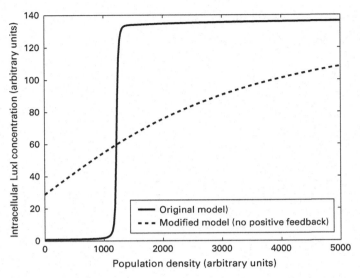

Figure 7.25
Dose-response curves for the model of quorum sensing in *V. fischeri* (solid curve). As the population density rises, switch-like activation of the quorum sensing mechanism occurs at a threshold value. This sigmoidal response is caused by the positive feedback loop involving LuxI and AHL. The dashed curve shows the dose response of a model variant in which the AHL production rate is fixed. In this case, the LuxI levels rise in a graded manner. Parameter values: $k_0 = 8 \times 10^{-4}$ (time^{-1}), $r = 0.6$ (time^{-1}), $R_T = 0.5$ (concentration), $k_1 = 0.5$ (time^{-1} · concentration^{-3}), $k_2 = 0.02$ (time^{-1} · concentration^{-3}), $a = 10$ (concentration · time^{-1}), $b = 0.07$ (time^{-1}), $K_M = 0.01$ (concentration), and $d = 1000$ (time^{-1}). In the modified model, the term $k_0 I(t)$ is replaced with $15k_0$, corresponding to a mid-range LuxI concentration. Units are arbitrary.

in diffusion into the extracellular environment at rate $p\, r(A(t) - A_{out}(t))$, where the parameter p accounts for the population density. The rate at which extracellular AHL diffuses away from the population (i.e., out of the system) is characterized by d. Parameter R_T is the total concentration of LuxR monomers (presumed fixed).

The solid curve in figure 7.25 shows the model's response (the LuxI concentration, which is proportional to the rate of light production) as a function of population density p. At low cell density, AHL diffuses out of the system, and there is no response in LuxI expression. As the density rises above a threshold value, the positive feedback causes a runaway increase in intracellular AHL and LuxI levels, culminating in maximal expression of LuxI (and luciferase). Figure 7.25 also shows a hypothetical graded response in which the positive feedback is absent (dashed curve), corresponding to a modified system in which AHL production is not LuxI-dependent.

Exercise 7.4.1 In the model (7.24), the LuxR concentration is held fixed (corresponding to constitutive expression and decay). In fact, LuxR expression is activated by AHL. Considering the dose-response curve in figure 7.25, would you expect this additional feedback on LuxR expression to make the response steeper or more shallow? ☐

7.4.2 Engineered Cell-to-Cell Communication

In the year 2000, Ron Weiss and Tom Knight published a paper describing two engineered strains of *E. coli* that demonstrate cell-to-cell communication (Weiss and Knight, 2001). Signals could be passed from cells of the first strain, called "sender cells," to cells of the second strain, called "receiver cells." They used the *Vibrio* autoinducer AHL as the intercellular signaling molecule. The two strains were created by splitting the *V. fischeri* quorum sensing network into separate sending and receiving modules: the sender cells host the *LuxI* gene, and hence can produce AHL; the receiver cells contain the *LuxR* gene and so respond to the presence of AHL (figure 7.26).

Weiss and Knight engineered the sender population so that AHL production could be controlled experimentally: they placed the *LuxI* gene under the control of a promoter that is repressed by TetR, and incorporated a constitutively expressed *tetR* gene in the cells. The addition of aTc (which inhibits TetR) induces expression of LuxI and hence generates AHL. Activity of the receiver cells is monitored via a *gfp* gene controlled by the LuxR-sensitive promoter. (As in the original network, LuxR expression was constitutive.)

Exercise 7.4.2 Modify the model of quorum sensing cells in (7.24) to arrive at a model of Weiss's receiver cell population as in figure 7.26. ☐

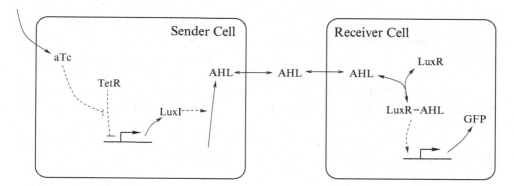

Figure 7.26
Engineered network for cell-to-cell communication. (Dashed arrows indicate regulation.) Addition of aTc to the sender cells induces expression of LuxI and hence leads to production of AHL. This chemical signal diffuses to the receiver cells, where it activates LuxR, leading to GFP expression. Adapted from figure 19 of Weiss et al. (2003).

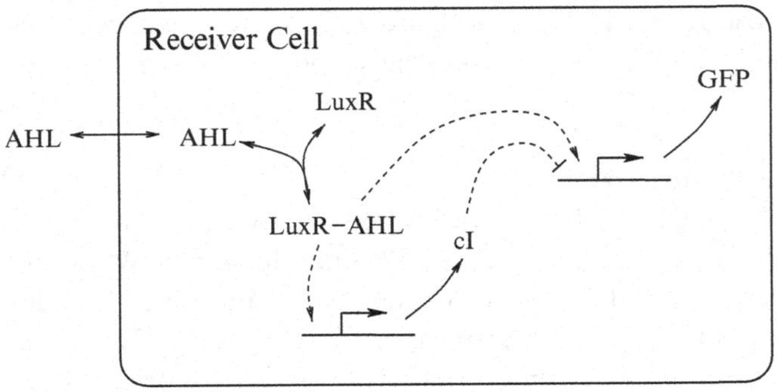

Figure 7.27
Engineered pulse generator. (Dashed arrows indicate regulation.) In this receiver cell, LuxR (constitutively expressed) binds AHL taken up from the environment. The LuxR–AHL complex activates expression of both cI and GFP. Production of GFP is repressed by cI, so only a transient pulse of fluorescence
is produced in response to an AHL signal. Adapted from figure 1 of Basu et al. (2004).

The Weiss group followed up on this design with elaborations of the receiver cell
network that result in spatio-temporal pattern formation. We next consider two of
their constructions.

Pulse Generation   As we saw in section 6.3.1, the bacterial chemotaxis signaling
network generates a transient response to a persistent stimulus. This behavior,
described as *pulse generation*, can be produced by a simple gene regulatory network
called a *feedforward loop*, in which a gene's activity is activated on a fast timescale
and inhibited on a slower timescale (see problem 7.8.16).

Subhayu Basu and colleagues adapted Weiss's cell-to-cell communication system
by engineering a new receiver population that responds to an AHL stimulus with a
transient pulse of GFP expression (Basu et al., 2004). The receiver cell network,
shown in figure 7.27, incorporates the *cI* gene from phage lambda into the original
receiver cell design. In this network, both *cI* and *gfp* are activated by the LuxR–AHL
complex, but GFP expression is also inhibited by cI. Exposure to AHL causes an
initial increase in GFP and cI levels, but once cI levels are sufficiently high, GFP
expression is repressed.

Basu and colleagues developed a model to explore the system's behavior. A simplified version of the model for receiver cell activity is

$$\frac{d}{dt}R*(t) = k_1 A^2 \left(R_T - 2R*(t)\right)^2 - k_2 R*(t)$$

$$\frac{d}{dt}C(t) = \frac{a_C \left(R*(t)/K_R\right)}{1 + \left(R*(t)/K_R\right)} - b_C C(t) \tag{7.25}$$

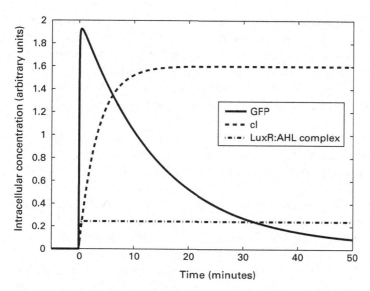

Figure 7.28
Pulse generation. At time zero, the AHL concentration rises from 0 to 10, causing a sudden increase in the LuxR–AHL abundance. This leads to a rapid rise in GFP and a slower rise in cI. As the cI concentration increases, repression of GFP leads to a drop in GFP abundance. Parameter values: $k_1 = 0.5$ (min^{-1} · concentration^{-3}), $k_2 = 0.02$ (min^{-1} · concentration^{-3}), $R_T = 0.5$ (concentration), $a_C = 0.5$ (concentration · min^{-1}) $K_R = 0.02$ (concentration), $b_C = 0.07$ (min^{-1}), $a_G = 80$ (concentration · min^{-1}), $K_C = 0.008$ (concentration), $b_G = 0.07$ (min^{-1}). Concentration units are arbitrary.

$$\frac{d}{dt}G(t) = \frac{a_G\left(R*(t)/K_R\right)}{1 + \left(R*(t)/K_R\right) + \left(C(t)/K_C\right)^2 + \left(R*(t)/K_R\right)\left(C(t)/K_C\right)^2} - b_G G(t),$$

where $R*$, C, and G are the concentrations of the AHL–LuxR complex, cI, and GFP, respectively. The AHL concentration A is taken as an input signal. Figure 7.28 shows the model behavior. AHL is introduced at time zero, after which both cI and GFP levels rise. Once cI levels are sufficiently high, the GFP abundance drops.

Basu and co-workers used their model as an aid to design. They selected destabilized versions of cI and GFP, a specific ribosome binding site for cI, and tuned the sensitivity of the GFP promoter to cI by introducing point mutations. Experiments confirmed the network's pulse-generating behavior.

Exercise 7.4.3 Verify that the GFP expression rate in equation (7.25) corresponds to the case that cI binds with strong cooperativity at two sites, LuxR–AHL binds at a single site, the binding events are independent, and expression occurs only when LuxR–AHL is bound and cI is not bound. □

Spatial Patterning In addition to temporal patterns like pulses, cell-to-cell communication can also produce steady spatial patterns. In another project, Basu and colleagues developed an AHL-receiver strain that acts as a *band detector* (Basu et al., 2005). These cells were engineered to respond only to a mid-range of inducer activity—no response is shown at low or high AHL concentrations. A population of these receiver cells surrounding a group of AHL-sender cells will then fluoresce in a bull's-eye pattern.

The gene network in the band detector cells is sketched in figure 7.29. The autoinducer AHL binds to LuxR, leading to expression of cI and the *lac* repressor, LacI. LacI is also expressed from a separate promoter that is repressed by cI. Finally, GFP expression is inhibited by LacI. The system's behavior can be understood in terms of the dose-response curves shown in figure 7.30. At high levels of AHL, the LuxR-induced levels of LacI are high, and GFP expression is repressed. At low AHL levels, cI is not expressed. Consequently, LacI is generated from the cI-repressible promoter and, again, GFP expression is repressed. At intermediate levels of AHL, moderate expression of cI and LacI occur. The system was tuned so that repression of LacI by cI is highly effective, whereas repression of GFP by LacI is not. Thus at these mid-range input levels, expression of LacI is sufficiently low that significant GFP expression occurs.

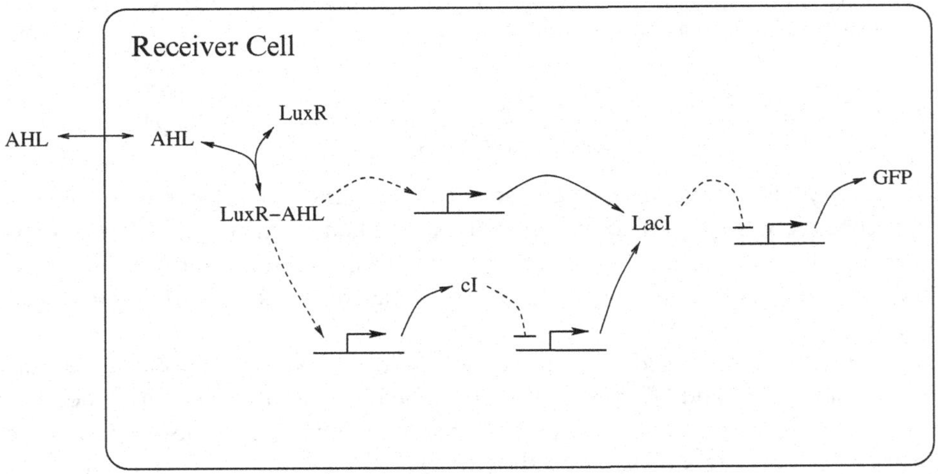

Figure 7.29
Engineered band detector. (Dashed arrows indicate regulation.) The AHL input binds LuxR, leading to expression of LacI and cI. LacI is also expressed from a separate cI-repressible promoter. GFP expression is repressed by LacI. At high and low levels of AHL, GFP expression is repressed. At intermediate AHL levels, a sufficiently low LacI level allows for GFP expression. Adapted from figure 1 of Basu et al. (2005).

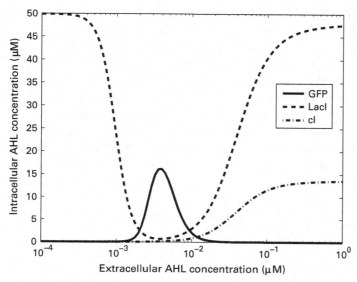

Figure 7.30
Dose-response curve for the band-detector circuit. The extracellular AHL concentration increases from left to right, so distance from the sender population increases from right to left. At high AHL levels, there is strong expression of LacI from the LuxR-induced promoter. At low AHL levels, there is strong expression of LacI from the cI-repressible promoter. In the intermediate range, LacI levels are sufficiently low to allow significant GFP expression. Parameter values: $k_1 = 0.5$ min^{-1} μM^{-3}, $k_2 = 0.02$ min^{-1} μM^{-3}, $R_T = 0.5$ μM, $a_{L1} = 1$ μM min^{-1}, $K_C = 0.008$ μM, $a_{L2} = 1$ μM min^{-1}, $K_R = 0.01$ μM, $b_L = 0.02$ min^{-1}, $a_C = 1$ μM min^{-1}, $b_C = 0.07$ min^{-1}, $a_G = 2$ μM min^{-1}, $b_G = 0.07$ min^{-1}, $K_L = 0.8$ μM. Adapted from figure 26 of Weiss et al. (2003).

The curves shown in figure 7.30 were generated from the following simple model of the band-detector network, which is a variant of the model developed by Basu and colleagues.

$$\frac{d}{dt}R^*(t) = k_1 A^2 \left(R_T - 2R^*(t)\right)^2 - k_2 R^*(t)$$

$$\frac{d}{dt}L(t) = \frac{a_{L1}}{1+(C(t)/K_C)^2} + \frac{a_{L2}R^*(t)}{K_R + R^*(t)} - b_L L(t)$$

$$\frac{d}{dt}C(t) = \frac{a_C R^*(t)}{K_R + R^*(t)} - b_C C(t) \tag{7.26}$$

$$\frac{d}{dt}G(t) = \frac{a_G}{1+(L(t)/K_L)^2} - b_G G(t).$$

The state variables $R^*, L, C,$ and G are the concentrations of the AHL–LuxR complex, LuxI, cI, and GFP, respectively. The AHL concentration A is taken as an input signal.

7.4.3 Synchronization of Oscillating Cells

Cell-to-cell communication allows intracellular oscillations to be synchronized across a population. This occurs in the circadian rhythm generators in animals. The engineered oscillatory networks discussed in section 7.3.3 can generate population-wide oscillations when cells are able to communicate their states to one another.

Synchronization behavior can be illustrated by a simple extension of the relaxation oscillator model of section 7.3.3, as follows. Consider a pair of identical cells each hosting a relaxation oscillator. Suppose further that the activation signal X can diffuse across the cell membranes and so can be shared between the two cells. Using a subscript $i = 1, 2$ to indicate concentrations in each cell, we can model the pair of networks as:

$$\frac{d}{dt} x_1(t) = \frac{1 + x_1(t)^2 + \alpha\sigma x_1(t)^4}{\left(1 + x_1(t)^2 + \sigma x_1(t)^4\right)\left(1 + y_1(t)^4\right)} - \gamma_x x_1(t) + D(x_2 - x_1)$$

$$\frac{d}{dt} y_1(t) = a_y \frac{1 + x_1(t)^2 + \alpha\sigma x_1(t)^4}{\left(1 + x_1(t)^2 + \sigma x_1(t)^4\right)\left(1 + y_1(t)^4\right)} - \gamma_y y_1(t),$$

$$\frac{d}{dt} x_2(t) = \frac{1 + x_2(t)^2 + \alpha\sigma x_2(t)^4}{\left(1 + x_2(t)^2 + \sigma x_2(t)^4\right)\left(1 + y_2(t)^4\right)} - \gamma_x x_2(t) + D(x_1 - x_2)$$

$$\frac{d}{dt} y_2(t) = a_y \frac{1 + x_2(t)^2 + \alpha\sigma x_2(t)^4}{\left(1 + x_2(t)^2 + \sigma x_2(t)^4\right)\left(1 + y_2(t)^4\right)} - \gamma_y y_2(t),$$

where D characterizes the rate at which activator X diffuses between the two cells. The behavior of this simple model is illustrated in figure 7.31, which shows the two cells beginning their oscillations out of phase, and then being drawn into synchrony by the shared activator.

The synchronization strategy used in this model cannot easily be applied to the relaxation oscillator design of Stricker and colleagues, as the activator (AraC) is a transcription factor protein (and so will not cross the cell membrane without a dedicated transporter). In 2010, Tal Danino and colleagues successfully demonstrated synchronization of intracellular relaxation oscillators: they used a design in which AHL acts as the intercellular signal (Danino et al., 2010). A synchronization scheme for the repressilator is addressed in problem 7.8.19.

7.5 Computation by Gene Regulatory Networks

The initial discovery of gene regulatory networks prompted an analogy to the human-made technology of electrical circuits and thus led to the term "genetic circuit." This analogy can be made explicit by treating promoter–transcription factor

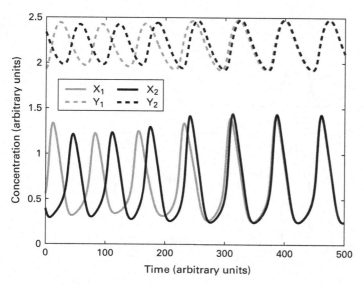

Figure 7.31
Synchronized relaxation oscillators. The cells each contain a relaxation oscillator. Initially, they are oscillating out of phase. The cells communicate their phase to one another through the shared activator: over time, this brings the cells into synchrony. Parameter values: $\alpha = 11$, $\sigma = 2$, $\gamma_x = 0.2$ (time^{-1}), $\gamma_y = 0.012$ (time^{-1}), $a_y = 0.2$ (concentration $\cdot$ time^{-1}), $D = 0.015$ (time^{-1}).

interactions (the building blocks of gene networks) as *logic gates* (the building blocks of computational electrical circuits). These ideas are reviewed in Weiss et al. (2003).

7.5.1   Promoters as Logic Gates

In digital electronics, a signal (e.g., voltage), is either considered HIGH (present) or LOW (absent), depending on whether a threshold has been passed. The same discretization process can be applied to the continuously varying concentrations of transcription factors in a gene network, as illustrated in figure 7.32. This abstraction results in a binary description of gene activity: at a give time point, each gene is either ON (expressing above threshold) or OFF (expressing below threshold). Dynamic models that describe two-state (ON/OFF) behaviors are called *Boolean* models. Boolean models are often used to describe gene regulatory networks and are particularly useful for addressing large networks. (See De Jong et al. (2002) for a review.)

Using the Boolean framework, all signals take either the value 1 (HIGH) or 0 (LOW). Applying this notion to the concentration of a transcription factor provides, as an example, an explicit comparison between repression of expression and a digital

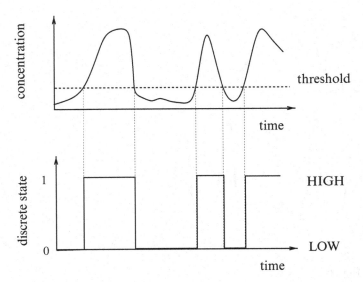

Figure 7.32
Discretization of gene activity. When the concentration of a gene's protein product is above a threshold, the gene is considered to be ON, and the protein concentration is assigned a value of 1 (HIGH). When the concentration drops below threshold, the gene is OFF, and the protein concentration is assigned a value of 0 (LOW).

inverter (figure 7.33). An inverter is a device that inverts a Boolean signal—a HIGH input yields a LOW output; a LOW input yields a HIGH output. Figure 7.33 also includes the *truth table* for the inverter, which summarizes its input–output behavior. The inverter is an example of a *logic gate*—a device that responds to a set of Boolean input variables (each equal to 0 or 1) and returns a Boolean output. The inverter is referred to as a NOT gate.

Promoters that are regulated by multiple transcription factors can be represented by multi-input logic gates. Figure 7.34 illustrates two promoters that are each regulated by two distinct activators. In panel A, the binding of *either* activator is sufficient to drive expression, so the promoter acts as an OR gate. Panel B shows the case in which binding of *both* activators is necessary to drive expression: this implements AND gate logic. Promoters that are regulated by two distinct repressors can be classified in a similar way: if either repressor suffices to inhibit repression, the promoter acts as a NOR (i.e., NOT-OR) gate, whereas if repression only occurs if both repressors are bound, then a NAND (i.e., NOT-AND) logic applies.

Exercise 7.5.1   Construct truth tables for the NOR and NAND logic gates. Verify that they can be constructed by connecting NOT gates downstream of OR and AND gates, respectively.   □

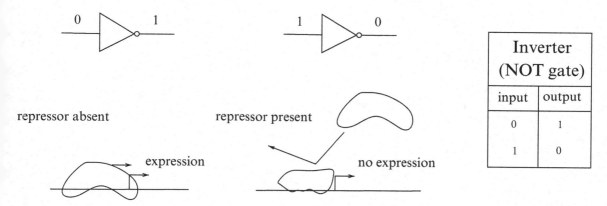

Figure 7.33
Repression of gene expression as an inverter. When the repressor input is absent (LOW; value 0), expression proceeds (the output is HIGH; value 1). Alternatively, when repressor is present (HIGH; value 1), expression does not occur (LOW; value 0). The promoter thus "inverts" the input signal (repressor concentration) to determine the output signal (expression rate, and thus concentration of protein product). This same behavior is exhibited by a digital inverter, or NOT gate, characterized by the truth table shown on the right. Adapted from figure 1 of Weiss et al. (2003).

Exercise 7.5.2   For promoters that are regulated by both an activator and a repressor, different cases arise depending on the priority of the inputs. The corresponding digital elements, called IMPLIES gates, can be built by combining an inverter and one of the two-input gates already considered (e.g., AND or OR). Referring to section 7.2.1, determine the truth table that corresponds to the *lac* operon, where the two inputs are the *lac* repressor and allolactose. Verify that the resulting IMPLIES logic can be constructed by combining a NOT gate and an OR gate.   □

7.5.2   Digital Representations of Gene Circuits

In digital electronics, elementary logic gates provide a foundation for the construction of complex computational devices. The same notions can be applied to the design of gene regulatory networks.

Consider, as an example, the repressilator circuit described in section 7.3.3. Because each promoter is repressed by the gene that precedes it in the loop, this device can be described as a set of three NOT gates strung together in a loop, as shown in figure 7.35. This type of network is known in digital electronics as a ring oscillator and is commonly used to generate periodic behavior.

The dynamic behavior of a ring oscillator can be simulated by supposing that all three of the elements update simultaneously at discrete time-steps. Although this rigid lock-step does not reflect the smooth variation of genetic processes, digital analogies often provide a useful abstraction of gene network behavior.

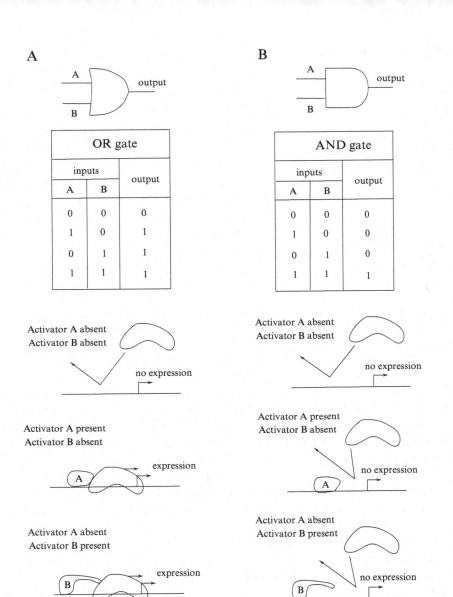

**Figure 7.34**
Dual-activator promoters as digital logic gates (A) If either transcription factor suffices to activate expression, the promoter exhibits an OR gate logic. (B) If expression only occurs when both activators are present, the promoter is represented by an AND gate.

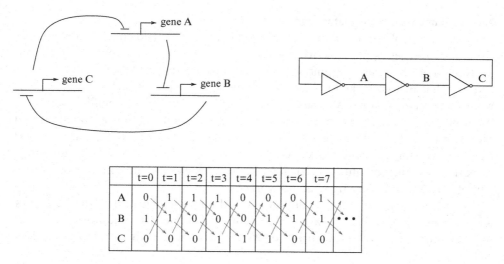

Figure 7.35

Digital representation of the repressilator circuit (see section 7.3.3). A loop of three NOT gates forms a ring oscillator. In the simulation shown, the signals are simultaneously updated at each time-step. At time $t = 0$, the three signals have values A = 0, B = 1, C = 0. At the next time-step ($t = 1$), the value of A is replaced with the new value NOT(C) = 1, while B is updated as NOT(A) = 1, and C takes the value NOT(B) = 0. (Because the updates occur simultaneously, the values of the signals at the previous time-step are used in the update computation.) The behavior is periodic: the state at the sixth time-step is identical to the initial state ($t = 0$). Adapted from figure 9 of Weiss et al. (2003).

The analogy between promoter activity and digital logic gates provides a useful framework for the design and analysis of gene regulatory networks. However, logic gates cannot provide a comprehensive description of gene circuit behavior. In addition to the abstraction introduced by discretization, the Boolean framework is not well suited to represent genes that exhibit multiple expression rates. This fact was vividly demonstrated in an experiment by Yaki Setty and colleagues, who mapped the response of the *lac* promoter to two inducers and found the resulting response to be a hybrid of OR and AND behaviors (Setty et al., 2003).

Another crucial distinction between electrical circuits and gene circuits is the manner in which the specificity of the interconnections is achieved. In an electrical circuit, all connections use the same signal (flow of charge). Undesired connections (short-circuits) are avoided by maintaining spatial separation between signals. In contrast, the signal carriers for gene circuits (transcription factors) are mixed together in a single compartment. Unwanted interconnections are avoided through chemical specificity (of the protein–DNA binding surfaces). This reliance on chemical specificity allows complex networks to operate on tiny spatial scales, but it means that each time a connection is added to an existing network, a chemically distinct promoter–regulator interaction must be introduced.

7.5.3   Complex Gene Regulatory Networks

The synthetic biology community is engaged in the design and construction of gene circuits of increasing complexity. Examples include a tunable version of a band detector, a gene cascade designed to display ultrasensitive responses, and a cellular counter that is able to keep track of a sequence of input events. (These projects were surveyed in a paper by Khalil and Collins (2010), which also highlights applications of synthetic circuits to biosensing, bioremediation, biofuel production, and biomanufacturing and delivery of drugs.) A broad range of computational and signal-processing gene networks has been proposed by the iGEM community.[1] Nevertheless, current attempts at gene circuit design pale in comparison to the complexity found in natural gene networks.

Natural gene circuits can be roughly divided into two classes: sensory networks, which mount a response to a cell's current environmental conditions; and cell-fate decision (or developmental) networks, which cause cells to adopt persistent states.

Sensory networks enhance a cell's survival by tailoring its behavior to suit the prevailing conditions. The *lac* operon of *E. coli* is a canonical example. Because they demand a timely response, sensory networks tend to be rather "shallow"—they do not usually involve long cascades of gene regulation between input and output. (Such cascades would introduce significant gene expression lags.)

Cell-fate decision networks do not normally act under tight time-constraints: they often involve long cascades of interacting genes and complex feedback loops, particularly positive feedback loops that "lock in" decisions. There are many known examples of bacterial cell-fate decision networks, such as the lysis–lysogeny switch in phage lambda (section 7.2.2) and the sporulation decision network in *B. subtilis*. These bacterial networks are typically simpler than the gene networks responsible for the development of multicellular organisms, which can involve dozens of individual genes, each typically regulated by several distinct transcription factors (reviewed in Stathopoulos and Levine (2005) and Davidson (2006)).

We conclude this section by introducing two well-studied examples of complex developmental gene networks.

The Segmentation Gene Network in the Fruit Fly *Drosophila melanogaster*   During their growth, *Drosophila* embryos develop a segmented body-plan. This spatial patterning is derived from maternal genes whose mRNA transcripts are placed in different regions of the egg. The gene regulatory network responsible for the segmentation process is sketched in figure 7.36. The temporal progression of activity in the network

---

1. International Genetically Engineered Machine (iGEM) competition: www.igem.org.

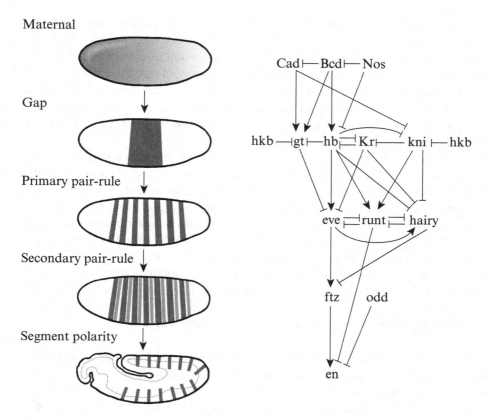

Figure 7.36
Segmentation gene network in *Drosophila*. Stages in embryonic development are illustrated. The shaded regions indicate spatial expression patterns for the indicated gene classes. Selected members of gene classes are depicted in the regulatory network that guides their behavior. Reproduced, with permission, from Carroll et al. (2005), figure 3.5.

corresponds to the steps in the segmentation process as shown. Because segmentation is a spatio-temporal process, ordinary differential equation (ODE)-based models are not directly applicable. An ODE model can be used if one supposes a compartmental structure across the embryo, but a more natural modeling approach is to make use of partial differential equations, as introduced in section 8.4. (An ODE model appears in Jaeger et al. (2004); a spatial model is presented in Perkins et al. (2006).)

The Endomesoderm Specification Network in *Strongylocentrotus purpuratus* Eric Davidson's lab has worked for many years on mapping the gene regulatory network that drives differentiation of cells in the early embryo of the sea urchin *Strongylocentrotus purpuratus* (figure 7.37). The behavior of this network begins with maternally specified asymmetries in the egg and leads to development of the endoderm (inside layer), skeletal, and mesoderm (middle layer) components of the embryo. A full kinetic characterization of the interactions in a network of this size is daunting, and so models are typically constructed using simpler methods, such as Boolean frameworks. (Appropriate modeling frameworks are reviewed in Bolouri and Davidson (2002).)

The study of complex gene regulatory networks has revealed an important insight into their structure: they often exhibit a *modular* architecture, meaning that the network is composed of subnetworks that play their role somewhat independently of one another. Modularity is a key aspect of human-engineered systems: it allows individual components to be designed, constructed, and tested independently of the entire system. Moreover, modularity allows the re-use of components in multiple systems—a feature that is likely of use in evolutionary "design." (Modularity is reviewed in Wagner et al. (2007); the challenges and opportunities that modular design presents to synthetic biology are discussed in Purnick and Weiss (2009).)

7.6* Stochastic Modeling of Biochemical and Genetic Networks

Chemical reactions result from collisions of individual molecules. Most molecular collisions do not cause reactions. On a molecular scale, reactions are thus rare events and are difficult to predict. In many cellular processes, this molecular randomness is averaged out over large numbers of reaction events, resulting in predictable system behavior. In contrast, processes that depend on small numbers of molecules can be strongly affected by the randomness of biochemical events. Gene expression often involves molecular species that are present in low numbers, and so gene regulatory networks can be subject to this random variation. (The effects of noise on developmental gene networks reveals itself in differences between genetically identical organisms, from bacteria to humans. Stochasticity in gene expression is discussed in Raj and van Oudenaarden (2008).)

Random variation is often considered an inconvenience that must be overcome: the fact that this randomness is usually referred to as "noise" suggests it is a nuisance. However, in some biological contexts, random behavior can be exploited for improved performance. An example is provided by the phenomenon of bacterial *persistence*, in which a genetically identical population gives rise to a small number of so-called persistent cells that exhibit antibiotic resistance at the cost of a reduced growth rate. In the absence of antibiotics, slow-growing persistent cells are quickly out-competed, but the presence of a handful of these cells ensures the population's survival when antibiotics are encountered.

Endomesoderm Specification to 30 Hours

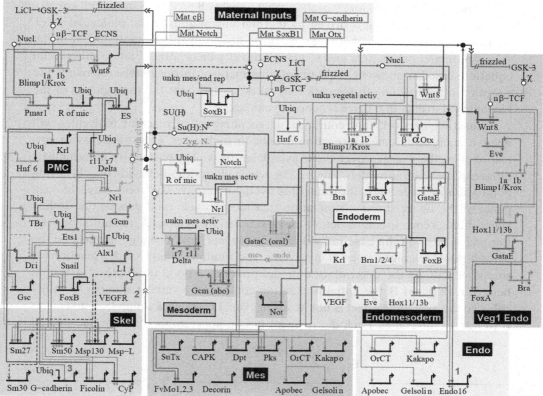

Figure 7.37
Endomesoderm specification network in the sea urchin *Strongylocentrotus purpuratus*. The genes are organized into boxes based on their function. Maternal inputs appear at the top; differentiation proteins are encoded by genes in the bottom boxes. The network describes events that occur in the 30 hours after fertilization. Reproduced, with permission, from Davidson (2006), figure 4.2.

At the cellular level, randomness can be partitioned into two categories: *extrinsic noise*, which refers to random variations that impact all processes in the cell equally, and *intrinsic noise*, which is driven by thermal fluctuations at the molecular level. In models of intracellular networks, extrinsic noise appears as randomness in the values of model parameters and so can be directly incorporated into a differential equation–based framework. In contrast, treatment of intrinsic noise demands the adoption of a modeling framework that takes into account the randomness of the biochemical events that drive reaction dynamics.

A reaction network that comprises large numbers of reactant molecules will involve many simultaneous reaction events. In such cases, network behavior corresponds to the average over these events and is well described by deterministic differential equation models. Figure 7.38, which shows the behavior of a decaying population of molecules, illustrates this averaging effect. The solid curve in each panel of the figure shows a simulation that incorporates randomness; the dashed curve shows a corresponding deterministic simulation. In panel A, the initial population size is large. In this case, individual decay events have a negligible effect on the overall pool. Averaged over many events, the random timing of the reactions is smoothed out, so the deterministic model provides a good description of system behavior. In panel B, the initial population consists of a smaller number of molecules, so the averaging effect is not as strong. Panel C shows a simulation that starts with just 10 molecules. Each decay event has an appreciable effect on the overall

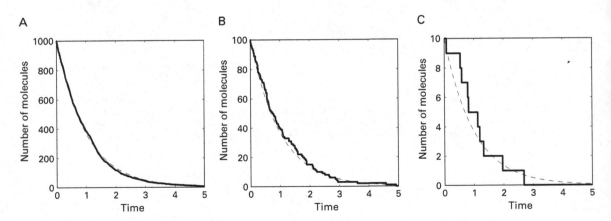

Figure 7.38
Simulations of constitutive decay. The solid curves show simulations that incorporate randomness (stochastic simulations). The dashed curves show the corresponding deterministic (differential equation–based) simulations. The initial pool sizes are 1000 (A), 100 (B), and 10 (C) molecules. For large pool size (A), the simulations agree. However, as the molecule count decreases (B, then C), random effects become more pronounced and are not well described by the deterministic model.

abundance. In this case, the system's discrete, random behavior is not well described by the deterministic simulation.

In this section, we introduce a *stochastic* modeling framework that is suitable for describing systems that involve small numbers of molecules. The term *stochastic* means "random": it is used to describe dynamic processes that have some element of randomness in their progression. (Appendix B contains a brief introduction to some basic concepts from probability.)

7.6.1   A Discrete Modeling Framework

In developing a stochastic modeling framework for chemical reaction networks, we will continue to assume spatial homogeneity and a fixed volume. The abundance of each chemical species will be described by the *number* of molecules in the reaction volume. The state of the system is then the vector $\mathbf{N}$ of molecule counts. (In contrast, the state of a differential equation model is the vector s of species *concentrations*, which change smoothly over time.) As the stochastic dynamics proceed, the molecule counts will change their values in discrete jumps.

We will characterize each reaction in the network by a *stoichiometry vector s* and a *propensity function a*. For each reaction, the stoichiometry vector indicates the identity and number of reactants and products: the *j*-th component of this vector is the net number of molecules of species *j* produced or consumed in the reaction. The propensity is a description of reaction rate.

To illustrate these ideas, consider the network composed of the two reactions

$$R_1: \quad A + B \xrightarrow{k_1} C \qquad\qquad R_2: \quad C \xrightarrow{k_2} .$$

The state of this system describes the numbers of molecules of species A, B, and C present at any given time. The stoichiometry vectors are

$$\mathbf{s}_1 = \begin{bmatrix} -1 \\ -1 \\ 1 \end{bmatrix} \begin{matrix} \leftarrow A \\ \leftarrow B \\ \leftarrow C \end{matrix} \qquad \text{and} \qquad \mathbf{s}_2 = \begin{bmatrix} 0 \\ 0 \\ -1 \end{bmatrix} \begin{matrix} \leftarrow A \\ \leftarrow B. \\ \leftarrow C \end{matrix}$$

When a reaction occurs, the state vector $\mathbf{N}$ is updated by addition of the corresponding stoichiometry vector. For example, suppose that at a given time the state is $\mathbf{N} = (N_A, N_B, N_C) = (12, 3, 4)$. If reaction R_1 were to occur, we would update the state by replacing $\mathbf{N}$ with $\mathbf{N} + \mathbf{s}_1 = (11, 2, 5)$.

The reaction propensities are functions of reactant abundance. We will assume that the probability of a reaction event is proportional to the product of the abundance of each reactant species (as in mass action). The propensities for this example are then

$$a_1(\mathbf{N}) = k_1 N_A N_B \qquad\qquad a_2(\mathbf{N}) = k_2 N_C.$$

Reaction propensities take the same form as mass-action rate laws, but differences appears when multiple copies of an individual reactant are involved.[2]

7.6.2 The Chemical Master Equation

We will build a stochastic modeling framework on the assumption that there are small time increments dt for which:

- At most one reaction event can occur during any time interval of length dt.

- The probability that reaction R_k occurs in any time interval $[t, t + dt]$ is the product of the reaction propensity at time t and the length of the interval: $a_k(\mathbf{N}(t))dt$.

Under these assumptions, the probability that no reactions occur during a time interval $[t, t + dt]$ is $1 - \sum_k a_k(\mathbf{N}(t))dt$, where the sum is taken over all reactions in the system.

Let $P(\mathbf{N}, t)$ denote the probability that the system is in state $\mathbf{N}$ at time t. This is called the *probability distribution* of the state (and is dependent on the initial condition—it is a conditional probability distribution). If the distribution $P(\mathbf{N}, t)$ is known at time t, we can use the assumptions above to describe the distribution at time $t + dt$:

$$P(\mathbf{N}, t + dt) = P(\mathbf{N}, t) \cdot \underbrace{\left(1 - \sum_k a_k(\mathbf{N})dt\right)}_{\text{Probability of no reactions firing}} + \sum_k \underbrace{P(\mathbf{N} - \mathbf{s}_k, t)a_k(\mathbf{N} - \mathbf{s}_k)dt.}_{\text{Probability of reaction } R_k \text{ occurring while in state } \mathbf{N} - \mathbf{s}_k}$$

$$(7.27)$$

This equation is called a *probability balance*. The first term is the probability of being in state $\mathbf{N}$ at time t and remaining in that state until time $t + dt$ (because no reaction events occur). The second term is the sum of the probabilities of transitioning into state $\mathbf{N}$ from another state (because reaction R_k causes a transition from $\mathbf{N} - \mathbf{s}_k$ to $(\mathbf{N} - \mathbf{s}_k) + \mathbf{s}_k = \mathbf{N}$).

As an example, consider the simple reaction chain in which species A is produced at zero order and degrades at first order:

$$R_1 : \xrightarrow{k_1} A \qquad\qquad R_2 : A \xrightarrow{k_2} .$$

The state of the system is the number of molecules of A (i.e., $N = N_A$). The reaction stoichiometries are $s_1 = [1]$, $s_2 = [-1]$. The reaction propensities are $a_1 = k_1$ and $a_2 = k_2 N_A$. The transitions between states follow the scheme in figure 7.39.

---

2. For instance, the propensity of the bimolecular reaction $A + A \xrightarrow{k} C$ is $k N_A(N_A - 1)/2$. This formula reflects the number of unique pairings of two A molecules.

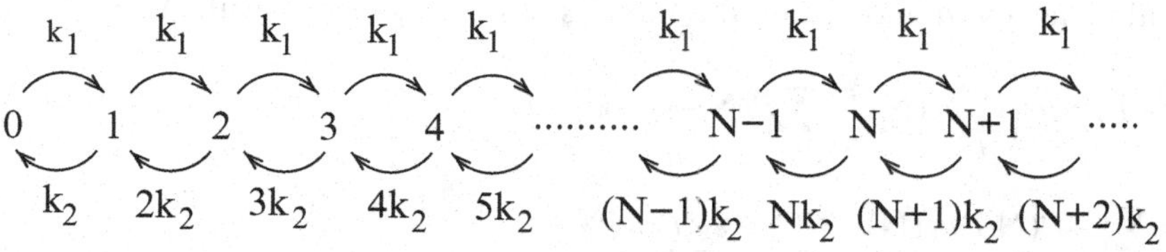

Figure 7.39
Transitions among states for the simple reaction chain $\xrightarrow{k_1} A \xrightarrow{k_2}$. The reaction propensities are indicated.

In this case, the probability balance reads:

$$P(0, t + dt) = P(0, t)(1 - k_1 dt) + P(1, t) \cdot k_2 dt$$

$$P(1, t + dt) = P(1, t)(1 - (k_1 + k_2)dt) + P(0, t) \cdot k_1 dt + P(2, t) \cdot 2k_2 dt$$

$$P(2, t + dt) = P(2, t)(1 - (k_1 + 2k_2)dt) + P(1, t) \cdot k_1 dt + P(3, t) \cdot 3k_2 dt$$

$$P(3, t + dt) = P(3, t)(1 - (k_1 + 3k_2)dt) + P(2, t) \cdot k_1 dt + P(4, t) \cdot 4k_2 dt$$

$$\vdots$$

$$P(N, t + dt) = P(N, t)(1 - (k_1 + Nk_2)dt) + P(N - 1, t) \cdot k_1 dt + P(N + 1, t) \cdot (N + 1)k_2 dt$$

$$\vdots$$

Exercise 7.6.1 Verify that the probability balance for the scheme:

$$R_1 : \quad \xrightarrow{k_1} A$$

$$R_2 : \quad \xrightarrow{k_2} B$$

$$R_3 : \quad A + B \xrightarrow{k_3}$$

is

$$P((N_A, N_B), t + dt) = P((N_A, N_B), t)(1 - (k_1 + k_2 + N_A N_B k_3)dt)$$

$$+ P((N_A - 1, N_B), t) \cdot k_1 dt + P((N_A, N_B - 1), t) \cdot k_2 dt$$

$$+ P((N_A + 1, N_B + 1), t) \cdot (N_A + 1)(N_B + 1)k_3 dt. \qquad \square$$

The probability balance (7.27) can be used to derive a differential equation describing the rate of change of the probability distribution, as follows. Subtracting $P(\mathbf{N}, t)$ from each side of equation (7.27) gives

$$P(\mathbf{N}, t + dt) - P(\mathbf{N}, t) = -P(\mathbf{N}, t) \left(\sum_k a_k(\mathbf{N}) dt \right) + \sum_k P(\mathbf{N} - \mathbf{s}_k, t) a_k(\mathbf{N} - \mathbf{s}_k) dt.$$

Dividing both sides by dt and taking the limit as dt tends to zero results in

$$\frac{d}{dt}P(\mathbf{N},t) = -P(\mathbf{N},t)\left(\sum_k a_k(\mathbf{N})\right) + \sum_k P(\mathbf{N}-\mathbf{s}_k,t)a_k(\mathbf{N}-\mathbf{s}_k)$$

$$= \sum_k \left(\underbrace{-P(\mathbf{N},t)a_k(\mathbf{N})}_{\text{Flow out of state } \mathbf{N}} + \underbrace{P(\mathbf{N}-\mathbf{s}_k,t)a_k(\mathbf{N}-\mathbf{s}_k)}_{\text{Flow into state } \mathbf{N}} \right)$$

This is called the *chemical master equation*. It is a system of differential equations describing the time-varying behavior of the probability distribution. The terms on the right-hand side account for probability flow out of, and into, the state $\mathbf{N}$ at time t. The master equation includes a differential equation for every state $\mathbf{N}$ that the system can adopt, and so typically involves an infinite number of equations.

For the simple reaction chain

$$\overset{k_1}{\rightarrow} A \overset{k_2}{\rightarrow}$$

described earlier, the master equation is

$$\frac{d}{dt}P(0,t) = -P(0,t)k_1 + P(1,t)k_2$$

$$\frac{d}{dt}P(1,t) = -P(1,t)(k_1+k_2) + P(0,t)k_1 + P(2,t)2k_2$$

$$\frac{d}{dt}P(2,t) = -P(2,t)(k_1+2k_2) + P(1,t)k_1 + P(3,t)3k_2 \qquad (7.28)$$

$$\vdots$$

$$\frac{d}{dt}P(N,t) = -P(N,t)(k_1+Nk_2) + P(N-1,t)k_1 + P(N+1,t)(N+1)k_2$$

$$\vdots$$

Exercise 7.6.2 Determine the chemical master equation for the system in exercise 7.6.1.   □

To illustrate the behavior of solutions of the master equation, we consider the closed reaction network:

$$R_1: A \xrightarrow{k_1} B \qquad\qquad R_2: B \xrightarrow{k_2} A.$$

To keep the analysis simple, we suppose that there are only two molecules present in the system. The system state $\mathbf{N} = (N_A, N_B)$ can then take only three possible values: $(2,0), (1,1)$, or $(0,2)$. The master equation is a system of three differential equations:

$$\frac{d}{dt} P((2,0),t) = -P((2,0),t)2k_1 + P((1,1),t)k_2$$

$$\frac{d}{dt} P((1,1),t) = -P((1,1),t)k_2 - P((1,1),t)k_1 + P((2,0),t)2k_1 + P((0,2),t)2k_2 \qquad (7.29)$$

$$\frac{d}{dt} P((0,2),t) = -P((0,2),t)2k_2 + P((1,1),t)k_1.$$

Note that the right-hand-sides sum to zero, as dictated by conservation of probability.

A simulation of system (7.29) is illustrated in figure 7.40, which shows plots of the probability distribution (histograms) at three time points.

For this network, the steady-state distribution $P^{ss}(N_A, N_B)$ can be found by setting the time rates of change to zero:

$$0 = -P^{ss}(2,0)2k_1 + P^{ss}(1,1)k_2$$

$$0 = -P^{ss}(1,1)k_2 - P^{ss}(1,1)k_1 + P^{ss}(2,0)2k_1 + P^{ss}(0,2)2k_2$$

$$0 = -P^{ss}(0,2)2k_2 + P^{ss}(1,1)k_1.$$

Solving these equations, along with the condition that probability is conserved ($P^{ss}(2,0) + P^{ss}(1,1) + P^{ss}(0,2) = 1$) yields the steady-state probability distribution:

$$P^{ss}(2,0) = \frac{k_2^2}{(k_1 + k_2)^2}, \qquad P^{ss}(1,1) = \frac{2k_1 k_2}{(k_1 + k_2)^2}, \qquad P^{ss}(0,2) = \frac{k_1^2}{(k_1 + k_2)^2}. \qquad (7.30)$$

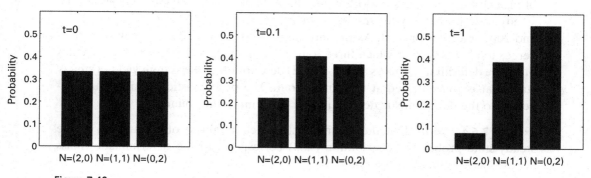

Figure 7.40
Evolution of probabilities for the closed reaction network (7.29). Probability distributions for $\mathbf{N} = (N_A, N_B)$ at times $t = 0$, $t = 0.1$, and $t = 1$ are shown. A uniform initial distribution is chosen, so that at time $t = 0$, all states are equally likely: $P((2,0),0) = P((1,1),0) = P((0,2),0) = 1/3$. Parameter values (in time^{-1}): $k_1 = 3$, $k_2 = 1$. Time units are arbitrary.

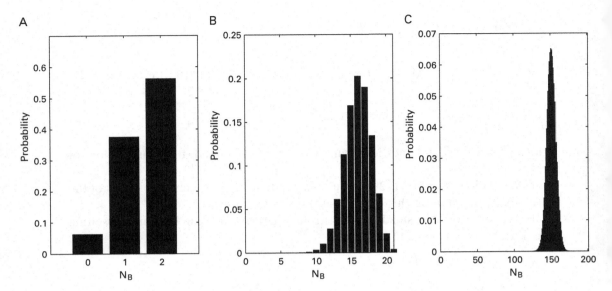

Figure 7.41
Steady-state probability distribution for N_B in reaction network (7.29) for a total molecule count of 2 (A), 20 (B), and 200 (C). As the molecule count grows, the distribution tends to a single peak for which $N_B = 3N_A$, which corresponds to the deterministic (mass-action) steady state.

Exercise 7.6.3 Verify equations (7.30). Does the simulation in figure 7.40 appear to have reached steady state by time $t = 1$? ☐

As the number of molecules in the system increases, the steady-state distribution of probabilities becomes smoother and more tightly peaked. Figure 7.41 shows the steady-state probability distributions for N_B in system (7.29) when there are 2, 20, and 200 molecules present. As the molecule count increases, the distribution converges to a tight peak at which three fourths of the total pool consists of molecules of B. The deterministic (mass action–based) description of the system yields a steady state that is concentrated at this single point. The probabilistic solution thus converges to the deterministic description for large molecule counts.

Exercise 7.6.4 Verify that the means (i.e., expected values) of N_A and N_B in the steady-state probability distribution (7.30) correspond to the deterministic (mass action–based) model of system (7.40). ☐

For most systems, the chemical master equation is intractable. (Simulations typically need to incorporate an infinite number of equations!) Consequently, a number of methods have been developed to provide alternative descriptions of stochastic behavior (reviewed in Khammash (2010)). These include the *linear noise approxi-*

mation, which generates differential equations whose solutions approximate the mean and variance of system behavior; *moment closure methods*, which allow calculation of approximate statistics for the probability distribution; and the *finite state projection*, which approximates the chemical master equation by a finite system of differential equations.

Rather than address these analytic approaches, we next consider a numerical method for generation of simulations of stochastic systems.

7.6.3 Gillespie's Stochastic Simulation Algorithm

Numerical algorithms that incorporate stochastic effects (by calling on random number generators) are called *Monte Carlo* methods. In 1977, Dan Gillespie published a Monte Carlo method for simulation of individual trajectories of chemical reaction networks characterized by the chemical master equation (reviewed in Gillespie (2007)). These trajectories, called *sample paths*, represent single elements drawn from a probability distribution generated by the system. Statistics of the trajectory distribution can be determined by generating a large collection of these sample paths (called an *ensemble*).

Gillespie's method, which he called the stochastic simulation algorithm (SSA), tracks each individual reaction event. The simulation does not proceed over a fixed time-grid but jumps forward in time from one reaction event to the next. After each reaction, the algorithm determines which reaction will occur next and how much time will elapse before it occurs.

The simulation algorithm depends on the properties of two *random variables*: the time T to the firing of the next reaction, and the reaction R that will occur next. We next consider how these two random variables are determined.

Determining the Next Reaction   The probability that a particular reaction will occur is proportional to the propensity of the reaction. Consider a network that involves three reactions, R_1, R_2, and R_3, with propensities a_1, a_2, and a_3. Let $P(R = R_i)$ denote the probability that R_i will be the next reaction to occur. Probability $P(R = R_i)$ is proportional to the propensity a_i of reaction R_i. Together, these probabilities sum to one. The probability distribution is

$$P(R = R_1) = \frac{a_1}{a_1 + a_2 + a_3}$$

$$P(R = R_2) = \frac{a_2}{a_1 + a_2 + a_3} \tag{7.31}$$

$$P(R = R_3) = \frac{a_3}{a_1 + a_2 + a_3}.$$

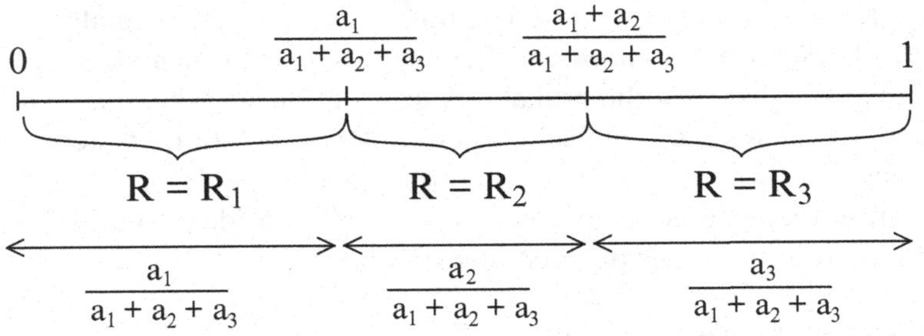

Figure 7.42
Selection of the next reaction. For a network with three reactions, the interval from zero to one is divided into three subintervals, whose lengths correspond to the probabilities of the reactions. A number sampled from the uniform zero-to-one distribution corresponds to a selection of the next reaction.

To implement a simulation of this network's behavior, we need to sample from this probability distribution. Most numerical software packages have built-in functions that generate random numbers drawn uniformly between zero and one. Samples from this uniform distribution can be converted to samples from the distribution (7.31), as follows. We divide the zero-to-one interval into three subintervals — one for each reaction — as in figure 7.42. The length of each subinterval is equal to the probability of the corresponding reaction. A number u that is drawn from the uniform distribution falls into one of these subintervals and thus corresponds to a particular reaction. This procedure can be formalized as follows:

If $\qquad 0 \le u \le \dfrac{a_1}{a_1 + a_2 + a_3}$, then we set $R = R_1$.

If $\qquad \dfrac{a_1}{a_1 + a_2 + a_3} < u \le \dfrac{a_1 + a_2}{a_1 + a_2 + a_3}$, then we set $R = R_2$. $\qquad\qquad$ (7.32)

If $\qquad \dfrac{a_1 + a_2}{a_1 + a_2 + a_3} < u \le \dfrac{a_1 + a_2 + a_3}{a_1 + a_2 + a_3} = 1$, then we set $R = R_3$.

Figure 7.43 provides a visualization of this process. Here, the uniform random number u is assigned to the vertical axis. The height of the staircase graph corresponds to the cumulative probabilities as used in algorithm (7.32). The next reaction is determined by selecting a number u from the uniform zero-to-one distribution and then extending a horizontal line to the staircase graph, as shown. This graph is called the *cumulative distribution function* for the random variable R.

Determining the Time to the Next Reaction The time T that elapses between reactions is also a random variable. Unlike R, it does not have a discrete value-set, but can

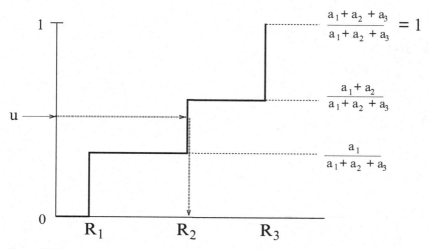

Figure 7.43
Cumulative distribution function for the random variable R. The height of the staircase graph corresponds to the cumulative probability as in algorithm (7.32). A reaction is chosen by selecting a number u from the uniform distribution on the vertical axis and then extending a horizontal line to the staircase graph.

take any nonnegative value. Because it can take infinitely many values, the probability of T having any particular value is vanishingly small. Thus, rather than frame our discussion in terms of point-wise probabilities, we will instead sample T directly from the cumulative distribution function, as we did for the random variable R in figure 7.43. The cumulative distribution function for T is given by

$$P(0 \leq T \leq t) = 1 - e^{-at}, \tag{7.33}$$

where a is the sum of the reaction propensities:

$a = a_1 + a_2 + a_3.$

Equation (7.33) characterizes T as an *exponential random variable*.

The cumulative distribution function for T is shown in figure 7.44. Most often, samples u from the uniform zero-to-one distribution will correspond to short wait-times between reactions: only occasionally (when u is chosen near 1) will a long time be selected. The steepness of the curve depends on a, the sum of the propensities. If this sum is large (many highly probable reactions), then the curve rises steeply, and waiting times are almost always short. If the sum is smaller, then the curve rises more slowly, and longer waiting times are more likely.

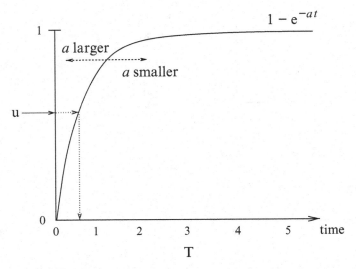

Figure 7.44
Cumulative distribution function for the waiting time T. The parameter a is the sum of the propensities. Waiting times T are determined by selecting numbers u from the uniform zero-to-one distribution and then extending a horizontal line to the graph, as shown. For large values of a, the curve rises sharply—most samples u correspond to short waiting times. For smaller a values, larger waiting times are more likely.

Gilliespie's algorithm can be summarized as follows:

Stochastic Simulation Algorithm (SSA)

1. Set the initial state $\mathbf{N}$. Initialize time t to zero.

2. Calculate the reaction propensities $a_k(\mathbf{N})$.

3. Draw a sample R_k from the random variable R (figure 7.43).

4. Draw a sample τ from the random variable T (figure 7.44).

5. Increment the simulation time $t \rightarrow t + \tau$ to account for the elapsed time.

6. Update the state vector $\mathbf{N} \rightarrow \mathbf{N} + \mathbf{s}_k$ to reflect the fact that reaction R_k has occurred.

7. Return to step 2.

The algorithm is usually continued until the simulation time t reaches the end of a specified time interval.

7.6.4   Examples
We conclude by using Gillespie's SSA to explore the behavior of some simple reaction networks.

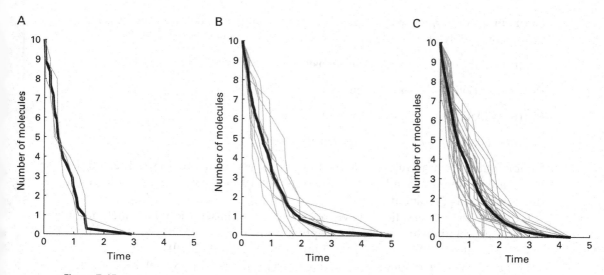

Figure 7.45
Ensembles of sample paths for the decay reaction. Each sample path begins with 10 molecules. Ensembles of 3 (A), 10 (B), and 30 (C) sample paths are shown (gray curves). The black lines show the ensemble average. This averaged behavior approaches the deterministic prediction (exponential decay) as the ensemble size grows. Parameter value: $k = 1$ (time^{-1}). Time units are arbitrary.

Constitutive Decay   Consider the decay reaction

$$A \xrightarrow{\ k\ } .$$

The behavior of this system was illustrated by the stochastic simulations in figure 7.38, which show that the trajectories are highly variable when the system consists of only a small number of molecules. Figure 7.45 shows ensembles of sample paths, each starting with only 10 molecules. Three ensembles are shown, along with the average behavior (solid line). Although the individual sample paths show significant variability, the average is more consistent. As the ensemble size increases, the averaged behavior converges to the solution of the deterministic model. By generating a large ensemble, a complete description of system behavior—including measures of the variability in the distribution of trajectories—can be reached (see problem 7.8.24).

In some cases, a very large number of sample paths is needed to guarantee confidence in these ensemble-derived results: generating a sufficiently large ensemble can be a time-consuming process. A number of refinements of the SSA have been proposed that aim to reduce the computational requirements for simulation (see problem 7.8.25 for an example).

Constitutive Gene Expression We next consider a simple model of unregulated gene expression, involving mRNA, M, and protein, P. The reaction network is

$$R_1 : \text{(transcription)} \quad \to M \quad \text{propensity}: k_r$$

$$R_2 : \text{(translation)} \quad \to P \quad \text{propensity}: k_p N_M$$

$$R_3 : \text{(degradation)} \quad M \to \quad \text{propensity}: \delta_r N_M$$

$$R_4 : \text{(degradation)} \quad P \to \quad \text{propensity}: \delta_p N_P .$$

(7.34)

Sample paths from a Gillespie simulation are shown in figure 7.46A. The mRNA traces are centered around an average of about 10 molecules. The protein count shows an average of about 60.

Experimental observations have revealed that transcription is sometimes a "bursty" process in which each transcription event leads to the production of multiple copies of mRNA (reviewed in Chubb and Liverpool (2010)).

This model can be modified to describe bursty transcription by replacing reaction R_1 with

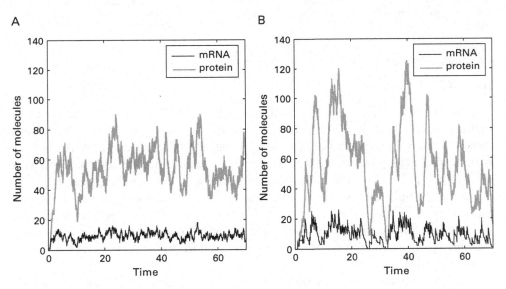

Figure 7.46
Stochastic simulations of constitutive gene expression. (A) Each transcription event produces a single mRNA transcript. (B) Transcription is modeled as "bursty": each transcription event produces five mRNA molecules. The propensity of the transcription reaction has been reduced by a factor of 5 to give the same average as in (A). Parameter values (in time^{-1}): $k_r = 10$, $k_p = 6$, $\delta_r = 1$, $\delta_p = 1$. Time units are arbitrary.

$$\tilde{R}_1 : \text{(bursty transcription)} \quad \rightarrow 5M \quad \text{propensity} : \frac{k_r}{5}.$$

In this modified model, each transcription event produces five mRNA molecules. To allow direct comparison with the original model, the propensity of this bursty transcription reaction has been reduced by a factor of 5, so that the time-averaged mRNA production rate is unchanged. Figure 7.46B shows simulations of this modified model. Although the mRNA and protein averages are the same in both models, the modified model exhibits considerably more variability. This difference in behavior could not be described by a mass action–based model: the deterministic versions of these two models are identical (in both cases, the transcription rate is k_r). Variability is an experimentally observable feature of system behavior that can only be captured in a stochastic modeling framework.

The Brusselator Our final example, the *Brusselator*, is a theoretical chemical system that exhibits sustained oscillations (see exercise 4.3.1 of chapter 4). The reaction network is

$$R_1 : \qquad \qquad \rightarrow X \qquad \text{propensity} : k_1$$
$$R_2 : \qquad X \rightarrow Y \qquad \text{propensity} : k_2 N_X$$
$$R_3 : \quad 2X + Y \rightarrow 3X \qquad \text{propensity} : \frac{k_3}{2} N_X (N_X - 1) N_Y$$
$$R_4 : \qquad X \rightarrow \qquad \text{propensity} : k_4 N_X.$$

A sample path is shown in figure 7.47, in both the time domain (panel A) and the phase space (panel B). The trajectories are somewhat jagged, but the oscillations are fairly regular. In contrast, some oscillatory stochastic systems exhibit considerable variability in the timing and shape of the cycles (see problem 7.8.27).

7.7 Suggestions for Further Reading

• **Modeling Gene Regulatory Networks** The book *An Introduction to Systems Biology: Design Principles of Biological Circuits* (Alon, 2007) surveys a number of models of gene regulatory networks. The text *Computational Modeling of Gene Regulatory Networks—A Primer* (Bolouri, 2008) addresses a wider range of modeling approaches than discussed in this chapter.

• **Phage Lambda** The book *A Genetic Switch: Phage Lambda Revisited* (Ptashne, 2004) provides a detailed description of the molecular genetics of the decision switch, including an accessible account of the experiments that led to these discoveries.

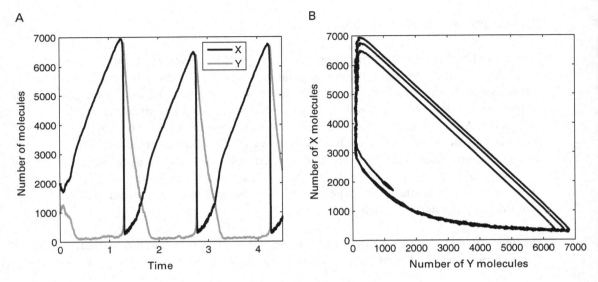

Figure 7.47
Stochastic simulation of the Brusselator. (A) Oscillations are evident in the time domain: a slow increase in Y is followed by a sudden crash when Y levels are sufficiently high. (B) This phase portrait shows the approximate limit cycle followed by the periodic trajectories. Initial conditions are $X = 1000$, $Y = 2000$. Parameter values (in time^{-1}): $k_1 = 5000$, $k_2 = 50$, $k_3 = 0.00005$, and $k_4 = 5$. Time units are arbitrary.

• **Synthetic Gene Circuits**   Discussions of modeling and design in synthetic biology are provided in the book chapter "Synthetic Gene Regulatory Systems" (Weiss and Kaern, 2006) and in *Engineering Gene Circuits* (Myers, 2010). The nontechnical book *Biology is Technology: The Promise, Peril, and New Business of Engineering Life* (Carlson, 2010) provides a thoughtful discussion of the potential impact of synthetic biology.

• **Stochastic Modeling in Systems Biology**   An introduction to stochastic modeling techniques in systems biology is provided in the book chapter "Modeling and Analysis of Stochastic Biochemical Networks" (Khammash, 2010). The book *Stochastic Modeling for Systems Biology* (Wilkinson, 2006) provides a detailed treatment of stochastic approaches.

7.8   Problem Set

7.8.1   Response Time of Autoinhibitory Genes
Consider expression from an unregulated gene as modeled in equation (7.2):

$$\frac{d}{dt}p(t) = \alpha_0 - \delta p(t) \, . \tag{7.35}$$

For comparison, consider an autoinhibitory gene whose expression can be modeled as in equation (7.8):

$$\frac{d}{dt}p(t) = \alpha \frac{1}{1+p(t)/K} - \delta p(t). \tag{7.36}$$

(a) Take $\delta = 1$ (time^{-1}) in models (7.35) and (7.36) and let $K = 1$ (concentration) for the autoinhibited gene. Verify that both genes generate the same steady-state protein concentration when $\alpha = \alpha_0(\alpha_0 + 1)$. (Hint: Substitute $p^{ss} = \alpha_0$ into the autoinhibited model.)

(b) Simulate the two models with $\alpha_0 = 5$ and $\alpha = 30$ (concentration $\cdot$ time^{-1}). Take the initial concentrations to be zero. Verify that as a result of having a higher maximal expression rate, the autoinhibited gene reaches steady state more quickly than the unregulated gene.

(c) How would you expect the response time to be affected by cooperative binding of multiple copies of the repressor? Verify your conjecture by comparing your results from part (b) with the model

$$\frac{d}{dt}p(t) = \alpha_2 \frac{1}{1+(p(t)/K)^2} - \delta p(t).$$

Take $\alpha_2 = 130$ (concentration $\cdot$ time^{-1}).

7.8.2   Robustness of Autoinhibitory Genes

(a) Verify that the steady states of the unregulated and autoinhibitory models in section 7.1 (equations 7.2 and 7.8) are given by

$$p_{unreg}^{ss} = \frac{\alpha}{\delta_p} \qquad \text{and} \qquad p_{reg}^{ss} = \frac{-1+\sqrt{1+4\alpha/(K\delta_p)}}{2/K},$$

where the unregulated model has expression rate α.

(b) Take derivatives with respect to α to verify the relative sensitivities:

$$\frac{\alpha}{p_{unreg}^{ss}} \frac{\partial p_{unreg}^{ss}}{\partial \alpha} = 1 \qquad \frac{\alpha}{p_{reg}^{ss}} \frac{\partial p_{reg}^{ss}}{\partial \alpha} = \frac{2\alpha/(K\delta_p)}{\left(\sqrt{1+4\alpha/(K\delta_p)}\right)\left(-1+\sqrt{1+4\alpha/(K\delta_p)}\right)}.$$

(c) Verify that the sensitivity of the autoinhibited protein is smaller than the sensitivity of the unregulated protein by showing that the expression for

$$\frac{\alpha}{p_{reg}^{ss}} \frac{\partial p_{reg}^{ss}}{\partial \alpha}$$

is always less than one, regardless of the parameter values. Hint: The formula depends on the quotient $4\alpha/(K\delta_p)$. Letting $x = 4\alpha/(K\delta_p)$, it is required to show that

$$\frac{x/2}{(\sqrt{1+x})(-1+\sqrt{1+x})} < 1$$

for any positive x value. You can convince yourself of this by plotting this function of x. For a mathematically rigorous proof, begin by expanding the denominator and multiplying by the conjugate $(1+x+\sqrt{1+x})$.

7.8.3 Alternative Regulatory Schemes
The models in this chapter focus on the regulation of transcriptional initiation (i.e., the binding of RNA polymerase to promoter regions). Other regulatory mechanisms include:

(a) *Antisense RNA.* Gene expression can be inhibited by the production of antisense RNA, which is complementary to a gene's mRNA. The antisense RNA binds tightly to its complementary partner (by base pairing) and so sequesters the mRNA away from the translation machinery. In this case, the transcription of the gene is unaffected, but translation of the resulting mRNA is inhibited. Extend the model of gene expression in equation (7.1) to incorporate inhibition by antisense RNA.

(b) *mRNA stability.* Protein production can be repressed by factors that target mRNA molecules for degradation. (An example is the protein β-tubulin, which destabilizes its own mRNA.) Develop a differential equation model that describes autoinhibition of protein expression in this manner.

7.8.4 Transcription Factor Multimerization
Many transcription factors function as multimers (e.g., dimers or tetramers). Consider a transcription factor P that dimerizes and binds an operator O. The reaction scheme is

$$P + P \underset{k_2}{\overset{k_1}{\rightleftharpoons}} P_2 \quad O + P_2 \underset{d}{\overset{a}{\rightleftharpoons}} OP_2.$$

(a) Letting $K_0 = k_2/k_1$ and $K_1 = d/a$, suppose that these binding events are in equilibrium and verify that the resulting promoter occupancy is

$$\text{Fraction of bound operators:} \quad \frac{[OP_2]}{[O]+[OP_2]} = \frac{[P]^2/K_0K_1}{1+[P]^2/K_0K_1}.$$

This is the same Hill function we used in section 7.1.2 to describe cooperativity in operator binding. However, when this function is used in a model of gene expression, additional nonlinearities appear.

(b) Ignoring mRNA dynamics, a model of an autoactivating gene whose product acts as a dimer takes the form:

$$\frac{d}{dt}p(t) = a\frac{p_2(t)/K}{1+p_2(t)/K} - 2k_1p^2(t) + 2k_2p_2(t) - \delta_1 p(t)$$

$$\frac{d}{dt}p_2(t) = k_1p^2(t) - k_2p_2(t) - \delta_2 p_2(t) \,,$$

where p is the monomer concentration, and p_2 is the dimer concentration. Verify that if $\delta_2 = 0$, then a quasi-steady-state approximation applied to p_2 will result in a reduced model that is equivalent to the cooperative-binding autoinhibitor model (7.10) (with $N = 2$). Thus, multimerization generates the same dynamics as cooperative DNA-binding if multimers are protected from degradation.

7.8.5   The *lac* Operon: Effect of Leak
Consider the model of the *lac* operon presented in section 7.2.1, with parameter values as in figure 7.7. The dose-response curve in figure 7.7B indicates that the system shows little response to external lactose levels below 55 μM. Modify the model by adding a small leak rate of transcription from the operon: add a constant term a_0 to the mRNA production rate in equation (7.11). Set $a_0 = 0.01$ molecules/min. Run simulations to determine how this change affects the triggering threshold. Explain your result in terms of the system behavior. How does the system behave when $a_0 = 0.1$?

7.8.6   The *lac* Operon: Role of Feedback
As presented in section 7.2.1, the model of the *lac* operon can be modified to explore the hypothetical situation in which there is no positive feedback from lactose to operon expression. In that case, the tasks of lactose uptake and lactose-to-allolactose conversion would be carried out by proteins that are present at fixed quantities. The model in section 7.2.1 can be modified to describe this hypothetical system by replacing equation (7.14) with

$$\frac{d}{dt}L(t) = \frac{4k_L E L_e}{K_{ML} + L_e} - \frac{k_g B(t)L(t)}{K_{Mg} + L(t)} - \frac{k_g E L(t)}{K_{Mg} + L(t)} - \delta_L L(t),$$

where E is fixed. (The factor 4 in the uptake rate has been included to preserve the ratio between the number of permease proteins and lactose-to-allolactose conversion enzymes.) In this scenario, β-galactosidase still metabolizes lactose, but it does not participate in the conversion of lactose to allolactose.

The dose-response curve for this modified model is shown in figure 7.7B. The graded response of this feedback-free system may be inefficient, but this modified system nevertheless has some advantages. Take $E = 40$ molecules and simulate this modified system's response to an abrupt introduction of external lactose. Comparing with figure 7.7A, comment on the speed of response for this hypothetical system. Explain your observations.

7.8.7 The *lac* Operon: CAP

Consider the model of the *lac* operon presented in section 7.2.1. With parameter values as in figure 7.7, extend the model to include the transcription factor CAP, which represses expression from the *lac* operon whenever glucose levels are sufficiently high, regardless of the lactose level.

7.8.8 The Goodwin Oscillator

Recall the generic model of an oscillating autoregulatory gene proposed by Goodwin (equation 7.22):

$$\frac{d}{dt} x(t) = \frac{a}{k^n + (z(t))^n} - bx(t)$$

$$\frac{d}{dt} y(t) = \alpha x(t) - \beta y(t)$$

$$\frac{d}{dt} z(t) = \gamma y(t) - \delta z(t).$$

This system exhibits limit-cycle oscillations provided the Hill coefficient n is sufficiently large. Unfortunately, for reasonable choices of the other parameter values, n has to be chosen very high (>8) to ensure oscillatory behavior. Modifications that generate oscillations with smaller Hill coefficients are as follows. (In exploring these models, make sure simulations run for sufficiently long that the asymptotic behavior is clear.)

(a) Taking parameter values as in figure 7.17, modify the model by adding a fourth step to the activation cascade. (Use dynamics identical to the third step.) Verify that the additional lag introduced by this fourth component allows the system to exhibit sustained oscillations with $n < 8$.

(b) Replace the term for degradation of Z by a Michaelis–Menten term: $-\delta z/(K_M + z)$. Verify that this modified system oscillates with no cooperativity (i.e., with $n = 1$). Take $a = 150$, $k = 1$, $b = \alpha = \beta = \gamma = 0.2$, $\delta = 15$, and $K_M = 1$. (Units as in figure 7.17.)

(c) Consider a one-state model in which the time delay caused by the cascade of molecular events is abstracted into an explicit time delay:

$$\frac{d}{dt}x(t) = \frac{a}{k^n + (x(t-\tau))^n} - bx(t).$$

Take parameter values $a = 10$, $k = 1$, $b = 0.5$, and $n = 4$. Verify that this one-dimensional model exhibits sustained oscillations when $\tau = 3$. Describe the effects of changing the delay by running simulations with $\tau = 2$, $\tau = 0.75$, and $\tau = 20$. (Units as in figure 7.17.) Details on simulation of delay equations can be found in appendix C.

Increased loop length was explored by Goodwin in his original paper (Goodwin, 1965); explicit delay and nonlinear degradation were considered in Bliss et al. (1982).

7.8.9 Circadian Rhythms: Goldbeter Model

Recall Goldbeter's model of a circadian oscillator from section 7.3.2.

(a) Using the parameter values in figure 7.19, run a simulation of the model and verify that the system oscillates with a period of roughly 24 hours.

(b) The oscillatory behavior of this model is crucially dependent on the level of cooperativity. Determine the minimum value of the Hill coefficient n for which this system exhibits oscillations. Does the period depend strongly on n?

(c) Modify the model so that the two phosphorylation steps are replaced by an explicit delay. (Details on simulation of delay equations can be found in appendix C.) What size delay is required to recover circadian (24-hour) oscillations?

(d) Returning to the original model formulation, verify Goldbeter's finding that the period can be shortened or lengthened by mutations to the *per* gene that affect the protein's degradation rate. Suggest an alternative effect of the mutation in the *per* gene that could also lead to changes in the period of the oscillation. Verify your conjecture by running simulations.

7.8.10 Circadian Rhythms: TIM

In addition to the PER protein described by the model in section 7.3.2, the circadian network in *Drosophila* also involves a protein called TIM (for "timeless"), expressed from the gene *tim*. John Tyson and colleagues published a simple model that incorporates the interaction between PER and TIM (Tyson et al., 1999). In their model, PER proteins form homodimers. These dimers then associate with two molecules of TIM into a PER_2–TIM_2 complex. These complexes migrate to the nucleus, where they inhibit expression of both PER and TIM. Degradation of both TIM and PER is constitutive, but PER is protected from degradation when in dimer form.

(a) Draw an interaction diagram describing the mechanism.

(b) Verify that this mechanism can explain the following experimental observations:

(i) Cells that lack the *tim* gene do not exhibit oscillatory behavior.

(ii) Circadian oscillations can be entrained to follow 24-hour light–dark cycles. Exposure to light enhances degradation of TIM.

(c) Develop a differential equation model of this system. Describe the features of your model that could enable oscillatory behavior. You may want to make use of the following reasonable assumptions: (i) mRNA dynamics occur quickly; (ii) dimerization events occur quickly; (iii) PER and TIM concentrations follow similar time courses and so the two species can be lumped into a single protein pool. As verified by Tyson and colleagues, a satisfactory model can involve as few as three state variables.

7.8.11 Repressilator: Ring Size
Recall the model of the repressilator in section 7.3.3.

(a) Taking parameter values as in figure 7.21, verify the oscillatory behavior of the system. Change the value of the Hill coefficient so that $n = 1.5$. Verify that the system does not oscillate in this case. Next, model an expanded network in which five genes participate in a ring of sequential repression. Using the same symmetric model framework and parameter values, verify that this expanded model can produce oscillations when the Hill coefficient is 1.5. Provide an explanation in terms of the lag in the negative feedback loop.

(b) Why is it that a ring of four sequential repressors would not be expected to oscillate?

7.8.12 Repressilator: Effect of Leak
Recall the model of the repressilator in section 7.3.3.

(a) Taking parameter values as in figure 7.21, verify that when the leak α_0 is increased to 2 (molecules per cell min^{-1}), the system no longer exhibits oscillations. Provide an intuitive interpretation of this finding: why does a persistent leak dampen the system's oscillatory behavior?

(b) With $\alpha_0 = 2$, can the oscillations be rescued by increasing the degree of cooperativity n? If so, how much of an increase in n is required?

7.8.13 Repressilator: IPTG Arrest
Recall the model of the repressilator in section 7.3.3.

(a) Modify the model to incorporate the effect of IPTG—an allolactose mimic—on the circuit. (Recall that allolactose deactivates the *lac* repressor, LacI, which is one of the three repressors in the loop.)

(b) Take parameter values as in figure 7.21 and choose parameter values for your model extension. Simulate the system's response to an extended pulse of IPTG. Your simulation should show the oscillations stopping when IPTG is introduced and resuming after it is removed. Comment on the protein concentrations in the arrested steady state.

7.8.14 Atkinson Oscillator

In 2003, Mariette Atkinson and colleagues described a synthetic genetic relaxation oscillator (Atkinson et al., 2003). Like the Stricker oscillator presented in section 7.3.3, Atkinson's circuit involves two genes and their protein products: a repressor and an activator. The activator induces expression of both genes (as in the Stricker oscillator), but the repressor only represses the activator. Model (7.23) can be modified to describe the Atkinson oscillator by removing the term $(1 + y(t)^4)$ from the denominator of the expression rate for Y (the repressor). Take parameter values as in figure 7.23, but with $\gamma_y = 1$, and verify that the resulting model displays damped oscillations, as were observed by Atkinson and her co-workers. Next, decrease a_y until the system exhibits sustained oscillations. Suggest a design feature that would provide control over the value of parameter a_y.

7.8.15 NF-κB Signaling

The transcription factor NF-κB is involved in a range of signaling pathways in eukaryotic cells (see section 1.6.2). A simple model of NF-κB activation, which captures the oscillations observed during signaling, was published in 2006 by Sandeep Krishna, Mogens Jensen, and Kim Sneppen (Krishna et al., 2006). Their model lumps the NF-κB inhibitors into a single pool, IκB, whose expression is activated by NF-κB. The network is shown in figure 7.48.

The model describes three state variables: nuclear NF-κB (concentration: N_n), IκB mRNA (I_m), and cytosolic IκB (I). The concentration scale is chosen so that the total NF-κB concentration is 1. Because NF-κB is conserved, the cytosolic NF-κB concentration is then $1 - N_n$. The model equations are

$$\frac{d}{dt} N_n(t) = A \frac{1 - N_n}{\varepsilon + I(t)} - B \frac{I(t)N_n(t)}{\delta + N_n(t)}$$

$$\frac{d}{dt} I_m(t) = N_n(t)^2 - I_m(t)$$

$$\frac{d}{dt} I(t) = I_m(t) - IKK(t) \frac{(1 - N_n(t))I(t)}{\varepsilon + I(t)}.$$

The input is the level of IκB kinase IKK. The units of time and concentration have been rescaled to reduce the number of model parameters.

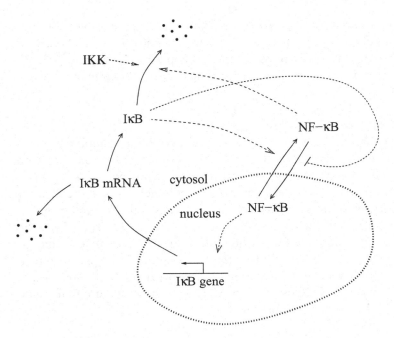

Figure 7.48
NF-κB signaling network for problem 7.8.15. (Dashed arrows indicate regulation.) Adapted from figure 1 of Krishna et al. (2006).

(a) Take parameter values (scaled to be dimensionless) $A = 0.007$, $B = 954.5$, $\delta = 0.029$, $\varepsilon = 0.00005$, and $IKK = 0.035$. Verify that the concentration of nuclear NF-κB (N_n) exhibits spike-like oscillations.

(b) The model time-units were scaled by the degradation rate of IκB mRNA, which is 0.017 min^{-1}. (That is, the model is expressed in time-units of $1/0.017$ minutes.) What is the period of the oscillations in NF-κB?

(c) Krishna and colleagues confirmed that the spiky oscillations in NF-κB are quite robust to variation in the parameter values. However, they observed that when IKK levels are increased, the oscillations become smoother. Confirm this result by running a simulation in which the IKK concentration is tripled.

(d) Recall from section 1.6.2 that there are three isoforms of IκB (IκBα, IκBβ, and IκBε) and that production of IκBβ and IκBε is not inhibited by NF-κB. These uninhibited isoforms can be included in the model by adding a constant production term for IκB mRNA (so that $dI_m / dt = c_0 + N_n^2 - I_m$). Verify that when $c_0 = 0.005$, the system exhibits sustained oscillations, but the amplitude of the spikes in nuclear NF-κB (N_n) is diminished. How does the system behave when c_0 is increased to 0.02?

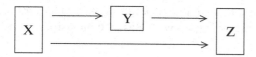

Figure 7.49
Coherent feed-forward loop (problem 7.8.16).

7.8.16   Feed-Forward Loops

Uri Alon has investigated the behavior of a three-gene network that is a common component of natural gene regulatory systems (Alon, 2007). The network, shown in figure 7.49, is activated at a single gene (X), which activates both the pathway target (Z) and an intermediary (Y). The intermediary coregulates the target. The network is called a *feed-forward loop*. If Z is activated by both X and Y, it is called *coherent*. (Alternatively, if Y represses Z, the loop is called *incoherent*: the pulse generator discussed in section 7.4.2 is an incoherent feed-forward loop.)

Consider a coherent feed-forward loop in which expression of Z requires activation by both X and Y (the Z promoter implements AND logic, as in section 7.5.1). This network provides signaling from X to Z that is robust to input noise, in the sense that brief input pulses will not lead to activation of the target gene Z.

(a) Using the formulation in section 7.1.2 for expression from promoters activated by single or multiple transcription factors, construct a differential equation model of a coherent feed-forward loop with AND logic. (Take protein X as an input to the pathway. You will need to develop differential equations describing the dynamics of protein products Y and Z. Treat mRNA dynamics in quasi–steady state.)

(b) Select parameter values and simulate the response of your system to an abrupt increase in X. You should observe an increase in Y, followed by a later increase in Z.

(c) Next, simulate the response of your system to a brief pulse in X (in which X jumps to a nonzero value for a short time and then jumps back to zero). Verify that for short pulses, no significant expression of Z is observed (compared to the direct response in Y). Provide an intuitive explanation of this behavior in terms of the network structure.

(d) Explain why a coherent feed-forward loop that implements an OR logic at the Z promoter would not exhibit the robustness property observed in part (c).

7.8.17   Band Detector: Sensitivity

Referring to the band-detector construction in section 7.4.2, Basu and colleagues used their model to determine that, compared to the other kinetic parameters in the model, the rate constant for LacI decay (b_L) had the most significant impact on the shape of the GFP band. Investigate the effects of perturbation in b_L and at least three other kinetic parameters. Do your findings agree with Basu's conclusion?

7.8.18 Band Detector: Design Variants

Referring to the band-detector circuit described in section 7.4.2, Basu and co-workers constructed variants of their design: one in which the LuxR concentration was reduced (by the use of a low-copy plasmid); and one in which the LuxR–DNA binding affinity was enhanced (by mutation). How would you expect these variations in LuxR activity to impact the system behavior? Implement changes in model (7.26) to mimic these variants, and run simulations to confirm your conjectures.

7.8.19 Synchronized Repressilators

In a 2004 paper, Jordi Garcia-Ojalvo, Michael Elowitz, and Steven Strogatz proposed an extension of the repressilator circuit (section 7.3.3) that could synchronize oscillations in a population of cells (Garcia-Ojalvo et al., 2004). In their proposed circuit, the *LuxI* gene, whose protein product catalyzes production of AHL (section 7.4.2), is placed under the control of the *lac* promoter. Additionally, a second copy of the *LacI* gene is introduced, under the control of the AHL–LuxR sensitive promoter from *V. fischeri*. Finally, a constitutively expressed copy of *LuxR* is added. The network is shown in figure 7.50.

   To explore the behavior of this circuit, consider a pair of identical cells, each hosting the network. AHL diffuses between the two cells and so can bring the two oscillators into synchrony. Garcia-Ojalvo and colleagues modeled the network as follows. The three repressilator genes (*tetR*, *cI*, and *LacI*, with mRNA concentrations a, b, and c) and their protein products (concentrations A, B, and C) are modeled as in the repressilator, except for an additional source of LuxI, which is induced by intracellular AHL (denoted S):

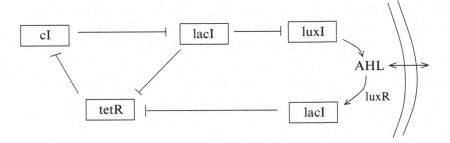

Figure 7.50
Extended repressilator circuit for synchronization (problem 7.8.19). Adapted from figure 1 of Garcia-Ojalvo et al. (2004).

$$\frac{d}{dt}a_i(t) = \frac{\alpha}{1+C_i(t)^n} - a_i(t) \qquad \frac{d}{dt}A_i(t) = \beta(a_i(t) - A_i(t))$$

$$\frac{d}{dt}b_i(t) = \frac{\alpha}{1+A_i(t)^n} - b_i(t) \qquad \frac{d}{dt}B_i(t) = \beta(b_i(t) - B_i(t))$$

$$\frac{d}{dt}c_i(t) = \frac{\alpha}{1+B_i(t)^n} + \frac{\kappa S_i}{1+S_i} - c_i(t) \qquad \frac{d}{dt}C_i(t) = \beta(c_i(t) - C_i(t)),$$

where the subscripts indicate concentrations in the two cells ($i = 1, 2$). The dynamics of LuxI are assumed to be identical to TetR, so the concentration of intracellular AHL can be described by

$$\frac{d}{dt}S_i(t) = k_{s1}A_i(t) - k_{s0}S_i(t) - \eta(S_i(t) - S_e(t)).$$

The first term in this equation describes LuxI-dependent production, the second term describes degradation/dilution, and the final term is transport, where S_e is the external AHL concentration. A simple formulation for the external AHL concentration is

$$S_e(t) = Q\frac{S_1(t) + S_2(t)}{2},$$

where the parameter Q characterizes diffusion of AHL away from the cells.

(a) Using parameter values $\alpha = 216$, $n = 2$, $\beta = 1$, $\kappa = 20$, $k_{s0} = 1$, $k_{s1} = 0.01$, $\eta = 2$, and $Q = 0.9$, simulate the system from an initial condition in which $A = 10$ in cell 1, $B = 10$ in cell 2, and all other concentrations start at zero. Confirm that the two cells synchronize after about 100 time units.

(b) What condition is described by setting $\kappa = 0$? Simulate the system under this condition, and provide a justification for the system behavior.

(c) With $\kappa = 20$, explore the effect of reducing Q. As Q decreases, AHL diffuses more quickly away from the cells. Do the cells synchronize when $Q = 0.2$? What is the effect on the synchronization speed? What happens when $Q = 0.02$?

7.8.20 Population Control

In a 2004 paper, Lingchong You and colleagues described a synthetic gene regulatory network that uses the *Vibrio fischeri* quorum sensing mechanism (section 7.4.1) to control the density of a population of *E. coli* cells (You et al., 2004). The quorum sensing network drives expression of a "killer" gene whose protein product causes cell death (figure 7.51).

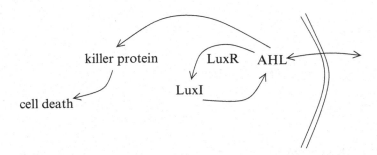

Figure 7.51
Population control network for problem 7.8.20(a). Adapted from figure 1 of You et al. (2004).

(a) You and colleagues used a simple model to describe a population of cells containing this circuit. The model describes the cell density (N) and the averaged intracellular concentrations of killer protein (E) and AHL (A):

$$\frac{d}{dt}N(t) = kN(t)(1 - N(t)/N_m) - d_N E(t)N(t)$$

$$\frac{d}{dt}E(t) = k_E A(t) - d_E E(t)$$

$$\frac{d}{dt}A(t) = v_A N(t) - d_A A(t) \,.$$

The population growth rate is $k(1 - N/N_m)$, which diminishes to zero as the population tends to N_m. The parameter N_m is called the *carrying capacity* of the environment. (This is a *logistic* growth model.)

(i) Take parameter values of $d_N = 0.004$ nM^{-1} hr^{-1}, $k_E = 5$ hr^{-1}, $d_E = 2$ hr^{-1}, $v_A = 4.8 \times 10^{-7}$ nM ml hr^{-1}, $k = 0.97$ hr^{-1}, $N_m = 1.24 \times 10^9$ CFU ml^{-1}, and $d_A = 0.639$ hr^{-1}. (Colony-forming units (CFU) is a standard measure of abundance of viable cells.) Simulate the model from initial condition $(N, E, A) = (1, 1, 1)$. Next, modify the model to mimic the control case in which no killer protein is produced, and compare the steady-state population densities (N) in the two cases.

(ii) Verify that the steady-state population size can be tuned by the value of k_E. Describe an intervention or design change that would alter this parameter value. What value of k_E would result in the population reaching half of the (no killer protein) control size?

(b) In a follow-up study, Frederick Balagaddé and co-workers constructed a two-species ecosystem using the density-dependent killing mechanism of part (a) (Bal-

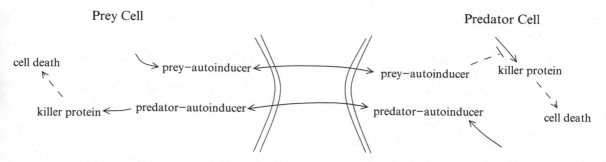

Figure 7.52
Predator–prey system for problem 7.8.20(b). Adapted from box 1 of Balagaddé et al. (2008).

agaddé et al., 2008). Their system involves two separate engineered strains of *E. coli*: a "prey" strain and a "predator" strain. The prey strain uses the *V. fischeri* quorum sensing mechanism to signal its population density. The predator strain uses an analogous quorum sensing mechanism from another bacterial species. Predation is mimicked by the prey's response to the predator-specific autoinducer—it activates expression of the killer protein in prey cells. The predator population is controlled by expression of killer protein, which is repressed by the prey-specific autoinducer (thus mimicking survival of predators when prey are abundant). The network is shown in figure 7.52.

Balagaddé and colleagues developed a model that describes the densities of predator (c_1) and prey (c_2) and the strain-specific autoinducer concentrations (a_1 and a_2, respectively):

$$\frac{d}{dt}c_1(t) = k_1 c_1(t)\left(1 - \frac{c_1(t) + c_2(t)}{c_m}\right) - d_1 c_1(t)\frac{K_1}{K_1 + a_2^\beta} - D c_1(t)$$

$$\frac{d}{dt}c_2(t) = k_2 c_2(t)\left(1 - \frac{c_1(t) + c_2(t)}{c_m}\right) - d_2 c_2(t)\frac{a_1^\beta}{K_2 + a_1^\beta} - D c_2(t)$$

$$\frac{d}{dt}a_1(t) = \gamma_1 c_1(t) - (\delta_1 + D)a_1(t)$$

$$\frac{d}{dt}a_2(t) = \gamma_2 c_2(t) - (\delta_2 + D)a_2(t).$$

As in part (a), a logistic growth model is used (with carrying capacity c_m). The killer protein abundance is in quasi–steady state, so growth inhibition is directly dependent on the autoinducer levels. For each autoinducer, the intracellular and extracellular concentrations are presumed identical. The parameters δ_1 and δ_2 are decay rates for the two autoinducers; the parameter D characterizes dilution of all species.

(i) Take parameter values $k_2 = 0.4$ hr^{-1}, $c_m = 10^5$ cells per nanoliter, $\beta = 2$, $d_1 = 1$ hr^{-1}, $d_2 = 0.3$ hr^{-1}, $K_1 = K_2 = 10$ nM, $\gamma_1 = \gamma_2 = 0.1$ nM ml hr^{-1}, $\delta_1 = 0.017$ hr^{-1}, $\delta_2 = 0.11$ hr^{-1}, and $D = 0.2$ hr^{-1}. Simulate the system for three cases: $k_1 = 0.2, 0.8$, and 1.4 hr^{-1}. Verify that in the first and last case, one population dominates over the other, whereas for $k_1 = 0.8$ hr^{-1}, the populations tend to a persistent oscillatory pattern. Explain the long-time system behavior in each case.

(ii) Using this model, Balagaddé and co-workers discovered that the system is more likely to exhibit oscillations when the predator death rate (d_1) is sufficiently large. They thus engineered a modified killer protein to reach a higher death rate. Confirm their finding by exploring the range of k_1 values over which the system oscillates at low ($d_1 = 0.2$) and high ($d_1 = 1$) predator death rates.

7.8.21   Genetic Logic Gate Design

(a) Develop a differential equation model of a regulated promoter that implements an OR logic gate, as discussed in section 7.5.1. Hint: Recall the expression rates for regulated promoters in section 7.1.2.

(b) Using genes that implement only OR, AND, and NOT gates, design a genetic circuit that implements an XOR (exclusive OR) gate. The output from an XOR gate is ON when exactly one of its two inputs is ON. Develop a differential equation model of your circuit.

7.8.22*   Chemical Master Equation: Closed System
Consider the reaction network

$$A \xrightarrow{k_1} B + C \qquad\qquad B + C \xrightarrow{k_{-1}} A.$$

(a) Suppose that the system starts with two molecules of A, one molecule of B, and no molecules of C; that is, $(N_A, N_B, N_C) = (2, 1, 0)$. Determine the set of possible states the system can adopt, and write the chemical master equation that describes the corresponding probability distribution.

(b) Take $k_1 = 1$ (time^{-1}) and $k_{-1} = 1$ (time^{-1}), and solve for the steady-state probability distribution.

7.8.23*   Chemical Master Equation: Open System
Consider the open system

$$\xrightarrow{k_1} A \xrightarrow{k_2}.$$

(a) Take $k_1 = 1$ (concentration $\cdot$ time^{-1}) and $k_2 = 1$ (time^{-1}). Referring to the corresponding master equation (7.28), verify that in steady state, the probabilities satisfy

$$P(N_A = n) = \frac{1}{n} P(N_A = n-1).$$

(b) Use the fact that $\sum_{n=0}^{\infty} (1/n!) = e$ (where the factorial $n! = n(n-1)(n-2) \cdots 3 \cdot 2 \cdot 1$, and e is Euler's number $e \approx 2.71828$) to derive the steady-state probability distribution:

$$P(N_A = n) = \frac{1/e}{n!},$$

for each $n = 0, 1, 2, \ldots$ (by convention $0! = 1$).

7.8.24* Statistics of an Ensemble of Sample Paths

Consider the simple model of unregulated gene expression in section 7.6.4:

R_1 : (transcription)	$\to M$	propensity : k_r
R_2 : (translation)	$\to P$	propensity : $k_p N_M$
R_3 : (degradation/dilution) $M \to$		propensity : $\delta_r N_M$
R_4 : (degradation/dilution) $P \to$		propensity : $\delta_p N_P$.

Take parameter values as in figure 7.46A. Simulate sample paths using the stochastic simulation algorithm. Analyze the statistics of your ensemble to verify that in steady state (so-called *stationary* behavior), the coefficient of variation (standard deviation divided by mean) for each species is

$$mRNA: \quad C_r = \left(\frac{\delta_r}{k_r}\right)^{1/2}, \qquad \text{protein}: \quad C_p = \left(\frac{\delta_r \delta_p}{k_r k_p}\right)^{1/2} \left(1 + \frac{k_p}{\delta_r + \delta_p}\right)^{1/2}.$$

(For a derivation of these formulas, see Khammash (2010).) Note that statistics can be gathered from an ensemble of sample paths or a single long simulation. XPPAUT users will have to export the data from the simulation and calculate the mean and variance using another program.

7.8.25* Stochastic Simulation: The First-Reaction Method

An alternative to the stochastic simulation algorithm presented in section 7.6.3 is the *first-reaction method*, which also involves stepping from reaction event to reaction event. However, rather than sample the next reaction and the waiting time separately (as in the SSA), the first-reaction algorithm samples a waiting time for each reaction in the network and then selects the shortest of this collection of times.

This selection specifies the identity of the next reaction and the elapsed time. Because these waiting times can often be re-used from one time-step to the next, this algorithm can be significantly more efficient than the SSA. (An implementation of the first-reaction method was presented in Gibson and Bruck (2000).)

Recall that the waiting time $T = T_{\text{wait}}$ in the SSA has a cumulative distribution function given by $P(0 \leq T_{\text{wait}} \leq t) = 1 - e^{-at}$, where a is the sum of the propensities for all of the reactions in the network. In the first-reaction algorithm, the waiting time T_{first} is the minimum of a collection of reaction-specific waiting times T_i, each of which is characterized by $P(0 \leq T_i \leq t) = 1 - e^{-a_i t}$, where a_i is the reaction propensity. Confirm that the cumulative distribution function for the first-reaction waiting time ($T_{\text{first}} = \min(T_1, T_2, \ldots, T_m)$) agrees with the distribution of $T = T_{\text{wait}}$ in the SSA. Hint: Verify that $P(T_{\text{wait}} > t) = e^{-at}$ and $P(T_i > t) = e^{-a_i t}$. Then use the fact that $P(T_{\text{first}} > t) = P((T_1 > t) \text{ and } (T_2 > t) \ldots \text{ and } (T_m > t))$ where the T_i are independent of one another.

7.8.26* Noisy Toggle Switch

Stochastic systems can exhibit a range of bistable-like behaviors, ranging from "true" bistability to frequent noise-induced transitions between two nominally stable states. To explore this behavior, consider a stochastic system that recapitulates the bistable toggle switch discussed in section 7.2.3:

$$R_1 : (\text{synthesis}) \qquad \rightarrow P_1 \qquad \text{propensity}: \frac{\alpha}{1 + N_2^{\beta}}$$

$$R_2 : (\text{synthesis}) \qquad \rightarrow P_2 \qquad \text{propensity}: \frac{\alpha}{1 + N_1^{\beta}}$$

$$R_3 : (\text{decay}) \qquad P_1 \rightarrow \qquad \text{propensity}: \delta N_1$$

$$R_4 : (\text{decay}) \qquad P_2 \rightarrow \qquad \text{propensity}: \delta N_2.$$

Here, N_1 and N_2 are the molecular counts for the two repressors. The Hill-type propensities for the synthesis reactions are not well-justified at the molecular level, but these expressions nevertheless provide a simple formulation of a bistable stochastic system. Take parameter values $\delta = 1$ and $\beta = 4$. The corresponding deterministic system (i.e., $dp_i / dt = \alpha / (1 + p_j^4) - p_i$) is bistable for any $\alpha > 1$. Run simulations of the stochastic system for $\alpha = 5, 50, 500,$ and 5000. Be sure to run the simulations sufficiently long so that the steady trend is clear (i.e., at least 10,000 reaction steps). Verify that for $\alpha = 5000$, the system exhibits bistability (with about 5000 molecules of the dominant species, in the long term). In contrast, verify that with $\alpha = 5$, noise dominates and the system shows no signs of bistability. What about at $\alpha = 50$ and 500? Comment on how the steady-state molecule abundance affects system behavior. (Note that it may be necessary to run multiple simulations to confirm your findings.)

7.8.27* Noise-Induced Oscillations

Stochastic systems can exhibit a range of oscillatory behaviors, ranging from near-perfect periodicity to erratic cycles. To explore this behavior, consider a stochastic relaxation oscillator studied by José Vilar and colleagues (Vilar et al., 2002). The system involves an activator and a repressor. The activator enhances expression of both proteins. The repressor acts by binding the activator, forming an inert complex. A simple model of the system is

R_1 : (activator synthesis) $\quad \rightarrow b_A A \quad$ propensity : $\dfrac{\gamma_A}{b_A} \dfrac{\alpha_0 + N_A / K_A}{1 + N_A / K_A}$

R_2 : (repressor synthesis) $\quad \rightarrow b_R R \quad$ propensity : $\dfrac{\gamma_R}{b_R} \dfrac{N_A / K_R}{1 + N_A / K_R}$

R_3 : (activator decay) $\quad A \rightarrow \quad$ propensity : $\delta_A N_A$

R_4 : (repressor decay) $\quad R \rightarrow \quad$ propensity : $\delta_R N_R$

R_5 : (association) $\quad A + R \rightarrow C \quad$ propensity : $k_C N_A N_R$

R_6 : (dissociation and decay) $\quad C \rightarrow R \quad$ propensity : $\delta_A N_C$

Here, N_A, N_R, and N_C are the molecular counts for the activator, repressor, and activator–repressor complex. The parameter b_A and b_R characterize the expression burst size. The Hill-type propensities for the synthesis reactions are not well-justified at the molecular level, but these expressions nevertheless provide a simple formulation of a stochastic relaxation oscillator.

(a) Take parameter values $\gamma_A = 250$, $b_A = 5$, $K_A = 0.5$, $\alpha_0 = 0.1$, $\delta_A = 1$, $\gamma_R = 50$, $b_R = 10$, $K_R = 1$, $k_C = 200$, and $\delta_R = 0.1$. Run simulations of this model and verify its quasi-periodic behavior.

(b) The deterministic version of this model is

$$\frac{d}{dt} a(t) = \gamma_A \frac{\alpha_0 + a(t) / K_A}{1 + a(t) / K_A} - k_C a(t) r(t) - \delta_A a(t)$$

$$\frac{d}{dt} r(t) = \gamma_R \frac{a(t) / K_R}{1 + a(t) / K_R} - k_C a(t) r(t) + \delta_A c(t) - \delta_R r(t)$$

$$\frac{d}{dt} c(t) = k_C a(t) r(t) - \delta_A c(t),$$

where a, r, and c are the concentrations of activator, repressor, and complex. Run a simulation with the same parameter values as in part (a). Does the system exhibit oscillations? How is the behavior different if you set $\delta_R = 0.2$?

(c) The contrast between the behavior of the models in parts (a) and (b), for $\delta_R = 0.1$, can be explained by the excitability of this relaxation oscillator. Run two simulations of the deterministic model ($\delta_R = 0.1$), one from initial conditions $(a, r, c) = (0, 10, 35)$ and another from initial conditions $(a, r, c) = (5, 10, 35)$. Verify that in the first case, the activator is quenched by the repressor, and the system remains at a low-activator steady state, whereas in the second case, this small quantity of activator is able to break free from the repressor and invoke a (single) spike in expression. Explain how noise in the activator abundance could cause repeated excitations by allowing the activator abundance to regularly cross this threshold. This is referred to as *noise-induced oscillation*.

8 Electrophysiology

The brain is the last and grandest biological frontier, the most complex thing we have yet discovered in our universe. It contains hundreds of billions of cells interlinked through trillions of connections. The brain boggles the mind.
—James Watson, foreword to S. Ackerman's *Discovering the Brain*

All cells actively maintain a difference in electrical charge across their cellular membranes. This difference in charge gives rise to a voltage difference, or *potential* (recall figure 1.8). In most cell types, the steady-state membrane potential, called the *resting potential*, is robust to perturbation. However, in some animal cell types (e.g., neurons and muscle cells), perturbations in the charge distribution can lead to sweeping changes in membrane potential, which are used to transmit signals. Such cells are called *excitable*.

Neurons, the primary cells of the animal nervous system, use their excitable cell membranes to communicate information. Neurons are built for communication. Each neuronal cell is composed of (i) a main cell body, called the *soma*; (ii) a tree of branched projections called *dendrites*; and (iii) a single extended projection called the *axon* (figure 8.1). Signals are received by the dendrites and are integrated at the soma. When the collection of incoming signals warrants a response, a pulse is generated in the transmembrane voltage at the soma. This pulse—called an *action potential*—propagates down the cell's axon and thus relays a signal to other neurons in the network.

This chapter will introduce models of the generation and propagation of action potentials. We will begin by developing a model of electrical activity at the membrane.

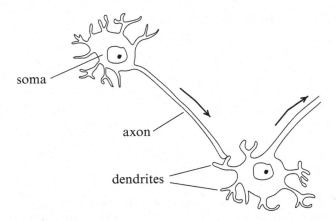

Figure 8.1
Neuronal communication. Projecting from each cell's body (called its soma) is a tree of dendrites and a long axon. Signals are sent down the axon and received at the dendrites.

8.1   Membrane Potential

Cells use ion pumps to maintain gradients of ion concentrations across their membranes. We will first consider the steady-state potential that arises from these ion gradients.

8.1.1   The Nernst Potential

We begin with a thought experiment (figure 8.2). Suppose that the intracellular and extracellular spaces each contain pools of cations, K^+, mixed with an equal number of anions, A^-. Each space is then electrically neutral (zero net charge). Suppose further that there is a higher concentration of ions in the intracellular space than in the extracellular space.

Next, imagine what happens when selective channels are opened in the membrane that allow K^+ ions to pass between the two spaces but do *not* allow A^- ions to pass. The initial effect will be diffusion of K^+ ions out of the cell (figure 8.2A). If the particles were not charged, diffusion would eventually lead to equal concentrations of K^+ in both spaces. However, because K^+ carries a positive charge, the migration of K^+ ions across the membrane changes the net charge of each reservoir (figure 8.2B): the intracellular space becomes negatively charged (because the A^- ions outnumber the K^+ ions), whereas the extracellular space becomes positively charged (K^+ outnumbers A^-). This buildup of charge causes the K^+ ions to be attracted back across the membrane to the negatively charged intracellular space. Eventually, the system will reach a steady state in which the electrically driven inward transport

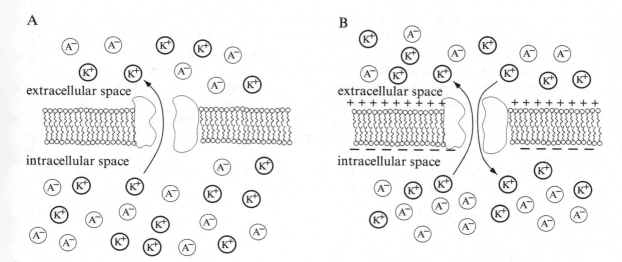

Figure 8.2
The Nernst potential. (A) In this thought experiment, the intracellular and extracellular spaces are initially electrically neutral. When a K^+-selective pore is opened in the membrane, the imbalance in $[K^+]$ drives diffusion of K^+ ions across the cell membrane. (B) Outward transport of K^+ leads to a charge difference between the two spaces, forcing K^+ ions back through the pore. The diffusion-driven and electrically driven transport rates come to equilibrium at the Nernst potential. Adapted from figure 1.1 of Ermentrout and Terman (2010).

rate balances the outward diffusion rate. The voltage difference at this equilibrium state is called the **Nernst potential** and is given, in volts, by

$$E_K = \frac{RT}{zF} \ln\left(\frac{[K^+]_e}{[K^+]_i}\right),$$ (8.1)

where R is the ideal gas constant, T is the temperature in degrees Kelvin, z is the charge on the migrating ion (its *valence*), and F is Faraday's constant.[1]

Exercise 8.1.1   Determine the Nernst potential for Na^+, K^+, and Ca^{2+} in mammalian skeletal muscle given the following steady-state ion concentrations:

Ion	Intracellular (mM)	Extracellular (mM)
Na^+	12	145
K^+	155	4
Ca^{2+}	10^{-4}	1.5

□

1. $R = 8.315$ J/(mol K), $F = 9.648 \times 10^4$ C/mol. At mammalian body temperature (37°C), $T = 310.15$ K, so $RT/F = 26.7 \times 10^{-3}$ J/C. One volt (V) is one joule per coulomb (J/C).

If a cell's membrane potential were set by a single ion type, the resting potential would be the corresponding Nernst potential. However, for most cells the resting potential is caused by transmembrane concentration differences in multiple ion species. In these cases, the resting membrane potential is a weighted average of the individual Nernst potentials. The weights are the ion-specific *conductances* of the membrane. Conductance is a measure of how easily an ion can cross the membrane. Membrane conductance is usually denoted g and is typically measured in millisiemens (mS) per area. (Siemens is the reciprocal of the standard unit of electrical resistance, the Ohm.)

As an example, if the voltage difference V across a membrane is determined by Na^+, K^+, and Cl^-, then the resting potential V^{ss} is

$$V^{ss} = \frac{E_{Na}g_{Na} + E_K g_K + E_{Cl}g_{Cl}}{g_{Na} + g_K + g_{Cl}}, \tag{8.2}$$

where E_i and g_i are the Nernst potential and membrane conductance for ion i.

Exercise 8.1.2   The membrane potential of the squid giant axon is set by three ions: K^+, Na^+, and Cl^-. (This is a long, thick axon in common squid, not a cell from a giant squid!) The corresponding Nernst potentials are (in millivolts; mV): $E_K = -75$, $E_{Na} = -54$, and $E_{Cl} = -59$. Determine the resting potential if the membrane conductances are related by $g_K = 25g_{Na} = 2g_{Cl}$.   □

8.1.2 The Membrane Model
We next consider the dynamic effects of ion transport on the membrane potential.

Single Ion Case   Consider first the case in which membrane potential is set by a single ion species, with corresponding Nernst potential E. We will assume that any local changes in ion concentration are negligible compared to the overall pool of ions, so will treat the resting potential E as constant.

The voltage difference across the membrane is caused by the difference in charge between the intracellular and extracellular spaces. The standard convention in electrophysiology is to define the transmembrane voltage difference, which is referred to simply as the *membrane voltage*, as the difference between the intracellular voltage and the extracellular voltage (so membrane voltage $V = V_{int} - V_{ext}$).

Let $I(t)$ denote the rate at which positive charge is transferred into the cell (across the membrane). We will refer to this as the electrical *current*. As current flows, the charge difference across the membrane changes, and thus the membrane voltage changes. The relationship between current and the rate of change of voltage is

$$C\frac{d}{dt}V(t) = I(t), \tag{8.3}$$

where C is the *capacitance* of the membrane. Capacitance is a measure of the membrane's ability (i.e., capacity) to store a charge difference. The capacitance of the membrane depends on its thickness: a thicker membrane has a smaller capacitance and thus does not hold charge as readily as a thin membrane. (In neurons, the membrane thickness can be increased by the process of *myelination*, in which a fatty sheath encases the axon.)

The current, $I(t)$, is driven by the difference between the membrane voltage and the Nernst potential. Ohm's law describes the current as

$$I(t) = g \cdot (E - V(t)), \tag{8.4}$$

where g is the membrane conductance.

Combining equations (8.3) and (8.4), we arrive at a differential equation describing the membrane voltage:

$$\frac{d}{dt}V(t) = \frac{g}{C} \cdot (E - V(t)). \tag{8.5}$$

The solution $V(t)$ relaxes exponentially to the resting potential E. (Compare equation 8.5 with equation 2.12 of chapter 2.) The rate of relaxation is dictated by the capacitance C and the conductance g.

Exercise 8.1.3 Verify that the solution to equation (8.5) with initial voltage V_0 is

$$V(t) = E - e^{-(g/C)t}(E - V_0). \qquad \square$$

Multiple Ion Case We next consider the case in which the membrane potential is set by multiple ion species. In this case, each ionic current contributes to changes in the membrane voltage, while each of these currents is driven by the difference between the membrane potential and the ion's individual Nernst potential.

As an example, suppose that Na^+, K^+, and Cl^- are responsible for setting the membrane potential. Then, equation (8.3) involves three separate currents:

$$C\frac{d}{dt}V(t) = I_{Na}(t) + I_K(t) + I_{Cl}(t). \tag{8.6}$$

As in equation (8.4), each ion-specific current is driven by the difference between the membrane potential $V(t)$ and the ion's Nernst potential,

$$I_{Na}(t) = g_{Na}(E_{Na} - V(t)), \qquad I_K(t) = g_K(E_K - V(t)), \qquad I_{Cl}(t) = g_{Cl}(E_{Cl} - V(t)), \tag{8.7}$$

where g_{Na}, g_K, and g_{Cl} are the ion-specific membrane conductances. Combining (8.6) and (8.7), we arrive at

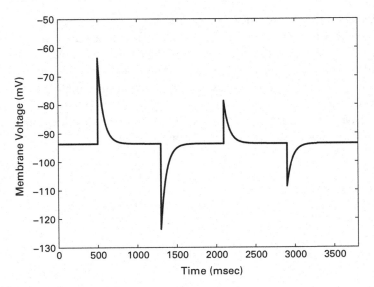

Figure 8.3
Response of a nonexcitable (passive) cell to perturbations in membrane voltage. Each time the membrane voltage is perturbed from the resting potential, it relaxes exponentially back to rest. Equation (8.5) is simulated with parameter values $C = 0.98$ microfarads per square centimeter (μF/cm^2), $g = 0.0144$ millisiemens per square centimeter (mS/cm^2), and $E = -93.6$ millivolts (mV).

$$\frac{d}{dt}V(t) = \frac{1}{C}(g_{Na} \cdot (E_{Na} - V(t)) + g_K \cdot (E_K - V(t)) + g_{Cl} \cdot (E_{Cl} - V(t))). \tag{8.8}$$

Exercise 8.1.4

(a) Verify that the steady-state voltage for equation (8.8) corresponds to the multiion resting potential given by equation (8.2).

(b) Verify that the solution to equation (8.8) decays exponentially to rest at rate $e^{-(g_T/C)t}$ where $g_T = g_{Na} + g_K + g_{Cl}$. $\square$

Equation (8.8) describes the behavior of the membrane voltage for nonexcitable cells (also called *passive* cells). Whenever the voltage is perturbed from its resting value, it relaxes exponentially to rest (figure 8.3).

8.2   Excitable Membranes

In developing the model (8.8), we treated the ion-specific conductances g_i as fixed parameters. We next consider membranes whose conductance is sensitive to changes in potential. This results in a feedback loop between current and voltage that is the foundation of excitable behavior.

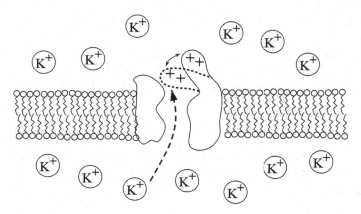

Figure 8.4
Voltage-gated ion channel. Charges in the protein are unequally distributed, so that a charge difference across the membrane induces a conformational shift, opening or closing the channel.

8.2.1 Voltage-Gated Ion Channels

The conductance of certain ion channels is sensitive to the voltage across the membrane. Changes in conductance of these *voltage-gated* ion channels are caused by voltage-induced conformational shifts in the channel proteins (figure 8.4).

In the simplest case, a voltage-gated ion channel can adopt two conformations: open and closed. If we let O denote the open conformation and C the closed, we can describe the conformational change by

$$C \underset{k_2(V)}{\overset{k_1(V)}{\rightleftharpoons}} O,$$

where the rate "constants" k_1 and k_2 depend on the membrane voltage V. To describe a population of channels, we let $n(t)$ denote the fraction of channels in the open state at time t. The fraction in the closed state is then $1 - n(t)$, so

$$\frac{d}{dt}n(t) = k_1(V)(1 - n(t)) - k_2(V)n(t). \tag{8.9}$$

For a fixed voltage V, this equation has steady state

$$n^{ss}(V) = n_\infty(V) = \frac{k_1(V)}{k_1(V) + k_2(V)},$$

where we have introduced the notation $n_\infty = n^{ss}$. Equation (8.9) can be rewritten as

$$\frac{d}{dt}n(t) = \frac{n_\infty(V) - n(t)}{\tau(V)}, \tag{8.10}$$

where $\tau(V) = 1/(k_1(V) + k_2(V))$ is the time constant for convergence of the channel population to its steady state.

For each ion, the conductance of the membrane is proportional to the number of channels that are open. We will express the conductance g_i as the product of a maximal conductance $\overline{g}_i$ and the fraction of ion-specific channels in the open state, so that

$$g_i(t) = \overline{g}_i\, n(t). \qquad (8.11)$$

In the next section, we will consider a model that incorporates two species of voltage-gated ion channels. In this model, feedback between ion current and membrane voltage results in excitable (action potential) behavior.

8.2.2   The Morris–Lecar Model

In 1981, Catherine Morris and Harold Lecar published a model of excitability in barnacle muscle (Morris and Lecar, 1981). They made use of the modeling formalism that had been established in the 1950s by Hodgkin and Huxley. (The pioneering work of Hodgkin and Huxley was introduced in chapter 1: their model of the squid giant axon is described in problem 8.6.4.)

Figure 8.5 shows the membrane activity of the Morris–Lecar model. Changes in voltage are dominated by two ionic currents: an inward calcium current and an outward potassium current. (Ion pumps maintain concentration gradients for both ions, with $[Ca^{2+}]$ high in the extracellular space and $[K^+]$ high in the cell.) Both the calcium and potassium channels are voltage-dependent.

The ion channels will be described using the formalism developed in the previous section. Let m represent the fraction of open Ca^{2+} channels and w the fraction of

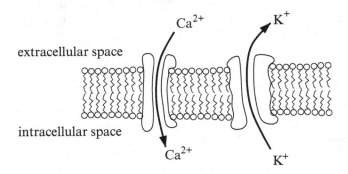

Figure 8.5
Ion exchange in the Morris–Lecar model of barnacle muscle. Voltage changes result from two ion currents: an inward calcium current and an outward potassium current. Both ion channels are voltage-gated.

open K^+ channels, with $m_\infty(V)$, $w_\infty(V)$, $\tau_m(V)$, and $\tau_w(V)$ the corresponding steady states and time constants, as in equation (8.10). Recalling the description of membrane voltage in equation (8.8) and the expression for voltage-gated conductance in equation (8.11), we can describe the electrical behavior at the membrane by

$$\text{Voltage}: \quad \frac{dV}{dt} = \frac{1}{C}(\bar{g}_{Ca}m(t)\cdot(E_{Ca} - V(t)) + \bar{g}_K w(t)\cdot(E_K - V(t)))$$

$$\text{Ca}^{2+} \text{ channels}: \frac{d}{dt}m(t) = \frac{m_\infty(V(t)) - m(t)}{\tau_m(V(t))}$$

$$\text{K}^+ \text{ channels}: \frac{d}{dt}w(t) = \frac{w_\infty(V(t)) - w(t)}{\tau_w(V(t))},$$

where $\bar{g}_{Ca}$ and $\bar{g}_K$ are the maximal membrane conductances.

Experimental observations of the system revealed that the calcium channels relax to steady state much more quickly than the potassium channels (i.e., $\tau_m(V) \ll \tau_w(V)$). Morris and Lecar used this fact to justify a quasi-steady-state approximation for the fraction of open calcium channels $m(t)$:

$$m(t) = m^{qss}(t) = m_\infty(V(t)).$$

As a further simplification, we will take the potassium time constant τ_w as a fixed parameter: it does not vary significantly over the voltage range of interest.

Besides calcium and potassium ions, Morris and Lecar included two other sources of current in their model. The first is a nonspecific "leak" current, which describes the background activity of other ion fluxes. This current is driven by the leak Nernst potential, E_{leak}, with fixed conductance g_{leak}. The second additional current source is an *applied current*, $I_{applied}$, which describes any current injected into the cell by an experimenter or generated in response to signals from other cells. The model is then

$$\frac{d}{dt}V(t) = \frac{1}{C}(\bar{g}_{Ca}m_\infty(V(t))\cdot(E_{Ca} - V(t)) + \bar{g}_K w(t)\cdot(E_K - V(t))$$

$$+ g_{leak}\cdot(E_{leak} - V(t)) + I_{applied}) \tag{8.12}$$

$$\frac{d}{dt}w(t) = \frac{w_\infty(V(t)) - w(t)}{\tau_w}.$$

The steady-state values for the gating variables m and w are given by

$$m_\infty(V) = \frac{1}{2}\left(1 + \tanh\left(\frac{V - V_m^*}{\theta_m}\right)\right) \quad w_\infty(V) = \frac{1}{2}\left(1 + \tanh\left(\frac{V - V_w^*}{\theta_w}\right)\right), \tag{8.13}$$

where V_m^*, V_w^*, θ_m, and θ_n are model parameters. Experimental measurements showed that the steady-state conductances vary in a switch-like manner between zero (all

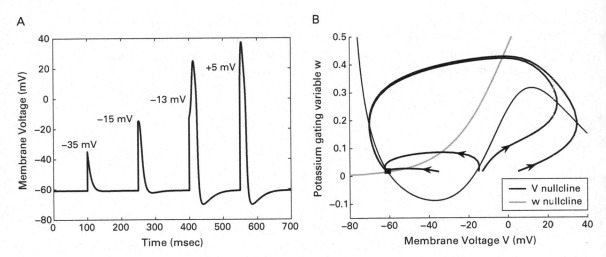

Figure 8.6
Excitability in the Morris–Lecar model. (A) Starting at rest (−61 mV), the membrane voltage V responds
to a series of perturbations, to (in mV) −35, −15, −13, and +5. The first two perturbations are subthreshold;
the response is an immediate return to rest. The latter two perturbations are above threshold; they result
in action potential responses. (B) The responses in (A) are shown in the phase plane, along with the
nullclines. Trajectories that start sufficiently to the right of the V nullcline increase in voltage before
eventually returning to rest. Parameter values: (in mV) $E_K = -84$, $E_{Ca} = 120$, $E_{leak} = -60$, $V_m^* = -1.2$, $V_w^* = 18$,
$\theta_m = 2$, $\theta_w = 30$; (in mS/cm²) $g_K = 8$, $g_{Ca} = 4.4$, $g_{leak} = 2$; $C = 20 \ \mu F/cm^2$, $\tau_w = 20$ milliseconds, $I_{applied} = 0$.

channels closed) and maximal values (all open). Morris and Lecar used the switch-
like function $(1/2)(1+\tanh(\cdot))$ to empirically fit the data.[2] From this formula, m_∞ is
zero (all calcium channels closed) when the membrane voltage V is significantly less
than V_m^*, and m_∞ is one (all calcium channels open) when the membrane voltage V is significantly greater
than V_m^*. The parameter θ_m characterizes how gradual the transition is. Likewise, in
steady state, the potassium channels are mostly closed ($w \approx 0$) for V less than V_w^*
and are mostly open ($w \approx 1$) for V greater than V_w^*.

The model's behavior is illustrated in figure 8.6. Panel A shows the response of
the membrane to a series of depolarizing perturbations (i.e., perturbations that
increase the membrane voltage toward zero). The membrane's resting potential is
about −60 mV, which is the Nernst potential for the leak current. At this value, both
$m_\infty(V)$ and $w_\infty(V)$ are near zero, and so the leak current dominates. The first two
disturbances (to −35 mV and −15 mV, respectively) elicit no response from the
voltage-gated channels: the membrane relaxes exponentially to rest as if it were
nonexcitable (compare with figure 8.3).

2. The function $\tanh(x) = (e^x - e^{-x})/(e^x + e^{-x})$ switches smoothly from −1 to 1 at $x = 0$; the function
$(1/2)(1+\tanh(\cdot))$ thus switches from 0 to 1.

The last two perturbations in figure 8.6A elicit excitations: the membrane potential responds to the perturbation by *increasing*, then dropping below rest before slowly climbing back to steady state. These are action potential responses. The below-rest interval is called the *refractory period*.

Note that an action potential is not triggered by a depolarization to −15 mV but is triggered by the slightly stronger depolarization to −13 mV. This difference in response reveals a threshold for invoking action potential behavior. This threshold can be seen more clearly in the phase plane in figure 8.6B, which shows the four responses from figure 8.6A. The perturbation threshold lies close to the V nullcline. For voltages below this threshold, trajectories converge rapidly to the steady state. When voltage is above threshold, trajectories exhibit an increase in voltage, which continues until the trajectory passes the nullcline's "elbow." Because these "excited" trajectories rapidly converge, the action potential profile is largely independent of the specific form of the initial triggering event.

Action potentials are caused by the activity of voltage-gated ion channels. In this model, super-threshold perturbations elicit significant changes in channel conductance. The calcium channels (m) respond immediately, resulting in an influx of Ca^{2+} and thus an increase in membrane voltage toward the Nernst potential for calcium ($E_{Ca} = 120$ mV). As voltage increases, more calcium channels are opened (positive feedback), leading to an increased rate of depolarization. However, before the voltage reaches E_{Ca}, the potassium channels—which lag the calcium channels in their response—begin to open. The maximal conductance of potassium is twice that of calcium, so once the potassium channels are open, they dominate and draw the membrane voltage toward the Nernst potential for potassium ($E_K = −84$ mV), which is lower than the resting potential. The potassium current drives the potential into the refractory condition, causing the calcium channels to close. (During the refractory period, the potassium channels are still closing, and so a second action potential cannot be triggered; see problem 8.6.2 for details.)

Sustained oscillations can readily be induced in the Morris–Lecar model. Figure 8.7 shows the model's behavior when a steady input current is applied. The simulations in panel A show that at a mid-range applied current, the system's action potential response is repeated indefinitely. The system's relaxation-oscillator limit cycle is shown in panel B. Comparing this phase portrait with figure 8.6B, in which there was no applied current, the V nullcline has shifted so that the "elbows" are on opposite sides of the w nullcline. This shift causes the steady state to be unstable, giving rise to the limit cycle.

Exercise 8.2.1 Consider the Morris–Lecar model in equations (8.12) and (8.13). Suppose that $E_{Ca} > 0 > E_K$.

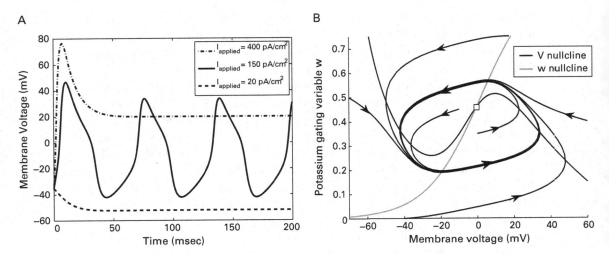

Figure 8.7
Oscillations in the Morris–Lecar model. (A) Starting from a subthreshold membrane voltage, steady-state behavior is observed for low or very high applied currents. For intermediate values, the applied current causes continual excitation, resulting in sustained oscillations. (B) This phase portrait shows the limit-cycle behavior for $I_{\text{applied}} = 150$ pA/cm^2. The trajectories follow the V nullcline until they pass an "elbow" after which the voltage changes abruptly. Parameter values are as for figure 8.6. Adapted from figure 2.9 of Fall et al. (2002).

(a) Noting that $0 < m_\infty < 1$, show that if this system admits an equilibrium $(V^{\text{ss}}, w^{\text{ss}})$, then

$$0 < w^{\text{ss}} < 1$$

$$\frac{g_{\text{leak}} E_{\text{leak}} + \bar{g}_{\text{K}} E_{\text{K}} + I_{\text{applied}}}{g_{\text{leak}} + \bar{g}_{\text{K}}} < V^{\text{ss}} < \frac{g_{\text{leak}} E_{\text{leak}} + \bar{g}_{\text{Ca}} E_{\text{Ca}} + I_{\text{applied}}}{g_{\text{leak}} + \bar{g}_{\text{Ca}}}.$$

(b) This model was reached by applying a quasi-steady-state assumption to the fraction of open calcium channels. If we were to apply a quasi-steady-state assumption to the fraction of open potassium channels as well, we would arrive at a one-dimensional model (for V). Why would this minimal model not be capable of displaying excitable (i.e., action potential) behavior?   □

8.3 Intercellular Communication

There are two primary mechanisms for passing electrical signals from cell to cell.

Neighboring cells whose membranes are in direct contact can exchange ions through *gap junctions*. These junctions are pores that span both cellular membranes, forming a "tunnel" that connects the cytoplasm of the two cells. The resulting ion exchange couples the electrical activity of the cells (as explored in problems 8.6.5 and 8.6.6).

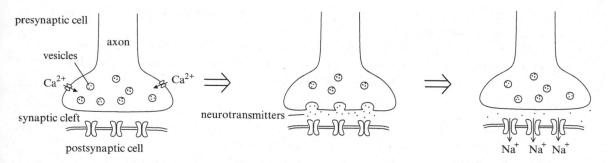

Figure 8.8
Synaptic transmission. When an action potential arrives at the end of an axon, it triggers an influx of calcium ions, which results in release of neurotransmitters into the synaptic cleft. These neurotransmitters diffuse across the cleft and activate receptors in the membrane of the postsynaptic cell, typically leading to current flow (shown as an inward sodium current). Adapted from figure 6.5 of Fall et al. (2002).

Gap junctional couplings tend to synchronize the electrical activities of the communicating cells. An alternative mechanism—which allows for one-way communication—is provided by *chemical synapses*.

8.3.1 Synaptic Transmission

When an action potential reaches the end of a neuron's axon, it activates voltage-gated calcium channels, resulting in an influx of Ca^{2+} across the cell's membrane (figure 8.8). These calcium ions cause the release of signaling molecules, called *neurotransmitters*, which diffuse through the extracellular space and are taken up by receptors in a dendrite of the target cell. The space between the axon of the signaling cell and the "postsynaptic" dendrite is called the *synaptic cleft*.

Because it uses chemical signals, synaptic transmission can implement a rich diversity of signal-response behaviors. The neurotransmitter receptors on the postsynaptic cell are typically complexed to ion channels. When triggered, these receptor–channel complexes can lead to excitation of the target cell membrane. When enough of these *excitatory* receptors are activated, an action potential will be triggered in the target cell, to be passed down its axon to other cells in the network. In contrast, some receptor–channel complexes are *inhibitory*: the ions that flow through these channels contribute to a return to the resting potential. Integration of these excitatory and inhibitory signals at the cell soma allows a wide variety of signaling "decisions" to be implemented. (See problem 8.6.8 for an example.) Additional diversity is provided by the dynamic behaviors of the receptor–channel complexes, which can respond to signals at different timescales and for different durations. (The presence of chemical intermediates in the otherwise electrical communication network allows neuronal behavior to be directly regulated by biochemical means. For example, the antidepressant Prozac increases the levels of the neurotransmitter serotonin; the drug nicotine mimics the neurotransmitter acetylcholine, causing persistent activation of acetylcholine receptors.)

We next develop a simple model of synaptic transmission. Suppose the postsynaptic receptor–channel complexes are open when bound to neurotransmitter but are otherwise closed. A simple kinetic model is

$$C + T \underset{\beta}{\overset{\alpha}{\rightleftharpoons}} O,$$

where C and O are closed and open channels, and T is the neurotransmitter. Letting s denote the fraction of channels in the open state and using T to denote the abundance of neurotransmitter molecules, we can write

$$\frac{d}{dt} s(t) = \alpha T(t) \cdot (1 - s(t)) - \beta s(t). \tag{8.14}$$

The time course for T is determined by the pre-synaptic cell. Presuming rapid neurotransmitter dynamics, we suppose that the transmitter is present in the cleft (at concentration T_{max}) whenever the voltage in the pre-synaptic cell (V_{pre}) is above a threshold voltage (V_{syn}^*). Thus,

$$T(t) = \begin{cases} 0 & \text{if } V_{pre} < V_{syn}^* \\ T_{max} & \text{if } V_{pre} \geq V_{syn}^* \end{cases}. \tag{8.15}$$

We can describe the neurotransmitter-induced current into the postsynaptic cell as

$$I_{syn}(t) = \overline{g}_{syn} s(t)(E_{syn} - V_{post}(t)),$$

where $\overline{g}_{syn}$ is the maximal conductance through the neurotransmitter-gated channels, E_{syn} is the Nernst potential for the corresponding ion, and V_{post} is the membrane voltage of the postsynaptic cell.

To illustrate synaptic dynamics, we use the Morris–Lecar model to describe two communicating neurons. (Although the Morris–Lecar model was developed to describe muscle cells, it provides a simple model of action-potential generation and so can be used as a generic model of neuronal activity.) Implementing the synaptic current as $I_{applied}$ in the postsynaptic cell, we arrive at a two-neuron model:

$$\frac{d}{dt} V_{pre}(t) = \frac{1}{C} (\overline{g}_{Ca} m_\infty(V_{pre}(t)) \cdot (E_{Ca} - V_{pre}(t)) + \overline{g}_K w_{pre}(t) \cdot (E_K - V_{pre}(t))$$
$$+ g_{leak} \cdot (E_{leak} - V_{pre}(t)))$$

$$\frac{d}{dt} w_{pre}(t) = \frac{w_\infty(V_{pre}(t)) - w_{pre}(t)}{\tau_w}$$

$$\frac{d}{dt}V_{\text{post}}(t) = \frac{1}{C}\left(\bar{g}_{\text{Ca}}m_{\infty}(V_{\text{post}}(t))\cdot(E_{\text{Ca}} - V_{\text{post}}(t)) + \bar{g}_{\text{K}}w_{\text{post}}(t)\cdot(E_{\text{K}} - V_{\text{post}}(t))\right.$$

$$\left. + g_{\text{leak}}\cdot(E_{\text{leak}} - V_{\text{post}}(t)) + \bar{g}_{\text{syn}}s(t)(E_{\text{syn}} - V_{\text{post}}(t))\right)$$

$$\frac{d}{dt}w_{\text{post}}(t) = \frac{w_{\infty}(V_{\text{post}}(t)) - w_{\text{post}}(t)}{\tau_w}$$

$$\frac{d}{dt}s(t) = \alpha T(t)\cdot(1 - s(t)) - \beta s(t),$$

where $T(t)$ is defined by equation (8.15). For simplicity, we have used identical parameters for the two cells. Figure 8.9A shows a simulation of this model in which an action potential in the pre-synaptic cell triggers a downstream action potential in the postsynaptic cell. Figure 8.9B shows the synaptic current, which is triggered by the action potential in the pre-synaptic cell.

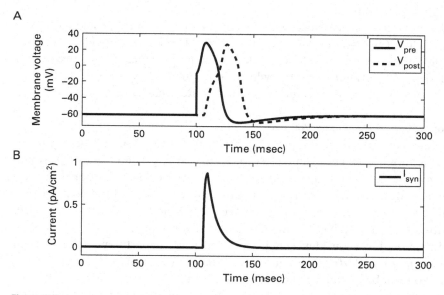

Figure 8.9
Synaptic transmission in the Morris–Lecar model. (A) An action potential is triggered in the pre-synaptic cell at time $t = 100$ milliseconds. A postsynaptic action potential follows after a short lag. (B) When the action potential reaches its peak, it causes channels to open in the postsynaptic cell, resulting in a spike in the synaptic current I_{syn}. This current induces the action potential in the postsynaptic cell. Parameter values: $\bar{g}_{\text{syn}} = 1$ mS/cm^2, $E_{\text{syn}} = 150$ mV, $\alpha = 1$ ms^{-1}, $\beta = 0.12$ ms^{-1}, $T_{\text{max}} = 1$, $V_{\text{syn}}^* = 27$ mV. Neurotransmitter units are arbitrary. The remaining parameter values are as for figure 8.6.

8.4*   Spatial Modeling

By using ordinary differential equations (ODEs) to model electrical activity at the membrane, we have implicitly assumed spatial homogeneity of membrane features and ion concentrations. This assumption is justified if attention is restricted to a small membrane patch or if electrical activity happens to be uniform over the entire cell membrane. However, ODE models cannot describe behavior over the extended projections of a neuron (the axon and dendrites).

In this section, we introduce *partial differential equation* (PDE) models, which describe dynamic behavior that is spatially distributed. We will use PDE models to describe the propagation of an action potential along the neuronal axon. (Appendix B contains some additional details on PDE modeling.) PDE models can also be used to describe biochemical and genetic networks, although their use in these domains is somewhat limited by the difficulty of collecting the relevant intracellular data. Problems 8.6.10 and 8.6.11 explore some PDE models of developmental signaling via morphogen diffusion.

Spatial effects can be described with ODE models, but only through compartmental modeling (as discussed in section 3.4). In this section, we will construct a PDE model by treating a spatially extended domain as a collection of small compartments. By considering the limit in which the compartment size shrinks to zero, we will arrive at a spatially continuous model of system behavior.

8.4.1   Propagation of Membrane Voltage

We will treat a cellular projection (i.e., an axon or dendrite) as a long, thin cylindrical tube. This allows us to keep our discussion of spatial effects simple, by describing behavior in only one spatial dimension. We will presume that membrane activity and ion concentrations are uniform in the *radial* direction (i.e., in the cross section of the tube) but vary in the *axial* direction (i.e., along the length of the tube). The axial position will be described by the independent variable x.

To build a model, we consider a chain of cylindrical compartments (slices), each with radius r and length dx, as in figure 8.10. Current (of positive charge) flowing along the axon will be denoted I_a. We will presume that the transmembrane (radial) current, denoted I_r, and the membrane voltage, V, are uniform in each compartment. (This assumption is justified provided the length dx is small.)

If the flow into a compartment does not match the flow out of the compartment, then charge builds up, causing a change in the local membrane potential. Consider the compartment at position x. With the direction of flow as indicated in figure 8.10, the axial flow into this compartment is $I_a(x)$, whereas the axial flow out of the compartment is $I_a(x + dx)$. To maintain consistency with the parameters in the previous section, we let I_r describe the per-area transmembrane current. The total transmem-

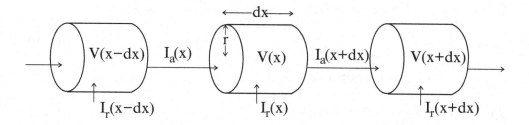

radial direction

axial direction

Figure 8.10
Compartmentalization of a long, thin cellular projection. The cellular "tube" is divided into small slices. Current flows along the tube (I_a) and across the membrane (I_r). The transmembrane current and voltage are presumed uniform over each cylindrical compartment of length dx.

brane current into the compartment is then $2\pi r I_r(x)dx$. (Recall that the surface area around a cylinder of length L and radius R is $2\pi RL$.) Likewise, if we let C denote the per-area capacitance of the membrane, then the capacitance of the compartment membrane is $2\pi r Cdx$. Charge balance then gives (compare with equation 8.3):

$$I_a(x) + 2\pi r I_r(x)dx - I_a(x+dx) = 2\pi r Cdx \frac{d}{dt} V(x,t). \tag{8.16}$$

We next consider the axial current, which is driven by voltage differences in the axial direction. Applying Ohm's law (compare with equation 8.4) gives

$$I_a(x) = g_a \cdot (V(x-dx) - V(x)), \tag{8.17}$$

where g_a is the conductivity of the compartment. This conductivity is proportional to the cross-sectional area of the compartment and inversely proportional to its length, giving

$$g_a = \bar{g}_a \frac{\pi r^2}{dx}, \tag{8.18}$$

where $\bar{g}_a$ is the per-length conductivity of the cell's cytoplasm.

To arrive at a differential equation, we consider the limit in which each compartment becomes vanishingly thin, that is $dx \to 0$. In this limit, equation (8.16) becomes

$$-\frac{d}{dx} I_a(x) = 2\pi r \left(C \frac{d}{dt} V(x) - I_r(x) \right). \tag{8.19}$$

In the same limit, equation (8.17), together with equation (8.18), gives

$$I_a(x) = -\overline{g}_a \pi r^2 \frac{d}{dx} V(x). \tag{8.20}$$

Differentiating equation (8.20) and substituting into equation (8.19) gives the PDE

$$\overline{g}_a r \frac{\partial^2}{\partial x^2} V(x,t) = 2\left(C \frac{\partial}{\partial t} V(x,t) - I_r(x,t) \right), \tag{8.21}$$

where we use partial derivative notation to distinguish rates of change with respect to time ($\partial/\partial t$) and position ($\partial/\partial x$).

Equation (8.21) is called the *cable equation*. To use this equation as a model for membrane voltage, we need to specify how the transmembrane current depends on the voltage. Before addressing the propagation of action potentials, we consider the simpler case of a nonexcitable membrane.

8.4.2 Passive Membrane

As discussed in section 8.1.2, current across a nonexcitable membrane is proportional to the difference between the membrane voltage and the resting potential. In the single-ion case,

$$I_r(x, t) = g_m \cdot (E - V(x, t)), \tag{8.22}$$

where g_m is the per-area membrane conductance. Substituting equation (8.22) into the cable equation (8.21) and rearranging gives

$$\frac{\partial}{\partial t} V(x,t) = \frac{1}{C}\left(\frac{\overline{g}_a r}{2} \frac{\partial^2}{\partial x^2} V(x,t) + g_m \cdot (E - V(x,t)) \right),$$

which is called the *linear cable equation*.

The simulation in figure 8.11A shows the response of this passive membrane model to a perturbation in voltage. The depolarizing perturbation is quickly smoothed out as charge diffuses through the membrane and along the length of the cell. (Compare with figure 8.3.) The simulation snapshots in figure 8.11B confirm that the entire length of membrane returns quickly to rest.

8.4.3 Excitable Membrane: Action Potential Propagation

Next, we consider propagation of voltage disturbances in an excitable membrane. We use the Morris–Lecar model for transmembrane current,

$$I_r(x,t) = \overline{g}_{Ca} m_\infty(V(x,t)) \cdot (E_{Ca} - V(x,t)) + \overline{g}_K w(t) \cdot (E_K - V(x,t)) + g_{leak} \cdot (E_{leak} - V(x,t)),$$

and arrive at a *nonlinear cable equation*

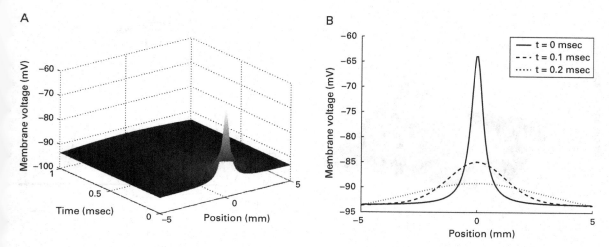

Figure 8.11
Passive membrane response. (A) The depolarizing perturbation at time zero quickly dissipates: charge diffuses both across the membrane and along the length of the cell. (B) Snapshots at different time points show the entire domain converging quickly to rest. (Boundary conditions are clamped at the resting potential on the domain shown.) Parameter values: $C = 0.98\ \mu F/cm^2$, $g = 0.0144\ mS/cm^2$, $E = -93.6\ mV$, $g_a = 30\ mS/cm$, and $r = 0.01\ cm$.

$$\frac{\partial}{\partial t}V(x,t) = \frac{1}{C}\left(\frac{\bar{g}_a r}{2}\frac{\partial^2}{\partial x^2}V(x,t) + \bar{g}_{Ca}m_\infty(V(x,t))\cdot(E_{Ca}-V(x,t)) \right.$$

$$\left. + \bar{g}_K w(t)\cdot(E_K - V(x,t)) + g_{leak}\cdot(E_{leak}-V(x,t)) \right)$$

$$\frac{d}{dt}w(x,t) = \frac{w_\infty(V(x,t))-w(x,t)}{\tau_w},$$

where w_∞ and m_∞ are given by equation (8.13).

Figure 8.12A shows the model behavior. An initial excitation triggers an action potential at position $x = 0$. This voltage profile then propagates down the length of the cell as a *traveling wave*. The time snapshots in figure 8.12B show the initial disturbance (at $t = 0$) and the characteristic action potential profile in two different positions along the cell length. Both the profile and the speed of propagation (called the *wave speed*) are independent of the form of the triggering disturbance (see problem 8.6.9).

8.5   Suggestions for Further Reading

• **Mathematical Neuroscience**   Mathematical models of neurons are introduced in the books *Computational Cell Biology* (Fall et al., 2002), *Mathematical Physiology*

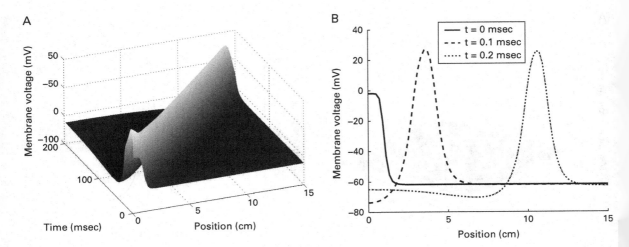

Figure 8.12
Propagation of an action potential in an excitable membrane. (A) The depolarizing perturbation at time zero triggers an action potential, which propagates along the length of the cell as a traveling wave. (B) Snapshots at different time points show the initial perturbation and the action potential profile at two different positions. (Boundary conditions are free, at $x = \pm 30$ cm.) Parameter values: $g_a = 30$ mS/cm, and $r = 0.05$ cm; other parameter values are as for figure 8.6 except $g_{leak} = 0.5$ mS/cm^2.

(Keener and Sneyd, 1998), and *Mathematical Foundations of Neuroscience* (Ermentrout and Terman, 2010). A survey of research in the field is presented in *Methods in Neural Modeling: From Ions to Networks* (Koch and Segev, 1998).

• **Electrophysiology**   The biophysics of electrophysiology is covered in *Biological Physics* (Nelson, 2004) and *Physical Biology of the Cell* (Phillips et al., 2009).

• **Spatial Modeling**   Spatial modeling in systems biology is reviewed in the book chapters "Spatial Modeling" (Iglesias, 2010) and "Kinetics in Spatially Extended Domains" (Kruse and Elf, 2006).

8.6   Problem Set

8.6.1   Passive Resting Potential

Consider a cell membrane that allows transfer of K$^+$ ions and Na$^+$ ions. Suppose the ion channels are not voltage dependent, so the membrane is not excitable. Let g_K and g_{Na} denote the membrane conductances for potassium and sodium ions, respectively, and let g_L denote the conductance of the membrane to a generic leak of ions (each in millisiemens per square centimeter; mS/cm^2). Let E_K, E_{Na}, and E_L denote the Nernst potentials (in mV) with respect to each of the currents, and suppose the membrane has capacitance C (microfarads per square centimeter; μF/cm^2).

(a) Use the membrane model (equation 8.8) to write a differential equation describing the voltage across the membrane.

(b) Suppose the parameter values are $C = 1$ μF cm^{-2}, $g_K = 36$ mS cm^{-2}, $g_{Na} = 82$ mS cm^{-2}, $g_L = 0.3$ mS cm^{-2}, $E_K = -72$ mV, $E_{Na} = 22$ mV, and $E_L = -49$ mV. What is the resting potential in this case?

(c) Suppose the membrane is treated with an agent that blocks 50% of the potassium channels. What is the resulting resting potential?

(d) Suppose that a stronger agent is added that blocks all potassium channels. Explain how a good approximation to the resulting resting potential can be directly identified from the parameter values given above, with no further calculation.

8.6.2 Morris–Lecar Model: Refractory Period
Consider the Morris–Lecar model (equations 8.12 and 8.13). Using the parameter values in figure 8.6, simulate the system to steady state. Run a second simulation that starts at steady state and introduces a 10-millisecond pulse of $I_{applied} = 150$ picoamperes per square centimeter (pA/cm^2), thus triggering an action potential. Next, augment your simulation by introducing a second, identical 10-millisecond burst of $I_{applied}$ that begins 100 milliseconds after the end of the first pulse. Verify that this triggers a second action potential that is identical to the first.

Next, explore what happens when less time elapses between the two triggering events. What is the response if the 10-millisecond pulses are separated by only 60 milliseconds? 30 milliseconds? For each case, plot the gating variable $w(t)$ as well as the voltage $V(t)$. Verify that even after the voltage has returned to near-rest levels, a second action potential cannot be triggered until the gating variable $w(t)$ has returned to its resting value. Provide an interpretation of this behavior.

8.6.3 Morris–Lecar Model: Bifurcation Diagram
Consider the Morris–Lecar model (equations 8.12 and 8.13). Using the parameter values in figure 8.6, produce a bifurcation diagram showing the steady-state voltage V as a function of the applied current $I_{applied}$ over the range 0–300 pA/cm^2. Determine the bifurcation points and verify that they are consistent with the behavior in figure 8.7A.

8.6.4 The Hodgkin–Huxley Model
The Morris–Lecar model (section 8.2.2) was built on a framework developed by Alan Hodgkin and Andrew Huxley (see section 1.6.4). The Hodgkin–Huxley model, which describes the excitable membrane of the squid giant axon, involves two ionic species: potassium (K$^+$) and sodium (Na$^+$).

The model formulation is similar to the Morris–Lecar model. The transmembrane voltage is described by

$$\frac{d}{dt}V(t) = \frac{1}{C}\left(g_{Na}(t)\cdot(E_{Na}-V(t)) + g_K(t)\cdot(E_K-V(t)) + g_{leak}\cdot(E_{leak}-V(t)) + I_{applied}\right).$$

The potassium conductance is given by

$$g_K(t) = \bar{g}_K(n(t))^4,$$

where $\bar{g}_K$ is the maximal conductance, and n is a gating variable. (The exponent characterizes the fact that four components of each channel must be in the open state in order for the channel to pass K^+ ions.) The sodium conductance is given by

$$g_{Na}(t) = \bar{g}_{Na}(m(t))^3 h(t),$$

where $\bar{g}_{Na}$ is the maximal conductance, m is an activation gating variable, and h is an inactivation gating variable.

The gating variable dynamics are given by

$$\frac{d}{dt}n(t) = \frac{n_\infty(V(t)) - n(t)}{\tau_n(V(t))}$$

$$\frac{d}{dt}m(t) = \frac{m_\infty(V(t)) - m(t)}{\tau_m(V(t))}$$

$$\frac{d}{dt}h(t) = \frac{h_\infty(V(t)) - h(t)}{\tau_h(V(t))},$$

where

$$n_\infty(V) = \frac{0.01(V+50)/(1-e^{-(V+50)/10})}{0.01(V+50)/(1-e^{-(V+50)/10}) + 0.125e^{-(V+60)/80}}$$

$$\tau_n(V) = \frac{1}{0.01(V+50)/(1-e^{-(V+50)/10}) + 0.125e^{-(V+60)/80}}$$

$$m_\infty(V) = \frac{0.1(V+35)/(1-e^{-(V+35)/10})}{0.1(V+35)/(1-e^{-(V+35)/10}) + 4e^{-(V+60)/18}}$$

$$\tau_m(V) = \frac{1}{0.1(V+35)/(1-e^{-(V+35)/10}) + 4e^{-(V+60)/18}}$$

$$h_\infty(V) = \frac{0.07e^{-(V+60)/20}}{0.07e^{-(V+60)/20} + 1/(1+e^{-(V+30)/10})}$$

$$\tau_h(V) = \frac{1}{0.07e^{-(V+60)/20} + 1/(1+e^{-(V+30)/10})}.$$

Model parameters are $\bar{g}_{Na} = 120$ mS/cm^2, $\bar{g}_K = 36$ mS/cm^2, $g_{leak} = 0.3$ mS/cm^2, $E_{Na} = 55$ mV, $E_K = -72$ mV, $E_{leak} = -49$ mV, and $C = 1$ μF/cm^2.

(a) Simulate the system with the given parameter values. Determine the resting potential and the rest values for the gating variables. Verify that an action potential occurs when the voltage is increased sufficiently from rest. What is the excitability threshold? That is, what is the initial voltage above which an action potential occurs (with the gating variables starting from their rest conditions)?

(b) Generate plots of the gating variables and the ionic currents during an action potential. Verify that the action potential begins with a sodium influx, followed by the flow of potassium out of the cell.

(c) When the external sodium concentration is reduced, the action potential behavior is diminished or eliminated. The Nernst potential for sodium is determined by equation (8.1) with $[Na^+]_{ext} = 440$ mM and $[Na^+]_{int} = 50$ mM. Consider an experimental condition in which the external sodium level is reduced to 44 mM. Determine the corresponding Nernst potential and run a simulation of the model in these conditions. How has the action potential behavior changed? Repeat for a condition in which $[Na^+]_{ext} = 4.4$ mM.

(d) The original development of this model relied heavily on *voltage clamp* experiments, in which feedback circuitry was used to keep the membrane voltage steady as the ionic currents were activated. Modify the model to allow simulation of a voltage clamp experiment. (That is, modify the model so that the voltage V is a fixed model parameter.) Simulate an experiment in which the voltage is increased abruptly from -65 mV to -9 mV. Plot the resulting ionic currents. Explain your results in terms of the dynamics in the gating variables. (Hodgkin and Huxley did the opposite—they used the results of voltage clamp experiments to determine the gating dynamics.)

Additional details on the development of the Hodgkin–Huxley model can be found in Guevara (2003).

8.6.5   Gap-Junctional Coupling: Passive Membranes

Consider the following model of two identical (nonexcitable) cells coupled by gap junctions:

$$\frac{d}{dt}V_1(t) = \frac{1}{C}(g_m(E - V_1(t)) + g_c(V_2(t) - V_1(t)))$$

$$\frac{d}{dt}V_2(t) = \frac{1}{C}(g_m(E - V_2(t)) + g_c(V_1(t) - V_2(t))).$$

In each cell, V_i is the membrane voltage. The capacitance, C, the membrane conductance, g_m, and the Nernst potential, E, are identical for both cells. The parameter g_c is the conductance of the gap-junctional coupling. The Ohmic gap junctional current flows from cell 1 to cell 2 at rate $g_c(V_1(t) - V_2(t))$. Consider the average potential $V_a = (1/2)(V_1 + V_2)$. Derive a differential equation for V_a and determine its solution. (Hint: Refer to section 2.1.3.) Repeat for the difference in potentials $V_d = V_1 - V_2$.

8.6.6 Gap-Junctional Coupling: Excitable Membranes

Consider a pair of excitable cells coupled by gap junctions. Build a model of this pair by describing each cell with the Morris–Lecar model (equations 8.12 and 8.13) and adding an Ohmic gap-junctional current from cell 1 to cell 2 at rate $g_c(V_1(t) - V_2(t))$. Take parameters as in figure 8.7B, so that each cell displays periodic oscillatory behavior. Take $g_c = 0.1$ mS/cm^2 and run simulations to verify that the coupling brings the oscillations in the two cells to synchrony. (Choose different initial conditions for the two cells so that they begin their oscillations out of phase.) How does the time required to reach synchrony change as the coupling strength g_c changes?

8.6.7 Synchronization by Synaptic Signaling

Consider two identical cells, each described by the Morris–Lecar model (equations 8.12 and 8.13), with membrane voltages V_1 and V_2. Take parameter values as in figure 8.7B, so that each cell exhibits periodic oscillatory behavior. Next, modify the model to describe mutual synaptic connections between the two cells. That is, add a current of the form

$$\bar{g}_{syn} s_1(t)(E_{syn} - V_1(t))$$

to cell 1, with

$$\frac{d}{dt} s_1(t) = \alpha T_2(t)(1 - s_1(t)) - \beta s_1(t)$$

$$T_2(t) = \begin{cases} 0 & \text{if } V_2 < V^*_{syn} \\ T_{max} & \text{if } V_2 \geq V^*_{syn} \end{cases},$$

and likewise for cell 2 (with the dynamics of s_2 driven by T_1). Take parameter values as in figure 8.9 for the two (symmetric) synaptic connections.

(a) Simulate the model. Choose different initial conditions for the two cells so that they begin their oscillations out of phase. Verify that for these parameter values, the mutual excitation brings the cells into a synchronous firing pattern.

(b) Explore the effect of reducing the activation timescale for the synaptic channels. (That is, explore the effects of reducing the parameter α.) What happens to the synchronization time as α is decreased? Verify that slower activation (e.g., α in the range 0.05–0.5 ms^{-1}) results in a longer lapse before synchrony is reached. What happens when α is reduced to 0.01 ms^{-1}?

(c) With α at its nominal value (1 ms^{-1}), explore the effect of changing the inactivation timescale for the synaptic channels. (That is, explore the effects of changing the parameter β.) How does the period change as β is decreased? What happens when β is increased to 4 ms^{-1}?

(d) The effect of the synaptic connections is inhibitory if the synaptic Nernst potential is close to the resting potential. Change E_{syn} to −50 mV and verify that (with the other parameter values as in figure 8.9), the cells fall into an *antiphase* behavior, in which they are exactly out of phase.

8.6.8 Computation in Neural Circuits

Consider a network in which two neurons both synapse onto a third. Suppose that neither presynaptic neuron is able individually to trigger a response in the target cell, but the combined input (both synapses firing) is sufficient to trigger an action potential. The target neuron then implements an AND-gate logic (see section 7.5.1). To simulate such a network, consider three identical Morris–Lecar neurons (modeled by equations 8.12 and 8.13). The synaptic current into the target cell can be described by

$$I_{syn}(t) = \overline{g}_{syn1} s_1(t)(V_3(t) - E_{syn1}) - \overline{g}_{syn2} s_2 (V_3(t) - E_{syn2}),$$

where V_3 is the voltage of the target cell, and $\overline{g}_{syni}$, s_i, and E_{syni} characterize the synaptic signal from the input cells ($i = 1, 2$). The synaptic dynamics are given by

$$\frac{d}{dt} s_1(t) = \alpha_1 T_1 (1 - s_1(t)) - \beta_1 s_1(t) \qquad \frac{d}{dt} s_2(t) = \alpha_2 T_2 (1 - s_2(t)) - \beta_2 s_2(t),$$

where

$$T_1(t) = \begin{cases} 0 & \text{if } V_1 < V_{syn}^* \\ T_{max} & \text{if } V_1 \geq V_{syn}^* \end{cases} \qquad T_2(t) = \begin{cases} 0 & \text{if } V_2 < V_{syn}^* \\ T_{max} & \text{if } V_2 \geq V_{syn}^* \end{cases}.$$

(a) Take parameter values for the cells as in figure 8.6 and treat the synapses as identical, with parameter values as in figure 8.9, but set $E_{syn1} = E_{syn2} = 60$ mV. Verify that the target cell does not respond significantly to an individual action potential in either input cell but produces an action potential when the input cells fire simultaneously.

(b) The AND gate network in part (a) can be viewed as an *incidence detector*, because it responds only to simultaneous action potentials in the input cells. Explore this incidence-detection property by determining how much time can elapse between the action potentials in cells 1 and 2 while still triggering a response from cell 3.

(c) How can you modify the model in part (a) to arrive at an OR-gate logic in the target neuron? Confirm your conjecture by simulation.

(d) If E_{syn2} is near the resting potential, cell 2 will repress activation of the target cell. Consider a model for which $E_{syn1} = 150$ mV and $E_{syn2} = -80$ mV. How does the target cell respond to a signal from cell 1? From cell 2? What about simultaneous signals from both cells?

8.6.9* Propagation of Action Potentials

Referring to section 8.4.3, the action potential shown in figure 8.12 was triggered by a narrowly peaked voltage pulse as shown in figure 8.12B. That initial pulse is given (in mV) by

$$V(x,0) = \frac{60}{1+x^8} - 61.85.$$

(Note: the resting potential is -61.85 mV. Position x is in cm.)

(a) Simulate the system using parameter values as in figure 8.12. Explore the dependence on initial conditions by simulating from initial voltage pulses of different shape or strength. Confirm that as long as the initial profile triggers an action potential, the long-term shape of the action potential profile—and the speed at which it propagates—are independent of the initial condition. (Appendix B provides some details on PDEs. Appendix C contains a brief tutorial on simulation of PDEs in MATLAB.)

(b) How would you expect the wave speed to change if the axial conductivity $\bar{g}_a$ is increased? Verify your conjecture via simulation.

8.6.10* Morphogen Diffusion

A morphogen is a chemical signal that induces cells to follow developmental pathways. As a simple model of developmental signaling, consider a morphogen that diffuses along a one-dimensional domain of length L. If the morphogen decays at rate k, the concentration of morphogen is described by the reaction–diffusion equation

$$\frac{\partial}{\partial t} c(x,t) = D \frac{\partial^2}{\partial x^2} c(x,t) - kc(x,t),$$

where D is the diffusion constant. Consider the case in which the morphogen is produced at one end of the domain ($x = 0$) and diffuses freely out of the domain at the other end ($x = L$). The boundary conditions are then

$$\left. \frac{\partial}{\partial x} c(x,t) \right|_{x=0} = -a \qquad \left. \frac{\partial}{\partial x} c(x,t) \right|_{x=L} = 0$$

(The parameter a is positive. The minus sign ensures that at $x = 0$, there is flux *into* the domain.)

(a) Take parameter values $k = 1\ \mathrm{s}^{-1}$, $D = 10\ \mu\mathrm{m}^2/\mathrm{s}$, $a = 1/10\ \mu\mathrm{M}/\mu\mathrm{m}$, and $L = 10\ \mu\mathrm{m}$. Simulate the system to steady state. (Appendix B provides some details on PDEs. Appendix C contains a brief tutorial on simulation of PDEs in MATLAB.)

(b) Explore the effect of changes in the diffusion coefficient D on the steady-state morphogen profile. Verify that as D decreases, the steady-state morphogen concentration at $x = L$ decreases, whereas the steady-state concentration at $x = 0$ increases.

(c) Explore the effect of changes in the domain length L on the steady-state morphogen profile. Verify that as the domain length increases, the steady-state concentration at both ends of the domain decreases, but that for large domains, the concentration at $x = 0$ does not drop below $a\sqrt{k/D}$ μM.

8.6.11* Pattern Formation in Development

In the 1950s, Alan Turing suggested that diffusion of morphogens could lead to spontaneous pattern formation, depending on the interactions between the morphogens. With two morphogens U and V diffusing in a one-dimensional domain, the dynamics in concentrations u and v can be described by

$$\frac{\partial}{\partial t}u(x,t) = D_u \frac{\partial^2}{\partial x^2}u(x,t) + F(u(x,t), v(x,t))$$

$$\frac{\partial}{\partial t}v(x,t) = D_v \frac{\partial^2}{\partial x^2}v(x,t) + G(u(x,t), v(x,t)).$$

The parameters D_u and D_v characterize the morphogen-specific diffusion rates; the functions F and G specify the interactions between the morphogens. Diffusion-driven patterns (called *Turing patterns*) can be generated as follows. (Appendix B provides some details on PDEs. Appendix C contains a brief tutorial on simulation of PDEs in MATLAB.)

(a) Suppose that species U activates production of both U and V, whereas species V inhibits production of U. The resulting *activator–inhibitor* system can exhibit spontaneous patterning if the inhibitor diffuses more readily than the activator (leading to local activation and long-range inhibition). Verify this result by simulating the system with

$$F(u,v) = \frac{\alpha u^2}{v} - \beta u \qquad \text{and} \qquad G(u,v) = \gamma u^2 - \delta v.$$

Take parameter values $D_u = 1\ \mu\mathrm{m}^2/\mathrm{s}$, $D_v = 10\ \mu\mathrm{m}^2/\mathrm{s}$, $\alpha = 1\ \mu\mathrm{M}/\mathrm{s}$, $\beta = 1/\mathrm{s}$, $\gamma = 1/\mu\mathrm{M}/\mathrm{s}$, and $\delta = 1.1/\mathrm{s}$. To observe spontaneous pattern formation, begin with a domain spanning $x = -20$ to $x = 20$ μm and take initial conditions (in μM) $u(x,0) = 1$, $v(x,t)$ $= 1/(x^2 + 1)$. Use free (zero-gradient) boundary conditions. Comment on the steady-state system behavior. Next, simulate over a larger domain, $x = -40$ to $x = 40$ μm. How has the steady-state pattern changed? The number of stripes is dependent on the domain size. If the domain is too small to support a single stripe, the homogeneous solution is stable, and no spatial pattern occurs. Simulate the system on the domain $x = -1$ to $x = 1$ μm to verify this result.

(b) Alternatively, suppose that species U activates its own production while inhibiting species V, and species V is self-inhibitory while activating U. This is called a *substrate-depletion* system: V is the substrate, U is the product. Set

$$F(u, v) = \alpha u + v - r_1 u v - r_2 u v^2 \quad \text{and} \quad G(u, v) = -\gamma u - \beta v + r_1 u v + r_2 u v^2.$$

Take parameter values $D_u = 0.45$ μm^2/s, $D_v = 6$ μm^2/s, $\alpha = \gamma = \beta = 0.9$/s, $r_1 = 0.2/\mu$M/s, and $r_2 = 0.02/\mu$M^2/s. To observe spontaneous pattern formation, begin with a domain spanning $x = -20$ to $x = 20$ μm and take initial conditions (in μM) $u(x,0) = 1$, $v(x,t)$ $= 1/(x^2 + 1)$. Use free (zero-gradient) boundary conditions. (This model describes displacements from a nominal concentration, so u and v can take negative values.) Comment on the steady-state behavior. Next, simulate over a larger domain, $x = -40$ to $x = 40$ μm. How has the steady-state pattern changed? The number of stripes is dependent on the domain size. As in part (a), if the domain is too small to support a single stripe, the homogeneous solution is stable, and no spatial pattern occurs. Simulate on the domain $x = -2$ to $x = 2$ μm to verify this result.

An accessible account of Turing patterns (and wave numbers, which dictate the number of stripes that appear in the simulation) can be found in Iglesias (2010).

Appendix A: Molecular Cell Biology

As living things ourselves, biology seems familiar to us, and we would like to find it easy to understand. But in many ways, rather than being the concrete science we like to think of it as, biology is more like mathematics, and thinking about life demands a great deal of mental work, much of it unfortunately very abstract. Besides, although living organisms are made up of a *finite* number of basic objects, there are still a great many of them. Unlike in physics or chemistry, where the numbers are small, in biology we have to deal with somewhere between a few hundred and several thousand objects, even if we simplify things as much as possible.
—Antoine Danchin, *The Delphic Boat*

This appendix introduces basic notions of molecular cell biology that are used in this book. The material presented here is covered in introductory courses on cell biology, biochemistry, and genetics. Standard references include *Molecular Biology of the Cell* (Alberts et al., 2007), *Lehninger Principles of Biochemistry* (Nelson and Cox, 2008), and *Genetics: From Genes to Genomes* (Hartwell et al., 2010). Concise introductions to molecular cell biology are presented in *New Biology for Engineers and Computer Scientists* (Tözeren and Byers, 2004) and *Systems Biology: A Textbook* (Klipp et al., 2009). There are many excellent online sources as well; Wikipedia is a great starting point for many topics.

A.1 Cells

The cell is the unit of life. All living organisms are composed of one or more cells. The models in this book address cellular physiology—the function and activity of cells.

Our human bodies are composed of about 50 trillion cells. Those cells grow from a single fertilized egg cell through the process of organism *development*. During development, progenitor cells *differentiate* (i.e., specialize) into about 200 different cell types. These cells are organized into *tissues*—groups of cells that together perform specific roles. Tissues are organized into organs, which make up our bodies. Cells themselves contain *organelles*, which are analogous to our own organs.

Cells are tiny. Animal cells are about 10 micrometers[1] long, with a volume of about 1 picoliter (10^{-12} L). Bacterial cells are about 1000-fold smaller (volume about 1 femtoliter (10^{-15} L), length about 1 micrometer). About 250,000 of these bacterial cells could fit comfortably on the period at the end of this sentence.

The interior of a cell is an aqueous environment, meaning that water is the primary solvent (making up about 70% of cell mass). However, cells are dense with macromolecules, so their interiors have the consistency of a gel, rather than a watery solution. Mixing by diffusion is much slower in cells than in water. In small cells, this mixing is nevertheless very fast. It takes about 10–100 milliseconds for a macromolecule to diffuse across a bacterial cell; smaller molecules move faster. Diffusion across an animal cell is about 1000-fold slower.

Cellular Organization

Cells can be classified into two primary cell types, as shown in figure A.1. Bacteria, which are unicellular organisms, are *prokaryotic*, whereas animals, plants, fungi, and single-celled protists are composed of *eukaryotic* cells.

Each cell is surrounded by a membrane. The material inside the membrane is called the *cytoplasm*. It contains, among other things, the ribosomes and DNA (encoding the genome). Eukaryotic cells have additional structure. They have a

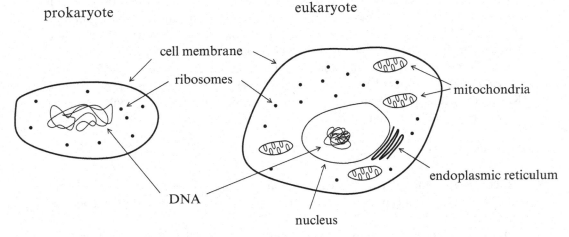

Figure A.1
Cell structure. Both prokaryotic and eukaryotic cells are surrounded by membranes. This sketch shows the ribosomes and DNA in the cytoplasm (cell interior). In prokaryotic cells, the DNA is organized into a *nucleoid*. Eukaryotic cells have additional structure. Their DNA is contained in the membrane-bound nucleus. Eukaryotic cells also contain membrane-bound organelles, such as the endoplasmic reticulum and mitochondria.

1. One micrometer (μm) = 10^{-6} meters. Micrometers are sometimes referred to as *microns*.

membrane-enclosed *nucleus*, which contains the DNA and keeps it separate from the rest of the cytoplasm. Eukaryotic cells also contain a number of membrane-bound organelles, such as mitochondria and, in plant cells, chloroplasts.

A.2 The Chemistry of Life

Although the periodic table of the elements contains more than 100 different entries, molecules that appear in living organisms (*biomolecules*) are composed almost exclusively of only six atoms: carbon (C), hydrogen (H), nitrogen (N), oxygen (O), phosphorus (P), and sulfur (S). Carbon is especially important to biomolecules: each carbon atom is capable of binding to four other atoms, with bonds that are flexible enough to support a wide variety of carbon chain structures. The molecular variety of carbon-based compounds is the basis for all organic chemistry.

A *molecule* is a group of atoms held together by strong chemical bonds. Molecules are classified as charged, polar, or nonpolar according to the distribution of charge among the atoms:

Nonpolar molecules, such as oxygen (O_2), have a symmetric distribution of charges.

Polar molecules, such as water (H_2O), have no net charge but exhibit an uneven charge distribution; part of the molecule is slightly positively charged, and part is slightly negative.

Charged molecules have unequal positive and negative charge content. A charged atom is called an *ion*. Many ions have important biological roles (e.g., Ca^{2+}, K^+, and Cl^-).

Chemical Bonds

Chemists distinguish a number of distinct chemical bond types. The two bonds that are most important in biochemistry are *covalent bonds* and *hydrogen bonds* (figure A.2). A covalent bond is formed when two neighboring atoms share electrons. The bonds among atoms within a molecule are covalent bonds. These strong bonds hold the atoms of a molecule together in a stable structure. In the cell, covalent bonds are formed and broken with the help of enzyme catalysts.

Hydrogen bonds are weak bonds formed by the attraction between a slightly positive hydrogen atom and a slightly negative oxygen or nitrogen atom. Hydrogen bonds are easily formed and broken. Atoms bound by hydrogen bonds are not considered part of the same molecule. Instead, biomolecules that are held together by one or more hydrogen bonds are said to be *associated*. Two or more molecules that are associated in this fashion are said to have formed a molecular *complex*. In the cell, the formation of hydrogen bonds is facilitated by the aqueous environment. Because water molecules are polar, they form hydrogen bonds with other polar

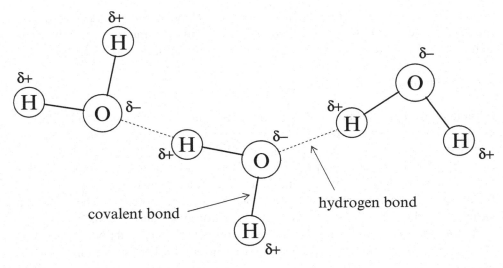

Figure A.2
Covalent bonds and hydrogen bonds. The atoms in each individual water molecule (H_2O) share electrons—they are covalently bound. Because hydrogen and oxygen atoms share electrons unequally, water
molecules are polar: the oxygen atoms have a partial negative charge, which is balanced by a partial
positive charge on the hydrogen atoms. This charge distribution allows the oxygen atom of one molecule
to attract a hydrogen atom from another, forming a hydrogen bond. Hydrogen bonds are weaker than
covalent bonds and are easily broken and reformed.

molecules. For that reason, polar molecules are also called *hydrophilic*; nonpolar
molecules, which do not form hydrogen bonds, are referred to as *hydrophobic* (from
Greek: *hydrophilic*, "water-loving"; *hydrophobic*, "water-fearing").

Small Molecules

In biochemistry, a distinction is made between *small molecules*, which contain at
most dozens of atoms, and *macromolecules*, which are composed of hundreds to
thousands of atoms.

Most cellular nutrients and waste products are small molecules. In addition, small
molecules are used to shuttle energy, protons, electrons, and carbon groups around
the cell. Two important examples are (figure A.3):

Glucose ($C_6H_{12}O_6$), a sugar that cells use as a source of energy and carbon.

Adenosine triphosphate (ATP; $C_{10}H_{16}N_5O_{13}P_3$), which is used as an "energy currency" within cells: most energy-requiring processes in the cell make use of ATP as
a chemical "power source."

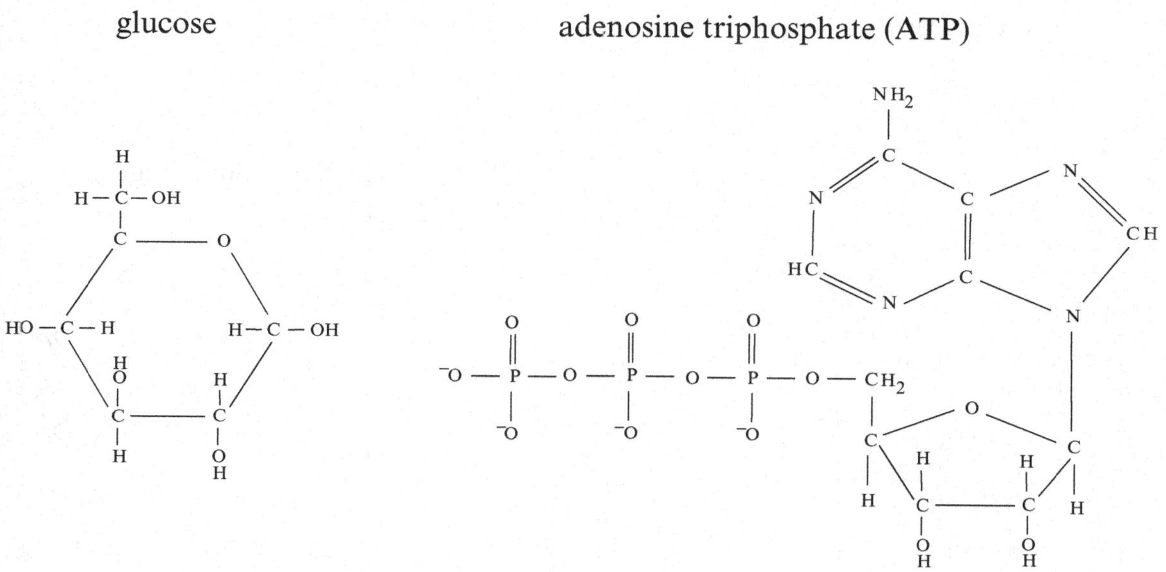

glucose

adenosine triphosphate (ATP)

Figure A.3
Small molecules. Glucose is a primary source of energy and carbon for cells, which harvest the energy in its covalent bonds. Much of that energy is used to produce adenosine triphosphate (ATP), which is the main energy currency in the cell. The cluster of negative charges in the three phosphate groups (PO_4^{3-}) is highly energetic. When the third phosphate group is removed, a significant amount of energy is released.

A.3   Macromolecules

Most cellular functions rely on macromolecules, which are typically *polymers*—molecules made up of similar (or identical) subunits called *monomers*. Biological macromolecules can be roughly classified into four group: carbohydrates, lipids, proteins, and nucleic acids.

Carbohydrates

Carbohydrates are built from simple sugars, such as glucose. Their general chemical composition is $C_m(H_2O)_n$. This composition includes multiple hydroxyl groups (O-H), which are polar. These hydroxyl groups cause carbohydrates to be hydrophilic—they readily form hydrogen bonds with the surrounding water molecules in the cell. Cells use carbohydrates to store chemical energy, for signaling, and as structural elements.

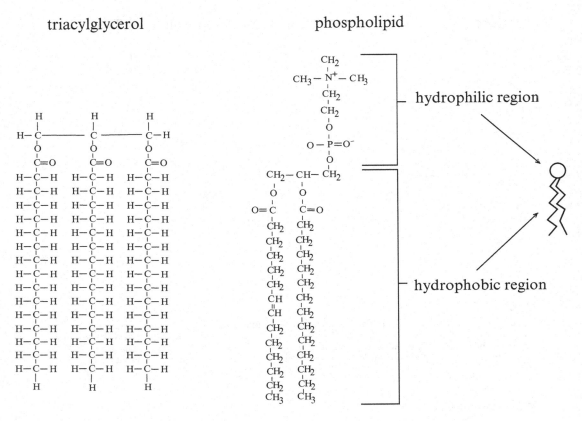

Figure A.4
Lipids. Triacylglycerol is a nonpolar fat. The structure of a phospholipid combines a polar "head" with nonpolar "tails."

Lipids

As illustrated in figure A.4, lipids consist primarily of carbon atoms surrounded by hydrogen atoms. This structure is nonpolar; lipids are hydrophobic. Lipids that consist solely of carbon chains are called fats. They are used primarily for energy storage. Other lipids consist of carbon chains covalently bound to a hydrophilic phosphate-containing group. These are called *phospholipids*. Because the carbon "tail" is nonpolar and the phosphate-containing "head" is polar, phospholipids are called *amphiphilic* (from Greek: "both-loving"; the term *amphipathic* ("both-suffering") is also used).

Because nonpolar carbon chains are hydrophobic, lipids tend to self-organize in the aqueous cellular environment. Fats form separate "islands" (or globules) that result in minimal contact with the surrounding water. Phospholipids self-organize

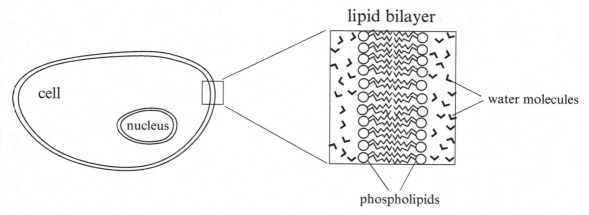

Figure A.5
Cross section of a bilipid membrane. In an aqueous environment, phospholipids spontaneously organize into bilayer sheets, with the nonpolar tails protected from interaction with the surrounding water. Cellular membranes are composed of these bilayer sheets.

into more interesting structures. It is energetically favorable for these molecules to have their polar heads interacting with the surrounding water while their nonpolar tails are sequestered from the aqueous surroundings. This arrangement can be reached by the formation of a *lipid bilayer*, in which the molecules arrange themselves into a double-layer sheet—the nonpolar tails are hidden inside the sheet while the polar heads lie on the exposed surfaces (figure A.5). When a bilayer sheet closes upon itself, the result is a *bilayer membrane*, also called a *bilipid membrane*. All cellular membranes—those that surround cells and those that surround intracellular organelles—are bilipid membranes. These membranes are fluid: the individual phospholipid molecules diffuse about in the two-dimensional membrane surface.

Bilipid membranes are *semipermeable*. Nonpolar molecules freely enter and leave the lipid core of the membrane. However, because the nonpolar interior provides no opportunity for hydrogen bond interactions, ions and polar molecules do not readily diffuse through the membrane.

Proteins

Most cellular functions are carried out by proteins: they catalyze metabolic reactions, transport molecules throughout the cell, transmit information, and much more.

Like carbohydrates and lipids, proteins are polymers. However, unlike carbohydrate and lipid polymers—for which the monomers show little variety—proteins are built from a diverse collection of 20 different monomer building blocks, called *amino acids*. These amino acid monomers share a common chemical backbone, by which they can be joined together via covalent *peptide bonds* (figure A.6). The resulting

amino acid   amino acid

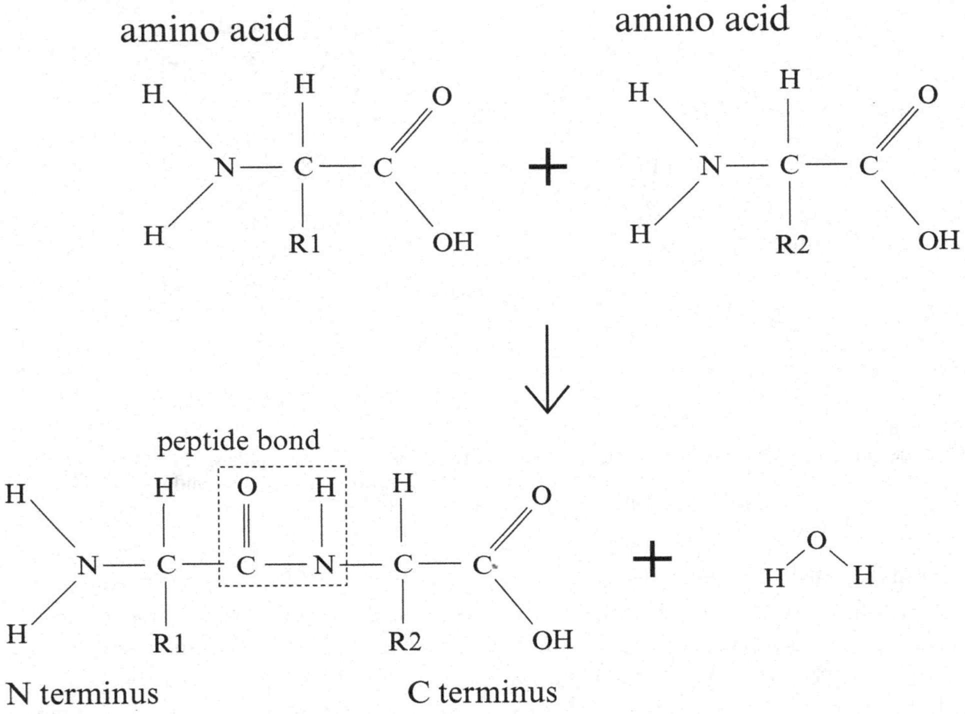

peptide bond

N terminus   C terminus

Figure A.6
Peptide bond formation. Two amino acids, with side chains R1 and R2, are covalently joined by a peptide bond. As more amino acids are added, the polypeptide chain grows. The two ends of the chain are named for the terminal atoms: the *N terminus* and the *C terminus*.

chain is referred to as a *polypeptide*. Each amino acid has a different functional group attached to its backbone. These functional groups, called *side chains*, exhibit a wide variety of sizes and charge placements (figure A.7).

The sequence of amino acid "beads" on a polypeptide "string" is called the *primary structure* of a protein (figure A.8). The peptide bonds are flexible and so allow the chain to fold back upon itself. The distribution of size and charge among the amino acids results in the polypeptide adopting a specific three-dimensional conformation, through the process of *protein folding*. The rough shape of most proteins results from nonpolar amino acids being pushed into the center of the molecule.

The final configuration is determined largely by hydrogen bonding among amino acids (which may or may not be neighbors in the peptide sequence). The conformation of the polypeptide chain is called the protein's *tertiary structure*. (*Secondary structure* refers to localized spatial patterns, such as α-helices and β-sheets.) Many

nonpolar

polar

charged

alanine

serine

histidine

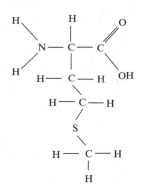

methionine

tyrosine

lysine

Figure A.7
Amino acids. Twenty different side chains appear on amino acids. The six shown here give some indication of their variability in size, polarity, and composition.

proteins also exhibit *quaternary structure*. These proteins are molecular complexes composed of multiple polypeptide chains held together by hydrogen bonds. The conformation of the entire complex is called the protein's quaternary structure. A complex of two polypeptide monomers is called a *dimer*; four monomers form a *tetramer*.

Proteins derive their function from their shape: specificity in conformation allows proteins to interact with specific molecules, catalyze specific reactions, and respond to specific chemical signals. Because protein structure is maintained by hydrogen bonds, the three-dimensional configuration is somewhat flexible. For some proteins, this flexibility is a crucial aspect of their function because it allows the protein to

primary structure   tertiary structure

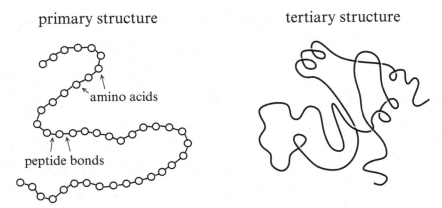

Figure A.8
Protein structure. The primary structure consists of the sequence of amino acid "beads" on the polypeptide "string." Together, the character of the amino acids results in the chain folding into a specific three-dimensional configuration, in which energy is minimized. This is the protein's tertiary structure.

adopt multiple conformations. For example, many proteins are nonfunctional in their native state but can be *activated* by an appropriate biochemical signal. Allosterically regulated proteins are activated by association with small molecules. Other proteins are activated by *phosphorylation*, a process that covalently attaches a phosphate group to a particular amino acid in the chain.

Nucleic Acids

Nucleic acids are polymers that store and transmit hereditary information in the cell. Their monomer subunits, called *nucleotides*, share a common backbone to which is attached one of four *bases*. Two classes of nucleic acids are found in cells: deoxyribonucleic acid (DNA) and ribonucleic acid (RNA). DNA polymers use the bases adenine (A), cytosine (C), guanine (G), and thymine (T) (figure A.9). These bases complement one another in size and charge distribution. As a result, hydrogen bonds form between bases when two DNA polymers are wound together, forming a double helix. The base-pair sequence of a cell's DNA encodes its *genome*—its complement of hereditary information. This sequence codes for the primary structure of the cell's proteins and RNAs.

RNA uses the same A, C, and G bases as DNA but substitutes uracil (U) in place of thymine. RNA polymers also exhibit complementary base pairing, but this more commonly occurs among the bases in a single RNA molecule, rather than between molecules. RNA molecules are integral to protein production and play a range of other roles in the cell.

DNA nucleotide (monomer)   DNA (polymer)   DNA double helix

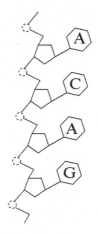

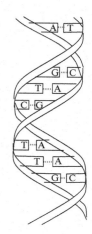

Figure A.9
DNA structure. Nucleotides consist of a pentose sugar bound to a phosphate group and one of four bases: adenine (A), cytosine (C), guanine (G), or thymine (T). These nucleotides can be strung together to form a DNA polymer. A double helix is formed when two complementary polymers anneal to each other by forming hydrogen bonds between complementary base pairs (A with T, C with G).

Production of Macromolecules

Carbohydrates and lipids are products of metabolism: specialized enzymes catalyze their formation from small-molecule building blocks.

DNA is only produced when a cell is preparing to divide. The process of duplicating the cell's DNA complement is facilitated by the double-stranded structure of DNA. A protein complex called *DNA polymerase* uses each of the strands as a template for constructing a new complementary strand.

Protein and RNA molecules are produced from DNA templates (figure A.10). A segment of DNA that codes for a protein's primary structure is called a *gene*. The production of a protein from this DNA recipe is called *gene expression*. In the first step of this process, called *transcription*, a protein complex called *RNA polymerase* produces an RNA copy of the gene. This RNA strand acts as a "message" (from the gene to the protein-production machinery), and so is called a *messenger RNA* (mRNA). The mRNA associates with a large protein–RNA complex called a *ribosome*, which processes the mRNA by "reading" its base-pair sequence and facilitates production of the corresponding polypeptide chain; this process is called *translation*. The flow of information from DNA to RNA to protein is called the "central dogma of molecular biology."

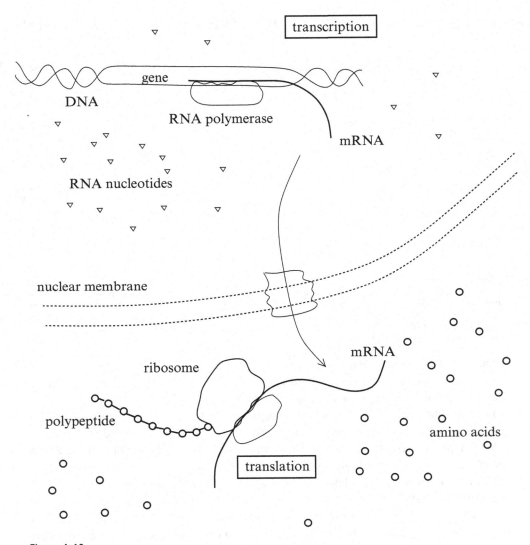

Figure A.10
Protein production (gene expression). The DNA of a gene is read by RNA polymerase and is transcribed into a messenger RNA (mRNA), which is produced from freely diffusing RNA nucleotides. In eukaryotes, the mRNA is then exported from the nucleus, as shown. In the cytosol, the mRNA is read by a ribosome, which produces the corresponding polypeptide chain from the pool of available amino acids: this process is called translation. (Amino acids are brought to the ribosome by specialized RNA molecules called *transfer RNAs* (tRNAs).) In prokaryotic cells, ribosomes have access to the mRNA as soon as it is produced and typically start translating an mRNA molecule before its transcription is complete.

A.4   Model Organisms

Much of molecular biological research is devoted to improving our understanding of the human body, especially in the study of health and disease. Other research areas involve the use of biological entities as technologies—in agriculture, manufacturing, energy production, environmental remediation, and other areas. Many of these biotechnological applications involve the use of bacteria or single-celled eukaryotes.

To consolidate the efforts of the biological community, research has focused on a manageable number of organisms. These *model organisms* serve as representatives of different biological classes. They span a range from bacteria to mammals: some represent specific organism or cell types, and others are used as models of specific biological processes.

Some model organisms that are referenced in this book are as follows.

• The bacterium *Escherichia coli* resides in the mammalian gut, where it aids in digestion. This bacterium has been the single most important model organism in molecular biology: many of the fundamental discoveries in molecular biology were based on studies of *E. coli*. In addition, *E. coli* is now the "workhorse" of molecular biology: it is routinely used as a biofactory to produce desired biomolecules and to aid in the manipulation of other cells.

• The single-celled yeast *Saccharomyces cerevisiae* (also known as budding yeast) has been used by humans for centuries for baking of bread and brewing of beer. *S. cerevisiae* has been adopted as a model eukaryote. Aspects of cell biology that are common across all eukaryotes, such as cell division, are frequently studied in this relatively "simple" organism.

• Studies of the fruit fly *Drosophila melanogaster* revealed many of the early insights into genetics. It is in wide use today for the study of genetics and development.

• The cress *Arabidopsis thaliana* has been adopted as the primary model organism for the study of plant biology.

• *Xenopus laevis* is an African frog. Xenopus embryos are large and can easily be manipulated. They are used to study the early development of fertilized egg cells.

Other model organisms include the roundworm *Caenorhabditis elegans*—a model of animal development—and the mouse *Mus musculus*, which is used to model mammalian biology and the progression of human disease.

Appendix B: Mathematical Fundamentals

The essential fact is that all the pictures which science now draws of nature, and which alone seem capable of according with observational facts, are mathematical pictures.
—James Jeans

B.1 Differential Calculus

The models in this book use *derivatives* to describe rates of change. Here, we review the notion of the derivative and summarize basic techniques for calculation of derivatives. Thorough introductions can be found in introductory calculus books such as *Modeling the Dynamics of Life* (Adler, 2004) and *Calculus for Biology and Medicine* (Neuhauser, 2004).

Functions

Some relationships that appear in this book are functions of a single variable. These functions relate the value of one variable to the value of another. The standard notation is $y = f(x)$, where x is the input variable (or *argument*), y is the output (the function value), and $f(\cdot)$ is the function itself. We also work with functions of multiple variables. Such functions each take a group of numbers as input and return a single number as output; for example, $y = g(x_1, x_2, x_3)$, where the input argument is the set of numbers x_1, x_2, and x_3.

The Derivative

The *derivative* of a function is the rate of change of the function value at a point (i.e., at a particular value of the input variable). To define the derivative of a function of a single variable, we first consider the average rate of change between two points (figure B.1). The rate of change of a function $f(\cdot)$ between the two points x_0 and x is the ratio of the change in the function value to the change in the argument, as follows:

Average rate of change: $\dfrac{f(x) - f(x_0)}{x - x_0}.$ \hfill (B.1)

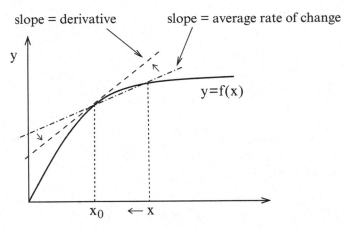

slope = derivative   slope = average rate of change

Figure B.1
Rates of change. The average rate of change corresponds to the slope of the line through two points on
the curve $y = f(x)$. The derivative is the slope of the tangent. The average rate of change converges to
the derivative as the two points come closer together.

As the point x is chosen closer and closer to x_0, the average rate of change converges to the rate of change at x_0. However, when $x = x_0$, equation (B.1) becomes
the undefined ratio 0/0. Consequently, the notion of a limit is needed to define the
rate of change at a point.

The *limit* of a function $h(\cdot)$ at a point x_0 is the value that $h(\cdot)$ would be expected
to take at x_0 based on the behavior of the function at arguments near x_0. When the
limit of a function $h(\cdot)$ at x_0 is equal to z, we write

$$z = \lim_{x \to x_0} h(x).$$

The derivative of a function $f(\cdot)$ at the point x_0 is defined as

$$\text{Derivative:} \quad \lim_{x \to x_0} \frac{f(x) - f(x_0)}{x - x_0}. \tag{B.2}$$

We use the notation

$$\frac{d}{dx} f(x) \Big|_{x = x_0}$$

to denote the derivative of $f(\cdot)$ at $x = x_0$. When we do not wish to specify the value
x_0, we write

$$\frac{d}{dx} f(x)$$

to denote the derivative of the function $f(\cdot)$ at the unspecified point x. A short-hand for this notation is

$$\frac{df}{dx}.$$

The derivative of a function is itself a function and has its own derivative, called the *second derivative*, denoted

$$\frac{d^2}{dx^2} f(x).$$

Differentiation Rules

The act of determining a function's derivative is called *differentiation*. Finding derivatives by using the definition in equation (B.2) is cumbersome. Fortunately, differentiation follows rules that can be applied to find derivatives of most functions:

1. **Constant function**

$$\frac{d}{dx} c = 0$$

2. **Identity function**

$$\frac{d}{dx} x = 1$$

3. **Power function**

$$\frac{d}{dx} x^n = nx^{n-1}$$

4. **Exponential function**

$$\frac{d}{dx} e^x = e^x$$

5. **Natural logarithmic function**

$$\frac{d}{dx} \ln(x) = \frac{1}{x}$$

6. **Scalar multiplication**

$$\frac{d}{dx} c f(x) = c \frac{d}{dx} f(x)$$

7. Addition

$$\frac{d}{dx}(f_1(x)+f_2(x)) = \frac{d}{dx}f_1(x) + \frac{d}{dx}f_2(x)$$

8. Product rule

$$\frac{d}{dx}(f_1(x)f_2(x)) = f_1(x)\left(\frac{d}{dx}f_2(x)\right) + f_2(x)\left(\frac{d}{dx}f_1(x)\right)$$

9. Quotient rule

$$\frac{d}{dx}\left(\frac{f_1(x)}{f_2(x)}\right) = \frac{f_2(x)\left(\frac{d}{dx}f_1(x)\right) - f_1(x)\left(\frac{d}{dx}f_2(x)\right)}{(f_2(x))^2}$$

10. Chain rule

$$\frac{d}{dx}f_1(f_2(x)) = \left(\frac{d}{dw}f_1(w)\Big|_{w=f_2(x)}\right)\frac{d}{dx}f_2(x).$$

Exercise B.1.1

(a) Determine the derivatives of the following functions:

(i) $f(x) = 2x + x^5$ (ii) $r(s) = \dfrac{2s}{s+4}$ (iii) $h(x) = x^3(e^x)$ (iv) $g(s) = \dfrac{3s^2}{0.5+s^4}$.

(b) Determine the derivatives of the following functions and evaluate the derivative at the specified point.

(i) $r(s) = s^2 + s^3$, $s = 2$ (ii) $f(x) = \dfrac{x}{x+1}$, $x = 0$

(iii) $g(s) = \dfrac{s^2}{1+s^2}$, $s = 1$ (iv) $h(x) = \dfrac{e^x}{1+x+x^2}$, $x = 2$. $\square$

Implicit Differentiation

When a function is specified by a formula $y = f(x)$, the function value y is defined explicitly in terms of the input argument x. In some cases, function values are defined *implicitly* by an equation involving the input argument and the function value; for example,

$$x + f(x) = \frac{1}{(1+f(x))^3}.$$

In these cases, evaluating the function value can be difficult. However, the derivative of the function can be found by the straightforward technique of *implicit differentiation*. To begin, both sides of the equation are differentiated. This yields an equation that can then be solved for the derivative.

As an example, we find the derivative of the function $f(\cdot)$ defined by

$$x + f(x) = \frac{1}{(1 + f(x))^3}.$$

We differentiate the equation:

$$\frac{d}{dx}(x + f(x)) = \frac{d}{dx}\left(\frac{1}{(1 + f(x))^3}\right)$$

$$1 + \frac{df}{dx} = \frac{-3(1 + f(x))^2 \dfrac{df}{dx}}{(1 + f(x))^6} = \frac{-3}{(1 + f(x))^4}\frac{df}{dx}.$$

We then solve this equation for df/dx, giving

$$\frac{df}{dx} = \frac{-(1 + f(x))^4}{3 + (1 + f(x))^4}.$$

Exercise B.1.2 Find the derivative of the function $y = y(x)$ defined implicitly by the following equation:

$$\frac{x}{y^2 + 1} = x^3.$$

$\square$

Partial Derivatives

When a function depends on multiple arguments (e.g., $y = g(x_1, x_2, x_3)$), it exhibits a separate rate of change with respect to each of the input variables. These rates of change are the *partial derivatives* of the function. Partial derivatives are defined by fixing the values of all but one of the input arguments—so that the function is treated as if it depends on a single variable. For instance,

First partial derivative: $\displaystyle\lim_{x_1 \to x_1^0} \frac{g(x_1, \bar{x}_2, \bar{x}_3) - g(x_1^0, \bar{x}_2, \bar{x}_3)}{x_1 - x_1^0}.$

We use a "curly d" notation for the partial derivative:

$$\frac{\partial}{\partial x_1} g(x_1, x_2, x_3).$$

Exercise B.1.3 Evaluate all partial derivatives of the following functions.

(i) $f(s_1, s_2) = \dfrac{3s_1 - s_2}{1 + \dfrac{s_1}{2} + \dfrac{s_2}{4}}$ (ii) $g(s, i) = \dfrac{2s^2}{i + 3s^2}$ $\square$

Partial Differential Equations

Section 8.4 introduces the use of partial differential equation (PDE) models for describing the propagation of action potentials. Here, we review some basic PDE concepts. A PDE involves derivatives in both time and space. We will restrict our discussion to models in a single spatial dimension, with position denoted by x. A comprehensive introduction to PDEs is provided in the book *Applied Partial Differential Equations* (Haberman, 2003).

The Diffusion Equation A common starting point for spatial models in biology is the *diffusion equation*:

$$\frac{\partial}{\partial t} c(x, t) = D \frac{\partial^2}{\partial x^2} c(x, t), \tag{B.3}$$

where $c(x, t)$ describes the state variable (e.g., species concentration). The parameter D is called the diffusion coefficient. This equation describes diffusion of the quantity c in a single spatial direction. (Because it was originally derived in the study of heat flow, this equation is often referred to as the *heat equation*.)

In some cases, the quantity c is involved in reactions as it diffuses. These cases can be described by a *reaction–diffusion equation*:

$$\frac{\partial}{\partial t} c(x, t) = D \frac{\partial^2}{\partial x^2} c(x, t) + f(c(x, t)),$$

where f describes the reaction effects. The simplest case is constitutive decay, for which $f(c(x, t)) = -kc(x, t)$.

Boundary Conditions Recall that in order to solve an ordinary differential equation (ODE), initial conditions must be specified. Solutions of PDEs depend on both initial conditions and *boundary conditions*. Before a PDE solution can be obtained, one must specify the spatial *domain* over which the solution should be defined. For equations with one spatial dimension, we can take the solution domain as an interval $[a, b]$. The initial condition must then be specified over the entire spatial domain, as a function of position: $c(x, 0) = c_0(x)$, for $a \leq x \leq b$.

Boundary conditions specify how the solution behaves at the two ends of the interval (i.e., $x = a$ and $x = b$). These conditions must be specified for all time (i.e., as functions of time). The simplest conditions, called *Dirichlet boundary conditions*, clamp the end-point values of the solution to given functions (e.g., $c(a, t) = c_1(t)$, $c(b, t) = c_2(t)$). An alternative is to allow the boundary values to be free but to specify the solution gradient at the end points. These are called *Neumann boundary conditions*. For example, if the gradient should be flat at the two end-points, we specify

$$\frac{\partial}{\partial x} c(x,t) \bigg|_{x=a} = 0, \ \frac{\partial}{\partial x} c(x,t) \bigg|_{x=b} = 0.$$

(Boundary conditions that specify both the value and gradient are called *Robin conditions*.)

Numerical Simulation Recall that the numerical simulation of an ODE is constructed on a mesh of time points. To simulate a PDE, a spatial mesh must be provided as well. Euler's method can be extended to PDEs by stepping forward in the time grid from the initial condition. At each time step, a spatial discretization of the PDE results in a system of algebraic equations. This system of equations, which depends on the boundary conditions, can be solved to determine the solution value at each point in the spatial grid. Details of PDE simulation in MATLAB are included in appendix C.

B.2   Linear Algebra

The material in section 5.4 depends on some basic notions from linear algebra, as described below. An accessible introduction to linear algebra is provided in the book *Linear Algebra and its Applications* (Strang, 2005).

Vectors and Matrices

A set of n numbers $x_1, x_2, \ldots, x_n$ can be organized into a *vector* of length n. For example

Column vector: $\mathbf{v} = \begin{bmatrix} 7 \\ 3 \\ 11 \\ -2 \end{bmatrix}$.

This is called a *column vector* because the numbers are stacked in a column. Alternatively, a *row vector* is formed when the numbers are placed side-by-side in a row:

Row vector: $\mathbf{w} = \begin{bmatrix} 7 & 3 & 11 & -2 \end{bmatrix}$.

A rectangular array of numbers is called a *matrix*:

Matrix: $\mathbf{M} = \begin{bmatrix} 3 & 1 & 0 \\ -1 & -1 & 6 \end{bmatrix}.$

When a matrix has n rows and m columns, we say it is an n-by-m matrix. (So $\mathbf{M}$ is a 2-by-3 matrix.) A column vector is an n-by-1 matrix, whereas a row vector is a 1-by-n matrix. A matrix that has the same number of rows as columns is called a *square* matrix.

The *transpose* of a matrix is constructed by "flipping" the matrix across the diagonal:

Matrix transpose: $\mathbf{M}^{\mathrm{T}} = \begin{bmatrix} 3 & -1 \\ 1 & -1 \\ 0 & 6 \end{bmatrix},$

where we have used the notation $(\cdot)^{\mathrm{T}}$ to denote the transpose. The transpose of an n-by-m matrix is an m-by-n matrix.

Vector and Matrix Multiplication   Two vectors of the same length can be multiplied together using the *inner product*. The result is a single number (which we call a *scalar* to distinguish it from a vector). The general procedure is

Inner product: $[x_1 \ \ x_2 \ \ \dots \ \ x_n] \cdot \begin{bmatrix} y_1 \\ y_2 \\ \vdots \\ y_n \end{bmatrix} = x_1 y_1 + x_2 y_2 + \cdots + x_n y_n.$

For example,

$$[1 \ \ 4 \ \ 7 \ \ -3] \cdot \begin{bmatrix} 2 \\ -5 \\ -4 \\ 1 \end{bmatrix} = (1)(2) + (4)(-5) + (7)(-4) + (-3)(1) = 2 - 20 - 28 - 3 = -49.$$

The inner product can be used to multiply a column vector by a matrix, treating the matrix as a stack of row vectors. For a column vector of length m, the matrix must have m columns. The product is a column vector whose length is the number of rows in the matrix:

$$\begin{bmatrix} x_{11} & x_{12} & \cdots & x_{1m} \\ x_{21} & x_{22} & \cdots & x_{2m} \\ \vdots & \vdots & \ddots & \vdots \\ x_{n1} & x_{n2} & \cdots & x_{nm} \end{bmatrix} \cdot \begin{bmatrix} y_1 \\ y_2 \\ \vdots \\ y_m \end{bmatrix} = \begin{bmatrix} x_{11} y_1 + x_{12} y_2 + \cdots + x_{1m} y_m \\ x_{21} y_1 + x_{22} y_2 + \cdots + x_{2m} y_m \\ \vdots \\ x_{n1} y_1 + x_{n2} y_2 + \cdots + x_{nm} y_m \end{bmatrix}. \tag{B.4}$$

For example

$$\begin{bmatrix} 1 & 4 & 7 & -3 \\ 0 & 3 & -1 & 1 \\ 2 & 2 & 0 & 0 \end{bmatrix} \cdot \begin{bmatrix} 2 \\ -5 \\ -4 \\ 1 \end{bmatrix} = \begin{bmatrix} (1)(2)+(4)(-5)+(7)(-4)+(-3)(1) \\ (0)(2)+(3)(-5)+(-1)(-4)+(1)(1) \\ (2)(2)+(2)(-5)+(0)(-4)+(0)(1) \end{bmatrix}$$

$$= \begin{bmatrix} 2-20-28-3 \\ 0-15+4+1 \\ 4-10+0+0 \end{bmatrix} = \begin{bmatrix} -49 \\ -10 \\ -6 \end{bmatrix}.$$

The product of two matrices can be defined by extending this procedure — treating the second matrix as a collection of column vectors. Again, the number of rows and columns must be aligned: the matrix product $\mathbf{M} \cdot \mathbf{N}$ is only defined if the number of columns in $\mathbf{M}$ equals the number of rows in $\mathbf{N}$. The general formula for the product can be written in terms of the column vectors that make up the matrix $\mathbf{N}$. Let

$$\mathbf{N} = [\mathbf{n}_1 \ \mathbf{n}_2 \ \cdots \ \mathbf{n}_r],$$

where each $\mathbf{n}_i$ is a column vector of length m. Then, if the matrix $\mathbf{M}$ has m columns, we can write

$$\mathbf{M} \cdot \mathbf{N} = \mathbf{M} \cdot [\mathbf{n}_1 \cdot \mathbf{n}_2 \ \cdots \ \mathbf{n}_r]$$

$$= [\mathbf{M} \cdot \mathbf{n}_1 \ \mathbf{M} \cdot \mathbf{n}_2 \ \cdots \ \mathbf{M} \cdot \mathbf{n}_r],$$

where each product $\mathbf{M} \cdot \mathbf{n}_i$ is a column vector as determined in equation (B.4).

Matrix multiplication is not *commutative*, meaning that the product $\mathbf{M} \cdot \mathbf{N}$ is not necessarily equal to the product $\mathbf{N} \cdot \mathbf{M}$ (even when both of these products are defined).

Exercise B.2.1 Evaluate the following products.

(a) $\mathbf{v} \cdot \mathbf{w}$, where

(i) $\mathbf{v} = [1 \ 2 \ 4]$ and $\mathbf{w} = \begin{bmatrix} 3 \\ 2 \\ -1 \end{bmatrix}$

(ii) $\mathbf{v} = [1 \ 1]$ and $\mathbf{w} = \begin{bmatrix} -2 \\ 2 \end{bmatrix}$.

(b) $\mathbf{M} \cdot \mathbf{w}$, where

(i) $\mathbf{M} = \begin{bmatrix} 1 & 0 & 1 \\ 1 & 2 & 3 \\ -1 & -2 & 0 \end{bmatrix}$ and $\mathbf{w} = \begin{bmatrix} -2 \\ -1 \\ 0 \end{bmatrix}$

(ii) $\mathbf{M} = \begin{bmatrix} -1 & 0 & -2 & 3 \\ -2 & 2 & 3 & 3 \end{bmatrix}$ and $\mathbf{w} = \begin{bmatrix} 1 \\ -2 \\ 1 \\ 1 \end{bmatrix}$.

(c) $\mathbf{M \cdot N}$, where

(i) $\mathbf{M} = \begin{bmatrix} 1 & 1 & 1 & 3 \\ -2 & 0 & 1 & -1 \end{bmatrix}$ and $\mathbf{N} = \begin{bmatrix} 2 & 2 \\ -1 & -1 \\ 1 & 4 \\ 0 & 5 \end{bmatrix}$

(ii) $\mathbf{M} = \begin{bmatrix} 1 & 0 & 1 \\ 2 & 2 & 3 \\ 2 & -4 & 1 \end{bmatrix}$ and $\mathbf{N} = \begin{bmatrix} 2 & 2 & 2 \\ -1 & -1 & 0 \\ 1 & 4 & 5 \end{bmatrix}$.

(d) Evaluate the product $\mathbf{N \cdot M}$ for the matrices in part (c)(ii), and verify that $\mathbf{N \cdot M} \neq \mathbf{M \cdot N}$. $\square$

The Identity Matrix   For scalars, the number 1 is called the *identity* element because multiplication of a number by 1 returns the original number. The set of square *n*-by-*n* matrices has an identity element, called the *identity matrix*, which has ones along the diagonal and zeros elsewhere. For example, the 3-by-3 identity matrix is

Identity matrix: $\mathbf{I}_3 = \begin{bmatrix} 1 & 0 & 0 \\ 0 & 1 & 0 \\ 0 & 0 & 1 \end{bmatrix}$.

Exercise B.2.2   Let

$\mathbf{M} = \begin{bmatrix} 2 & -1 & 3 \\ 1 & 1 & 4 \\ -1 & -2 & 2 \end{bmatrix}$.

Verify that $\mathbf{M \cdot I}_3 = \mathbf{I}_3 \cdot \mathbf{M} = \mathbf{M}$. $\square$

The Matrix Inverse   For a scalar x, the number $x^{-1} = 1/x$ is called the *multiplicative inverse* of x because the product $x \cdot (1/x)$ is the identity element 1. The *inverse* of a matrix is defined the same way. For an *n*-by-*n* matrix $\mathbf{M}$, the inverse matrix, denoted $\mathbf{M}^{-1}$, satisfies

$\mathbf{M} \cdot \mathbf{M}^{-1} = \mathbf{I}_n = \mathbf{M}^{-1} \cdot \mathbf{M}$.

Only square matrices can have inverses, and many square matrices have no inverse.

Exercise B.2.3

(a) Verify the following matrix–inverse pair

$$\mathbf{M} = \begin{bmatrix} 1 & 0 \\ 2 & 2 \end{bmatrix}, \qquad \mathbf{M}^{-1} = \begin{bmatrix} 1 & 0 \\ -1 & \dfrac{1}{2} \end{bmatrix}.$$

(b) Show that each of the following matrices does not have an inverse:

(i) $\quad \mathbf{M}_1 = \begin{bmatrix} 1 & -2 & -3 \\ 2 & 2 & 1 \\ 0 & 0 & 0 \end{bmatrix}$ (ii) $\quad \mathbf{M}_2 = \begin{bmatrix} 2 & 2 \\ -1 & -1 \end{bmatrix}.$

Hint: For $\mathbf{M}_1$, consider the last row of any product $\mathbf{M}_1 \cdot \mathbf{N}$; for $\mathbf{M}_2$, consider the general product $\mathbf{M}_2 \cdot \mathbf{M}$ and verify that it cannot take the form of the identity matrix. □

The Nullspace Vector multiplication has the property that the product of two nonzero elements can be zero. For example

$$[1 \quad -1] \cdot \begin{bmatrix} 2 \\ 2 \end{bmatrix} = (1)(2) + (-1)(2) = 0.$$

When the product of a matrix $\mathbf{M}$ and vector $\mathbf{v}$ is the zero vector (all elements zero), then $\mathbf{v}$ is said to be in the *nullspace* of the matrix $\mathbf{M}$.

If a vector $\mathbf{v}$ is in the nullspace of a matrix $\mathbf{M}$, then for any constant c, the vector $c\mathbf{v}$ is also in the nullspace of $\mathbf{M}$. Similarly, if vectors $\mathbf{v}$ and $\mathbf{w}$ are both in the nullspace of a matrix $\mathbf{M}$, then any sum of the form $c_1\mathbf{v} + c_2\mathbf{w}$ (called a *linear combination* of $\mathbf{v}$ and $\mathbf{w}$) is also in the nullspace of $\mathbf{M}$. (Scalar multiplication and vector addition are carried out component-wise, so that $c[3\ 5] = [3c\ 5c]$ and $[3\ 5] + [1\ 7] = [4\ 12]$.)

Exercise B.2.4 Verify that the vectors $\mathbf{v}$ and $\mathbf{w}$ are in the nullspace of the matrix $\mathbf{M}$, where

$$\mathbf{M} = \begin{bmatrix} 1 & -1 & 0 & 0 \\ 0 & -1 & -2 & -1 \end{bmatrix}, \qquad \mathbf{v} = \begin{bmatrix} 2 \\ 2 \\ 1 \\ 0 \end{bmatrix}, \qquad \text{and} \qquad \mathbf{w} = \begin{bmatrix} 1 \\ 1 \\ 0 \\ -1 \end{bmatrix}. \qquad □$$

B.3   Probability

Section 7.6 addresses the role of randomness in systems biology models. Here, we briefly review some fundamental notions in probability theory. A thorough introduction to this material can be found in the book *Introduction to Probability Models* (Ross, 2007).

An experiment can be called *random* if we cannot predict its outcome with absolute certainty. For such an experiment, we define the *sample space* as the set of all possible outcomes. Subsets of the sample space are called *events*. For instance, if the experiment involves flipping two coins, the sample space is

$$S = \{(H, H), (H, T), (T, H), (T, T)\},$$

where H and T signify heads and tails, respectively. The outcome "both heads," (H, H), is an event, as is the outcome "at least one heads," which corresponds to the subset $\{(H, H), (H, T), (T, H)\}$.

To each event E we assign a probability, $P(E)$, so that

(i) The probability of any event lies between zero and one ($0 \leq P(E) \leq 1$).

(ii) The probability of the entire sample space is one ($P(S) = 1$).

(iii) When two events are mutually exclusive, their probabilities sum to the probability that one of the two events occurs (i.e., $P(E_1) + P(E_2) = P(E_1 \text{ or } E_2)$).

In the coin-tossing example, we have

$$P((H,H)) = \frac{1}{4} \quad \text{and} \quad P(\{(H,H),(H,T),(T,H)\}) = \frac{3}{4}.$$

Discrete Random Variables

A *random variable* is a function defined on the set of events. For instance, if we assign heads a value of 1 and tails a values of 2, then we can define a random variable as the sum that results from flipping two coins. Calling this random variable X, we have

$$P(X = 2) = P((H,H)) = \frac{1}{4}$$

$$P(X = 3) = P((H,T)) + P((T,H)) = \frac{1}{2} \tag{B.5}$$

$$P(X = 4) = P((T,T)) = \frac{1}{4}.$$

This random variable is called *discrete* because it can take only a finite set of values. The equations in (B.5) specify its (discrete) *probability distribution*. This distribution can be used to determine the *cumulative distribution function*, which describes the probability that the random variable takes a value less than or equal to a given number. Letting $F(b)$ denote the cumulative distribution function, we have, from (B.5):

$$F(b) = P(X \le b) = \begin{cases} 0 & \text{for} \quad b < 2 & \text{(as } X \text{ is never less than 2)} \\ 1/4 & \text{for} \quad 2 \le b < 3 & \text{(as } X < 3 \text{ only for (H,H))} \\ 3/4 & \text{for} \quad 3 \le b < 4 & \text{(as } X < 4 \text{ for (H,H), (H,T), or (T,H))} \\ 1 & \text{for} \quad 4 \le b & \text{(as } X \text{ is always less than or equal to 4)} \end{cases}$$

The *expected value* of a random variable is the average over many experiments. Denoting the expected value of X by $E[X]$, we can determine the expected value as a weighted average over all values of X:

$$E[X] = \sum_{X_i = 2,3,4} X_i \cdot P(X = X_i) = 2 \cdot \frac{1}{4} + 3 \cdot \frac{1}{2} + 4 \cdot \frac{1}{4} = 3.$$

Exercise B.3.1 Consider an experiment that involves flipping two weighted coins, for which the probability of heads is only 1/3 (so tails has probability 2/3). Assign a value of 1 to heads and 3 to tails. Define a discrete random variable X as the sum that results from flipping the two coins. Determine the discrete probability distribution for X, its cumulative distribution function, and its expected value. □

Continuous Random Variables

Random variables that take values over a continuum are called *continuous*. For most continuous random variables, the probability of any particular value is vanishingly small, and so we cannot define a probability distribution as in (B.5). Instead, we define a *probability density function*, which describes the probability of the variable taking values in specified intervals. Letting X denote a continuous random variable, the probability density function $f(\cdot)$ satisfies

$$P(a \le X \le b) = \int_a^b f(x)\,dx.$$

The cumulative distribution function $F(\cdot)$ is then given by

$$F(b) = P(X \le b) = \int_{-\infty}^b f(x)\,dx.$$

The expected value is the weighted sum of the probabilities over all values:

$$E[X] = \int_{-\infty}^{\infty} x \cdot f(x)\,dx.$$

Standard examples of continuous random variables are the uniform random variable and the exponential random variable, as described next.

The *uniform* random variable on the interval [0, 1] has equal probability at each point in the interval and so has probability density function $f(x)$ given by

$$f(x) = \begin{cases} 1 & \text{for } 0 \leq x \leq 1 \\ 0 & \text{otherwise} \end{cases}.$$

The cumulative distribution function is

$$F(b) = P(X \leq b) = \begin{cases} 0 & \text{for} \quad b < 0 \\ b & \text{for} \quad 0 \leq b < 1. \\ 1 & \text{for} \quad 1 \leq b \end{cases}$$

The expected value is

$$E[X] = \int_{-\infty}^{\infty} x f(x)\, dx = \int_0^1 x \cdot 1\, dx = \frac{1}{2}.$$

The *exponential* random variable describes the time that elapses between events that occur continuously at a constant average rate. The probability density function depends on a parameter λ and is given by

$$f(x) = \begin{cases} 0 & \text{for } x < 0 \\ \lambda e^{-\lambda x} & \text{for } x \geq 0 \end{cases}.$$

The cumulative distribution function is

$$F(b) = P(X \leq b) = \begin{cases} 0 & \text{for } b < 0 \\ 1 - e^{-\lambda b} & \text{for } b \geq 0 \end{cases}.$$

Appendix C: Computational Software

A computer does not substitute for judgment any more than a pencil substitutes for literacy. But writing without a pencil is no particular advantage.
—Robert McNamara

The mathematical techniques covered in this text can be implemented in a number of computational software packages. This appendix introduces two such programs: XPPAUT and MATLAB. Other commonly used software packages include Copasi, Mathematica, Maple, and Berkeley Madonna.[1] The Systems Biology Markup Language (SBML) facilitates the transfer of models between different packages by providing a common model-description platform.[2]

XPPAUT is recommended for users who have limited experience with computational software packages. This program, which is freely available, was designed specifically for the simulation and analysis of dynamic mathematical models. The XPPAUT interface is primarily menu-based—analysis and simulation are carried out through a graphical user interface (GUI).

MATLAB is a general computational environment. Many researchers consider it to be the standard tool for the analysis introduced in this text. MATLAB is a commercial product and so may not be available to all readers. (The free programs Octave and Scilab provide functionality that is similar to that of MATLAB.[3]) Although MATLAB's interface is command-based, an add-on package called the Systems Biology Toolbox (described later) can be used to facilitate a GUI approach to model analysis and simulation.

---

1. Copasi: www.copasi.org. Mathematica: www.wolfram.com/mathematica. Maple: www.maplesoft.com. Berkeley Madonna: www.berkeleymadonna.com.

2. Information on SBML and tools for translating between formats can be found at www.SBML.org.

3. Octave: www.gnu.org/software/octave. Scilab: www.scilab.org.

C.1 XPPAUT

XPPAUT was created and is maintained by G. Bard Ermentrout. The name of the program incorporates three key features: it runs in the X-windows environment; phase plane analysis is one of its primary uses; and it employs a program called AUTO for bifurcation analysis. (AUTO was created by E. Doedel.)

The XPPAUT software and documentation are available online.[4] In addition to the tutorial at the XPPAUT Web site, Ermentrout has written a book that details the use of the program—*Simulating, Analyzing, and Animating Dynamical Systems: A Guide to XPPAUT for Researchers and Students* (Ermentrout, 2002). A more concise introduction, also by Ermentrout, is provided in *Computational Cell Biology* (Fall et al., 2002).

Input (.ode) File

The XPPAUT program relies on user-created text files—with file extension .ode—to describe models. An .ode file specifies a set of differential equations, model parameters, and initial conditions.

We will illustrate with an .ode file that describes a two-species reaction chain ($\rightarrow s_1 \rightarrow s_2 \rightarrow$) with irreversible mass-action kinetics. Lines that begin with # are comments and are ignored by the program. The file input is case insensitive.

```
# File chain1.ode

# Model of two-species reaction chain (-> s1 -> s2 ->)

#

# Declaration of model parameters

par v0=5

par k1=3, k2=2

#

# Declaration of model equations (the notation s1' can also be
used)

ds1/dt = v0 - k1*s1

ds2/dt = k1*s1 - k2*s2
```

---

4. See www.math.pitt.edu/~bard/xpp/xpp.html. The Web site www.math.uwaterloo.ca/~bingalls/MMSB contains XPPAUT code for figures in this book, as well as code to support the end-of-chapter problem sets.

```
#
# Set initial conditions (optional; default is zero)
init s1=1, s2=2
#
# The file must end with the following command
done
#
```

A specific syntax must be followed in the .ode file. In assigning parameter values and initial conditions, spaces are not allowed around the equals (=) sign. Variable names are limited to 10 characters and can only include alphanumeric characters and underscores.

The .ode file can also include lines to declare auxiliary output variables (e.g., aux R2rate=k1*s1), functions (e.g., f(s1,s2)=k1*s1−k2*s2), and "fixed variables" (e.g., r=k2*s2). Functions and fixed variables can be referred to within the file but will not be accessible once the program is running. Auxiliary output variables will be available for graphing. Fixed variables can be used to describe conserved moieties.

The function heav(·) can be used to trigger events at certain times. This is the *Heaviside function*, it returns 0 for negative arguments and 1 for positive arguments. Thus, the statement input=K*heav(t-t0) will define a fixed variable input that will transition from value 0 to value K at time t=t0.

Lines in the .ode file that begin with @ are used to set internal computational parameters. These parameters, which can also be set in the GUI, allow the user to set the length of time to be simulated (e.g., @ total=5; the default is 20), the bound on the state variable size (e.g., @ bound=1000; the default is a bound of ±100), and to allocate additional memory for storage of the simulation data (e.g., @ maxstor=10000; the default is space for 5000 data points). The variables to be plotted and the size of the plotting window can also be set (e.g., @ xp=t, yp=s1, xlo=0, xhi=5, ylo=0, yhi=3; the default window is $[0, 20] \times [-1, 1]$). A space should always be included after the @ symbol.

The XPPAUT GUI

When the program is opened, the user is prompted to select an .ode file, after which the XPPAUT GUI appears. The main window is used to display simulations and phase plane trajectories. On the left-hand side of the window is a menu of commands for visualization and analysis. These commands can be selected with the mouse or by pressing the hot-key shortcut, indicated by the capitalized letter in the command name. (The lowercase letter is the hot-key shortcut.) The escape key (Esc) is used to exit submenus. The program can be shut down by selecting **File|Quit|Yes** (hotkeys: **F Q Y**).

Across the top of the window are triggers that open additional windows. **ICs** allows initial conditions to be set. **Param** allows modification of the model parameter values. **Eqns** displays the model equations. (These cannot be altered in the GUI.) **Data** displays the numerical output from simulations.

Model Simulations

To simulate the model, select **Initialconds|Go** (hotkeys: **I G**). This will run a simulation from the initial conditions specified in the .ode file (or in the **ICs** window). New initial conditions can be set by selecting **Initialconds|New** and entering values at the prompts. Selecting **Initialconds|Last** will start a new simulation (at time $t=0$) starting from the final state of the previous simulation. The length of the simulation can be set by selecting **Numerics|Total** and entering the final time at the prompt. Selecting **Continue** allows the previous simulation to be continued over a longer time-interval.

The simulation tracks all state and auxiliary variables in the model, as can be verified in the **Data** window. By default, the time course of only one state variable is plotted—the variable described by the first equation in the .ode file. The display window can be adjusted to fit the trajectory by selecting **Window/zoom|Fit** or by selecting **Window/zoom|Window** and then specifying ranges for the horizontal (X) and vertical (Y) axes. Multiple simulations can be run in sequence (with, e.g., different initial conditions or parameter values). Each time a new simulation is run, the data in the **Data** window are written over. Plots of previous simulations may remain in the display window but will disappear when the display is redrawn. The user can ensure that these curves are retained in the display (and in any exported graphics) by selecting **Graphic stuff|Freeze|On freeze**.

To plot a different variable, select **Xi vs t** and enter the name of the variable at the prompt. To plot multiple variables in a single plot, select **Graphic stuff|Add curve**, and specify the name of a state or auxiliary variable in the Y-axis field. You can alter the line style and color by changing the numbers in the specified fields. The display can be exported as a postscript (.ps) graphics file (by selecting **Graphic stuff|Postscript** and specifying a file name), or as a .gif file (by selecting **Kinescope|Capture**, then **Kinescope|Save**, and entering 2 to select .gif at the prompt). Another export option is simply to use a printscreen command. Alternatively, data can be exported directly by selecting **Write** from the data browser and selecting a filename. These data can then be plotted in, for instance, a spreadsheet program.

Phase Plane Analysis

Trajectories can be visualized in the phase plane by selecting **Viewaxes|2D** and specifying which variables should be displayed on the two axes (the window size can also be set). Trajectories can be generated with **Initialconds|Go**. There is also a more direct method for setting initial conditions in the phase plane: selecting **Initialconds|Mouse** allows the user to specify an initial condition by clicking a location in the display window. This command can be accessed multiple times by selecting **Initialconds|mIce**. (Press Esc to quit back to the main menu.)

A direction field can be generated by selecting **Dir.field/flow** and then selecting either **Direct field** or **Scaled Dir.Fld** and specifying a grid size. In an unscaled field, the length of each arrow is proportional to the speed of motion. A scaled field, in which all arrows have the same length, sometimes provides a cleaner picture.

Nullclines can be plotted by selecting **Nullcline|New**. The option **Nullcline|Save** allows the user to export these curves in a data file.

Linearized Stability Analysis

Recall that the intersections of the nullclines are the steady states of the system (see section 4.2). The location of these points can be determined by holding the mouse button down as you move the cursor across the display window—the (x, y) coordinates of the cursor appear at the bottom of the GUI. Precise values for steady states can be determined by selecting **Sing pts|Mouse** and then clicking near a steady state of interest. The numerical value of the state variables at the nearby steady state will be reported, along with the results of a stability analysis. The tally of eigenvalues in each of the following classes is reported: **c+**, **c−**, **im**, **r+**, **r−**, corresponding to eigenvalues that are complex-valued with positive or negative real part, purely imaginary, positive real, or negative real. (Recall that a steady state is stable if all eigenvalues have negative real part.) If the chosen steady state is unstable, the user is also given the option to plot any *invariant sets*. These are the trajectories that connect this unstable point with other stable points (so called *heteroclinic* trajectories), and the separatrices—the balanced trajectories that converge to the unstable state and form the boundary between basins of attraction. (Recall the trajectories that are balanced on the 'mountain ridge' as they approach the unstable saddle point in figure 4.11.)

Continuation and Bifurcation Analysis

Continuation and bifurcation analysis in XPPAUT is handled by the embedded program AUTO, which can be opened by selecting **File|Auto**. The AUTO window contains a graphical display and a list of commands. The small window on the lower left displays the eigenvalues of the Jacobian (transformed in such a way that eigenvalues with negative real part appear inside the displayed circle). The program displays coordinates and classification of points at the bottom of the main display window.

AUTO should only be called once a trajectory has been simulated to steady state in XPPAUT. The last point of this simulation is passed to AUTO as a starting point for the continuation analysis. (The user can ensure that steady state has been reached by re-running a simulation from its final point with **Initialconds|Last**.)

Once the AUTO window is open, some internal parameters must be set before a continuation curve can be plotted. The **Parameter** command opens a window where the user can choose which parameter should be varied to produce the curve. This parameter should be entered as **Par1**. Next, the display window can be set by selecting **Axes|hI-lo** and specifying which state variable should be plotted (on the **Y-axis**) and setting the window size. (The **X-axis** will correspond to the **Main Parm**, which is the same as **Par1**.) Each run of the continuation routine will vary the parameter in one direction (up or down). The direction of motion can be controlled in the **Numerics** window. By default, the step-size, **Ds**, has a positive value (0.02), so the parameter value will increase as the analysis proceeds. Specifying a negative step-size (e.g., −0.02) will generate a curve in the opposite direction. In the **Numerics** window, the user can also specify the range over which the parameter can be altered by adjusting **Par Min** and **Par Max** (the default range is 0 to 2).

Once these conditions are properly set, the analysis can be carried out by selecting **Run|Steady state**. This should generate a continuation curve. Stable steady states are shown with a thick curve; unstable states with a thin curve. The points on the curve can be explored by clicking **Grab**, which generates a cross-shaped cursor on the curve. This cursor can be moved along the curve by pressing the left and right arrow keys. The coordinates of the point are displayed in the box below the main display; the eigenvalues of the Jacobian are displayed in the lower left window. Pressing Tab jumps to the 'special' points that AUTO has labeled with small crosses and numbers. These are bifurcation points or end points of the analysis run. These points each have a 'type' (**Ty**) as indicated in the bottom window: end points are labeled **EP**, saddle-node bifurcations as **LP** (for limit point), and Hopf bifurcations as **HB**. The user can exit the **Grab** setting by hitting Esc or by pressing Enter, which loads the currently selected point into memory. If a special point is grabbed in this way, a new continuation run can be performed (e.g., to continue the curve after an end point was reached.) When a Hopf bifurcation point is grabbed, the user can select **Run|Periodic** to generate the envelope (i.e., upper and lower limits) for the resulting limit cycle. Two-parameter bifurcation diagrams, such as in figure 7.15 of chapter 7, can be generated by grabbing a bifurcation point, setting **Two param** in the **Axis** window, and clicking **Run** again. The points on the continuation curve can be exported by selecting **File|Write pts**.

To illustrate, we consider two examples from chapter 4. The bifurcation plot in figure 4.19B (for the bistable model (4.2)) can be generated with the following .ode file.

```
# File bistable_example1.ode

# Bistable network in Figure 4.6

par k1=20, k3=5, k2=20, k4=5, n1=2, n2=2

S1'=k1/(1+S2^n1) - k3*S1

S2'=k2/(1+S1^n2) - k4*S2

init S1=1,S2=0.1

done
```

With the parameter values as specified, this system is bistable (as can be verified by plotting the nullclines in the phase plane). To generate a bifurcation diagram against parameter k1, we begin by shifting the value of this parameter so that the system is monostable. In the **Param** window, set the value of k1 to 35 and then press **I G**. The resulting trajectory shows a high steady state in s1. Press **I L** a few times to ensure that the steady state has been reached. Next, hit **F A** to bring up the AUTO window. Tap **Parameter** and confirm that k1 is in the **Par1** position. Next, click on **Axes|hi-lo**. Confirm that the axis assignments are k1 (**Main Parm**) and s1 (**Y-axis**). Set the window size to $[5, 35] \times [0, 7]$. Next, open the **Numerics** window. Change the range of allowed parameter values to $[0, 40]$ (i.e., Par Min = 0, Par Max = 40). Also, add a negative sign to **Ds** (so it is −0.02). Finally, return to the main menu and choose **Run|Steady state** to generate the bifurcation curve. You can explore the curve with **Grab**. The Tab key will bring the cursor to the bifurcation points at k1 = 16.15 and 28.98.

As a second example, the bifurcation plot in figure 4.21 (for the oscillatory model (4.10)) can be generated from the following file.

```
# File oscillatory_example1.ode

# Oscillatory network in Figure 4.14

par n=2.5, k0=8, k1=1, k2=5, K=1

S1'=k0 - k1*S1*(1+(S2/K)^n)

S2'=k1*S1*(1+(S2/K)^n) - k2*S2

init S1=1,S2=1

done
```

The given parameter values lead to oscillations, as can be verified by simulation. To generate a bifurcation diagram, begin by adjusting n so that steady-state behavior results, as follows. In the **Param** window, set n to 1.5. Run a simulation and hit I L a few times to ensure that steady state has been reached. Next, open the AUTO window (with **F A**). By default, the main parameter is n, and steady states of S1 are plotted. Set the window size (in **Axes|hI-lo**) to [1.5, 3.5] × [0, 4]. In the **Numerics** window, set the parameter range to [0, 4]. Then select **Run|Steady state**. A continuation curve should appear, showing the change in stability at n = 2.407 (at the Hopf bifurcation, HB). Click **Grab** and navigate by pressing Tab until you reach the bifurcation point. Then press Enter to load that point into memory as the starting point for the next continuation analysis. Finally, select **Run|Periodic** to see the upper and lower limit of the oscillatory trajectory.

Local Sensitivity Analysis

XPPAUT does not have an explicit feature for computation of sensitivity coefficients. These can be determined by running simulations for two nearby parameter values and then manually comparing the differences in parameter values and the resulting steady states (see equation 4.15 of chapter 4). This procedure can be facilitated by assigning parameters to the 'sliders' below the display window. Clicking on the Par/Var? box for one of these sliders opens a window that allows a model parameter to be assigned to the slider, along with minimum and maximum values. The parameter value can then be modified by dragging the slider.

Beyond Ordinary Differential Equations

Delay Differential Equations Some differential equation models incorporate an explicit delay in the effect of one variable on another (e.g., problem 7.8.9 of chapter 7). Such models involve terms of the form $s(t - \tau)$ in the description of rates of change; for example,

$$\frac{d}{dt}s_1(t) = -k_1 s_1(t) + k_2 s_2(t - \tau).$$

In this case, the effect of s_2 is only felt after a lag of τ time units.

XPPAUT has an internal parameter called delay that specifies the maximum allowed delay. The default value is zero, so it must be assigned a new value before a delayed equation can be simulated. This value can be set in the GUI by selecting **nUmerics|dElay** or in an .ode file with the @ delay command (e.g., @ delay=10). The XPPAUT syntax for $s_2(t - \tau)$ is delay(s2, tau). A delay equation model needs a 'pre-initial' condition to refer to until time $t=\tau$, which can be set in the **Delay** window. A constant initial condition can be set with init in the .ode file.

Gillespie's Stochastic Simulation Algorithm Gillespie's stochastic simulation algorithm (SSA) can be implemented in XPPAUT by a discrete-time dynamical system that steps from one reaction event to the next. The update statements for this system track the elapsed time and the changes in species abundance. Discrete update steps are implemented in XPPAUT using the same equation structure as differential equations (i.e., $x'=\ldots$) along with the declaration @ meth=discrete. The length of the simulation is specified in terms of the number of steps (i.e., reaction events), not the elapsed time.

As an example, the following file simulates the two-species chain $\rightarrow s_1 \rightarrow s_2 \rightarrow$. The state variable tr will be used to track the elapsed time because the internal variable t will be tracking the number of steps that have occurred. The simulation time tr follows an update law of the form tr(t+1)=tr(t)+log(1/ran(1))/a, where a is the sum of the reaction propensities, and ran(1) is a uniform random variable on the interval $[0, 1]$. In the .ode file, this will appear in the form of a differential equation, following XPPAUT convention for discrete-time systems: tr'=tr+log(1/ran(1))/a.

The state variable is also updated in a differential equation–like statement once the next reaction has been selected. Logic statements (e.g., a<b) are used to draw a sample from the cumulative distribution for the next reaction. In XPPAUT, these logic statements are assigned a value of 1 or 0 depending on whether the statement is true or false.

```
# File gillespie1.ode

# Gillespie stochastic simulation for two-species reaction chain
# (-> s1 -> s2 ->)

# Set kinetic parameters

par v0=20, k1=1, k2=2

# Initialize concentrations and simulation time tr

init s1=0, s2=0, tr=0

# Set reaction propensities

a0=v0

a1=k1*s1

a2=k2*s2

# Define the total of all propensities

a_total=a0+a1+a2
```

```
# Set the update rule for tr

tr'=tr+log(1/ran(1))/a_total

# Determine cumulative reaction probabilities (note, p2=1)

p0=a0/a_total

p1=(a0+a1)/a_total

p2=(a0+a1+a2)/a_total

# Select next reaction to occur:

# First generate a random number between 0 and 1

s=ran(1)

# Next, compare this number with the cumulative distribution

z0=(0<=s)&(s<p0)

z1=(p0<=s)&(s<p1)

z2=(p1<=s)&(s<=p2)

# The single non-zero z variable corresponds to the next
# reaction

# (Note, the statements (0<=s) and (s<=p2) are always true)

# Update the molecule counts

s1'=s1 + z0 - z1

s2'=s2 + z1 - z2

# Set the total number of steps to be taken

@ total=100

# Set the model as discrete-time (update rule-based)

@ meth=discrete

# Set the plot to be s1 against tr

@ xp=tr,yp=s1,xlo=0,xhi=10,ylo=0,yhi=100

done
```

Because every reaction event is stored, the stochastic simulation algorithm sometimes generates far more output data points than needed. To reduce the amount of data points that are reported, select **nUmerics|nOutput** and set n_out to a number larger than one. The program will only record one data point for every n_out reaction steps. The @ maxstor line can be used to ensure that sufficient memory is allocated.

C.2   MATLAB

MATLAB is a product of The MathWorks, Inc. The software can be purchased directly from the company Web site[5] or readers may be able to access an institutional license. The program name is a portmanteau of Matrix Laboratory—the software is especially well suited to matrix calculations.

There are many MATLAB tutorials available online, and much documentation is provided at the MathWorks Web site. There are several texts that introduce the use of MATLAB, including *MATLAB: An Introduction with Applications* (Gilat, 2008), which provides a basic introduction, and *Mastering MATLAB 7* (Hanselman and Littlefield, 2004), which is a more comprehensive treatment.

The main MATLAB interface is command-line driven. Most analysis is carried out through user-coded scripts. The following introduction covers basic MATLAB commands for the techniques addressed in this book and describes some add-on packages that allow certain analyses to be carried out via graphical user interface (GUI).

Simulation

MATLAB files end with the file extension .m. They are referred to as *m-files*. When describing a model for simulation, a MATLAB function must be written that specifies the differential equations. In addition, a script must be written that generates the simulation and produces the desired graphical output. As an example, the following m-file generates a simulation of a two-species reaction chain ($\rightarrow s_1 \rightarrow s_2 \rightarrow$) with irreversible mass-action kinetics. Lines in the code that begin with % are comments that are ignored by the program. MATLAB is case-sensitive. The end of a command is indicated with a semicolon.

```
% File chain1.m

% Model of two-species reaction chain (-> s1 -> s2 ->)
```

---

5. See www.mathworks.com/matlab. The Web site www.math.uwaterloo.ca/~bingalls/MMSB contains MATLAB code for figures in this book, as well as code to support the end-of-chapter problem sets.

```
%

% This function generates the simulation and produces a plot

function chain1

%

% Declaration of model parameters

% Global declarations allow parameters to be shared between
% functions

global v0; global k1; global k2;

%

% Declaration of model equations (function chain1dtt is defined
% below)

ODEFUN=@chain1ddt;

%

% Assignment of parameter values

v0=5; k1=3; k2=2;

%

% Set initial conditions (as a vector)

s0=[1,0];

%

% Specify the final time; the simulation interval is [0, Tend]

Tend = 5;

%

% Generate the simulation.

[t,s]= ode45(ODEFUN, [0,Tend], s0);

% The vector t is the mesh of simulation time-points

% The rows of the matrix s are the vectors [s1(t),s2(t)]

%
```

```
% Open a plotting window

figure(1);

%

% Plot the time-course for s1 and s2

plot(t, s(:,1), t, s(:,2));

% The colon notation (:) denotes a submatrix

%

% Specify the window size; add axis labels and a legend

axis([0 Tend 0 5]); xlabel('Time'); ylabel('Concentration');
legend('s_1', 's_2');

%

% Close the function chain1

end

%

%%%%%%%%%%%%%%%%%%%%%%%%%%%%%%%%%%%%%

% The second function defines the model equations

% The inputs (t,s) are the time and state at a given point

% The output ds is the right-hand-side of the ODE

%

function ds=chain1ddt(t,s)

%

global v0; global k1; global k2;

%

% Assignment of state variables

s1=s(1); s2=s(2);

%

% Declaration of model equations.
```

```
ds1 = v0 - k1*s1;

ds2 = k1*s1 - k2*s2;

%

% Assign the output vector

ds = [ds1;ds2];

%

% Close the function chain1ddt

end
```

This m-file makes use of the command ode45 to generate the simulation. This is one of many simulation routines that are available in MATLAB. For models that involve significant differences in timescales (that have not been resolved by model reduction), the command ode23s may perform better.

The first function in the file above has no input and so could instead be written as a *script* (without the initial function declaration or the end statement). Multiple functions can be written in a single m-file. Each script needs to be in its own m-file, and so writing this simulation code as a script would require two separate m-files.

To run the file, open MATLAB and set the current directory (shown at the top of the window) to the directory where your m-file was saved. MATLAB has a comprehensive help system that can be accessed by entering help command_name at the command line. Type the name of the file (without the extension, e.g., chain1) to execute your function. Alternatively, the file can be executed interactively within the MATLAB editor, as follows. Choose **File|Open** and select the m-file. Along the top of the editing window are controls for the interactive debugging environment. The "play" button executes the file. Clicking anywhere along the left-hand side of the file (to the right of the line numbers) introduces a stop-sign *breakpoint*. When breakpoints are set, file execution stops at each of these points. The controls next to the "play" icon allow the user to resume execution or run the file line-by-line from the break. While stopped at a breakpoint, the current values of the model variables can be queried by hovering the cursor over the variable in the editor or by entering the variable name in the command window.

Figures generated by the m-file can be saved in a number of graphics formats (e.g., .jpg, .eps). The command dlmwrite('mydata.dat', M, ',') can be used to export the data in matrix M. This will generate a comma-delimited .txt file. (The simulation data can be stored in M by assigning M = [t,s].)

Phase Plane Analysis

Trajectories can be plotted in the phase plane using the command `plot`. For instance, in the function `chain1`, we could add the lines:

```
figure(2);

plot(s(:,2),s(:,1));
```

A collection of trajectories can be plotted by running multiple simulations and plotting them together. (The command `hold on` specifies that previously plotted curves should be retained.)

A direction field can be produced with the command `quiver`, as follows. The command `meshgrid` generates a mesh of points at which the direction arrows can be placed. The length of the arrows in the x and y directions can then be assigned by evaluating the right-hand side of the differential equations. A scaled direction field results from dividing each arrow by its length. (To generate an unscaled field, remove `L` in the code below.) In function `chain1`:

```
% generate grid on [0,3]x[0,6], with step-sizes 0.1 and 0.25

[xx,yy] = meshgrid(0:0.1:3, 0:0.25:6);

s1dot=v0 - k1*xx;

s2dot=k1*xx - k2*yy;

L = sqrt(s1dot.^2 + s2dot.^2);

quiver(xx,yy,s1dot./L,s2dot./L,0.5); % 0.5 is a scaling factor
```

Nullclines can be generated with the command `ezplot`. Although the nullclines in `chain1.m` can be solved explicitly, we nevertheless use this model to illustrate the general procedure. The `ezplot` command is used to plot the solution to implicit equations. These equations can be defined as *anonymous functions* in MATLAB, as follows.

```
ncfun1 = @(x,y) v0-k1*x;

ncfun2 = @(x,y) k1*x-k2*y;

figure(3); hold on;

ezplot(@(x,y) ncfun1(x,y), [0 3 0 6]);

ezplot(@(x,y) ncfun2(x,y), [0 3 0 6]);
```

PPLANE: Phase Plane Analysis   The program PPLANE, by John Polking, provides a GUI for phase plane analysis in MATLAB.[6] The interface allows the user to specify model equations and display a direction field or nullclines. Clicking on a point in the display window produces the trajectory that passes through the selected point.

Matcont: Continuation and Bifurcation Analysis

MATLAB does not have built-in functions for continuation and bifurcation analysis. The package Matcont, developed under the supervision of W. Govaerts and Yu. A. Kuznetsov, provides a GUI for bifurcation analysis. A comprehensive tutorial is provided at the developer's Web site.[7]

To run Matcont, direct MATLAB to the **Matcont** folder and enter `matcont` at the MATLAB command line. This brings up several windows. To enter a system of equations, choose **Select|System|New**. The user can then give the system a name, specify the state variables and the model parameters (in comma-separated lists—no spaces), and then enter the system equations, following the syntax `X'=-2*X-Y;`. (The order of the equations should correspond to the preceding list of state variables.) When complete, click **OK**. (You can later load this system from the list of saved systems in **Select|System|Edit/Load**.)

To generate a simulation, select **Type|Initial Point|Point** and **Type|Curve|Orbit**. Then enter the initial condition and parameter values in the **Starter** window. The length of the simulation can be adjusted in the **Integrator** window (in the **Interval** field). To visualize the simulation, select **Window|Graphic|2Dplot** and assign variables to the x-axis (abscissa) and the y-axis (ordinate). (The state variables are classified as coordinates.) The axes can also be assigned by selecting **Layout|Variables on axes** from the menu at the top of the graphics window. In the same menu, **Layout|Plotting region** allows the plotting region to be specified. To run the simulation, select **Compute|Forward** from the **Matcont** window.

To generate a continuation/bifurcation curve, begin by running a simulation to steady state. (Make sure the **Interval** is sufficiently long that steady state has been reached.) To set up a continuation from this point, select this steady state as the initial point for the analysis by following **Select|Initial point** and then selecting the line "This is the last point on the curve." This will open the **Starter** and **Continuer** windows. Select the desired bifurcation parameter by tapping the corresponding button in the **Starter** window. In the **Continuer** window, you can change the number of points to be calculated along the curve (**MaxNumPoints**). Set the display window

---

6. The program uses the file `pplane8.m`, available at www.math.rice.edu/~dfield. A Java applet running pplane can be found at www.math.rice.edu/~dfield/dfpp.

7. The program is available at www.matcont.ugent.be/matcont.html. To run the package, MATLAB must have access to a C compiler (e.g., gcc), which you may have to install separately.

to show the desired state variable plotted against the bifurcation parameter, and select **Compute|Forward** (or **Compute|Backward**) to generate the curve. The computation will stop at bifurcation points, where you can choose to **Resume**. The bifurcation values can be observed in the **Numeric** window (**Window|Numeric**).

Local Sensitivity Analysis

Parametric sensitivities can be approximated numerically in MATLAB (see equation 4.15 of chapter 4). A template script is

```
nom_par=5;                          % set nominal parameter value

par=nom_par;                        % assign nominal parameter value

[t_nom, s_nom]=ode45(ODEFUN,[0,Tend],s0);    % nominal trajectory

s_nom_ss = s_nom(length(t_nom));    % steady-state concentration

delta=0.05;                         % set deviation (5 percent)

par=par*(1+delta);                  % perturbed parameter value

[t_pert,s_pert]=ode45(ODEFUN,[0,Tend],s0);   % perturbed trajectory

s_pert_ss=s_pert(length(t_pert));   % steady-state concentration

abs_sens=(s_pert_ss-s_nom_ss)/(delta*nom_par);      % absolute
                                                    % sensitivity

rel_sens=abs_sens*(nom_par_value/s_nom_ss); % relative sensitivity
```

Beyond Ordinary Differential Equations

Delay Differential Equations   Delay differential equations can be simulated with the command `dde23`, which requires arguments that specify the lag time `tau` and the history `s_hist` (which replaces the initial condition). The syntax is

```
sol = dde23(ODEFUN, [tau], s_hist, [0, Tend])
```

A constant history can be described by giving `s_hist` a single value.

Gillespie's Stochastic Simulation Algorithm   Gillespie's stochastic simulation algorithm (SSA) can be implemented directly in MATLAB. The simulation can be run either for a specific number of reaction events (in a `for` loop) or for a specific time interval (in a `while` loop). In the latter case, the number of reaction events is not prespecified, so the state and time vectors must be initialized as sufficiently long to ensure the entire simulation will be recorded.

As an example, consider a two-species reaction chain ($\rightarrow s_1 \rightarrow s_2 \rightarrow$). Using a `for` loop, simulation is carried out by the following m-file:

```
% File gillespie.m

% Stochastic simulation of -> s1 -> s2 ->

% Set parameter values

v0=5; k1=3; k2=2;

% Set number of reaction events

Tend=10000;

% Initialize the state and time vectors

s=zeros(Tend,2); t=zeros(Tend,1)

% Set initial conditions

s(1,:)=[10,0]; t(1)=0;

% Run the simulation

for j=1:Tend

%       Calculate propensities

        a1=v0;

        a2=k1*s(j,1);

        a3=k2*s(j,2);

        asum=a1+a2+a3;

%       Select the next reaction:

        u = rand(1);

        z0=0; z1=0; z2=0;

        if 0 <= u && u < a1/asum

            z0=1;

            else if a1/asum <= u && u < (a1+a2)/asum

                    z1=1;

                else
```

```
                    z2=1;

                end

            end

        end

%       Update time

        t(j+1)=t(j)+log(1/rand(1))/asum;

%       Update the state

        s(j+1,1)=s(j,1)+z0-z1;

        s(j+1,2)=s(j,2)+z1-z2;

%       Update reaction counter

        j=j+1;

end

% Plot the simulation

figure(1)

stairs(t, s(:,1), t, s(:,2))

axis([0 70 0 140]);

ylabel('Number of Molecules'); xlabel('Time');
legend('s_1', 's_2')

end
```

Partial Differential Equations

MATLAB's built-in PDE solver is called pdepe. Simulation of a PDE requires four functions—one to call the solver and one each to specify the equations, the initial condition, and the boundary conditions.

The equation is specified in the general form:

$$c\left(x,t,u,\frac{\partial u}{\partial x}\right) \cdot \frac{\partial u}{\partial t} = x^{-m}\frac{\partial}{\partial x}\left(x^m b\left(x,t,u,\frac{\partial u}{\partial x}\right)\right) + s\left(x,t,u,\frac{\partial u}{\partial x}\right)$$

for functions $c(\cdot)$, $b(\cdot)$, and $s(\cdot)$. (The exponent m specifies certain spatial symmetries.)

Boundary conditions are specified in the form

$$p_l(x_l,t,u(x_l)) + q_l(x_l,t) \cdot b\left(x_l,t,u(x_l),\frac{\partial u}{\partial x}(x_l)\right) = 0$$

$$p_r(x_r,t,u(x_r)) + q_r(x_r,t) \cdot b\left(x_r,t,u(x_r),\frac{\partial u}{\partial x}(x_r)\right) = 0$$

where x_l and x_r are the left and right end-points of the interval. The function $b(\cdot)$ is the same as in the PDE.

We will illustrate the PDE solver with the reaction–diffusion equation

$$\frac{\partial}{\partial t}u(x,t) = 5\frac{\partial^2}{\partial x^2}u(x,t) - 3u(x,t)$$

on the interval [0, 1] with boundary conditions

$$u(0,t) = 1, \frac{\partial}{\partial x}u(1,t) = 0$$

and initial condition $u(x, 0) = 2 - (x - 1)^2$. The equation is specified by the function:

```
function [c,b,s] = eqn(x,t,u,DuDx)

% Specification of reaction-diffusion equation (for m=0)

c=1;

b=5*DuDx;

s=-3*u;

end
```

Here, DuDx indicates $\partial u/\partial x$. The initial condition is specified by

```
function value=initial(x)

% Specification of initial condition

value=2-(x-1)^2;

end
```

The boundary conditions are specified by

```
function [pl,ql,pr,qr]=bc(xl,ul,xr,ur)
```

```
% Specification of boundary conditions

pl=ul-1;

ql=0;

pr=0;

qr=1;

end
```

The following function creates a grid of x and t values, calls the PDE solver, and plots the solution.

```
% function pdesolver

m=0;

x=linspace(0,1,30); t=linspace(0,0.5,20);

% Call to the solver

sol=pdepe(m, @eqn, @initial, @bc, x, t); u = sol(:,:,1);

%Plot solution surface

figure(1); surf(x,t,u); xlabel('Position'); ylabel('Time');

% Plot solution slices

figure(2); plot(x,u(1,:),x,u(5,:),x,u(10,:));
```

Linear Algebra

MATLAB is an excellent tool for linear algebraic analysis. Vectors and matrices are defined using the following syntax: $v = [1 \ 2 \ 3]$ (row vector), $w = [4;5;6]$ (column vector), $M = [1 \ 2; \ 3 \ 4]$ (2-by-2 matrix). Matrix and vector multiplication are carried out with $*$. Commands relevant to this text are the following: M' (transpose), inv(M) (matrix inverse), eye(n) ($n \times n$ identity matrix), null(M) (nullspace), null(M,'r') (nullspace specified with rational, i.e., fractional, values), eig(M) (eigenvalues).

Frequency Response Analysis

As discussed in section 6.6, a linear input–output system can be specified as a set of four matrices: A, B, C, and D. Once these are specified in MATLAB, the frequency response can easily be generated, and the corresponding Bode plots can be produced. The command [num den]=ss2tf(A,B,C,D) determines the transfer function for the system, specified in terms of the coefficients of the numerator and denominator polynomials. A MATLAB transfer function can then be created with the command `sys=tf(num,den)`, from which Bode plots can be produced with `bode(sys)`.

The MATLAB Systems Biology Toolbox

The MATLAB Systems Biology Toolbox, developed by Henning Schmidt, provides a GUI for a range of simulation and analysis tools that are relevant to biological models.[8] All toolbox commands can be queried using MATLAB's `help command_name` function.

Systems Biology Toolbox: Model Simulation   The toolbox provides a GUI for specification of models. To access this GUI, enter `model=SBmodel()` at the MATLAB command line, and then enter `SBedit(model)`. The model data can be entered in the window that appears for each component. As an example, to model a two-species chain ($\rightarrow s_1 \rightarrow s_2 \rightarrow$), click **States** and enter:

```
d/dt(s1)=r0-r1
```

```
d/dt(s2)=r1-r2
```

```
s1(0)=2
```

```
s2(0)=1
```

Click **Parameters** and enter:

```
k0=5
```

```
k1=3
```

```
k2=2
```

Define the reaction rates in **Reactions** as:

```
r0=k0
```

```
r1=k1*s1
```

```
r2=k2*s2
```

The model can be saved by exporting as text. To reload, enter `model=SBmodel('model.txt')` at the MATLAB command line. Models can also be

---

8. The package is available at www.sbtoolbox.org. Once downloaded, direct MATLAB to the folder SBTOOLBOX and run the command `installSB`.

exported in XPPAUT or SBML formats. Click **Simulate** to run a simulation. This opens a window where you can select which variables are plotted.

Systems Biology Toolbox: Model Analysis Analysis begins with loading a model (e.g., `model = SBmodel('model.txt')`).

The command `ss=SBsteadystate(model)` will return the steady state nearest the initial condition. The system Jacobian at the steady state `ss` can be calculated with `J = SBJacobian(model, ss)`. The MATLAB command `eig(J)` calculates the eigenvalues.

Local parametric sensitivity analysis is carried out in two steps. For steady-state sensitivities, first type `output = SBsensdatastat(model)` at the command line. Sensitivity coefficients can then be generated with the command `SBsensstat(output)`, which plots the sensitivities as a bar graph. The user can choose which sensitivity coefficients to display, whether to display sensitivities in their absolute (unscaled) or scaled (normalized) form, and what sort of plot should be shown. (Plots that involve multiple parameters will display statistics of a collection of sensitivity coefficients.)

The toolbox also provides functions to calculate sensitivity coefficients for the amplitude and period of limit-cycle oscillations, with the command `output = SBsensdataosc(model,[Tend nTr])` followed by `SBsensperiod(output)` and `SBsensamplitude(output)`. Here, the length of the interval to be simulated is `Tend`, and `nTr` is the number of intervals of length `Tend` that should be treated as the transient.

The toolbox has functions to support stoichiometric network analysis. The stoichiometry matrix can be produced with `M = SBstoichiometry(model)`. Moiety conservations can be identified with the command `SBMoietyconservations(model)`. The toolbox can also be used for parameter estimation: the main command is `SBparameterestimation`.

Bibliography

Ackerman, S. (1992). *Discovering the Brain: Symposium on the Decade of the Brain: Papers*. New York: John Wiley & Sons.

Adler, F. R. (2004). *Modeling the Dynamics of Life* (2nd ed.). Pacific Grove, CA: Brooks/Cole.

Albert, M. A., Haanstra, J. R., Hannaert, V., Van Roy, J., Opperdoes, F. R., Bakker, B. M., et al. (2005). Experimental and *in silico* analyses of glycolytic flux control in bloodstream form *Trypanosoma brucei*. *Journal of Biological Chemistry, 280*, 28306–28315.

Alberts, B., Johnson, A., Lewis, J., Raff, M., Roberts, K., & Walter, P. (2007). *Molecular Biology of the Cell* (5th ed.). New York: Garland Science.

Alon, U. (2007). *An Introduction to Systems Biology: Design Principles of Biological Circuits*. London: CRC Press.

Ang, J., Ingalls, B., & McMillen, D. (2011). Probing the input-output behaviour of biochemical and genetic systems: system identification methods from control theory. *Methods in Enzymology, 487*, 279–317.

Aström, K., & Murray, R. (2008). *Feedback Systems: An Introduction for Scientists and Engineers*. Princeton, NJ: Princeton University Press.

Atkinson, M. R., Savageau, M. A., Myers, J. T., & Ninfa, A. J. (2003). Development of genetic circuitry exhibiting toggle switch or oscillatory behaviour in *Escherichia coli*. *Cell, 113*, 597–607.

Bakker, B. M., Michels, P. A. M., Opperdoes, F. R., & Westerhoff, H. V. (1999). What controls glycolysis in bloodstream form *Trypanosoma brucei*? *Journal of Biological Chemistry, 274*, 14551–14559.

Balagaddé, F. K., Song, H., Ozaki, J., Collins, C. H., Barnet, M., Arnold, F. H., et al. (2008). A synthetic *Escherichia coli* predator-prey ecosystem. *Molecular Systems Biology, 4*. doi:10.1038/msb.2008.24.

Barkai, N., & Leibler, S. (1997). Robustness in simple biochemical networks. *Nature, 387*, 913–917.

Basu, S., Mehreja, R., Thiberge, S., Chen, M.-T., & Weiss, R. (2004). Spatiotemporal control of gene expression with pulse-generating networks. *Proceedings of the National Academy of Sciences of the United States of America, 101*, 6355–6360.

Basu, S., Gerchman, Y., Collins, C. H., Arnold, F. H., & Weiss, R. (2005). A synthetic multicellular system for programmed pattern formation. *Nature, 434*, 1130–1134.

Berg, H. (2004). *E. coli in Motion*. New York: Springer.

Bliss, R. D., Painter, P. R., & Marr, A. G. (1982). Role of feedback inhibition in stabilizing the classical operon. *Journal of Theoretical Biology, 97*, 177–193.

Bolouri, H. (2008). *Computational Modeling of Gene Regulatory Networks — A Primer*. London: Imperial College Press.

Bolouri, H., & Davidson, E. H. (2002). Modeling transcriptional regulatory networks. *BioEssays, 24*, 1118–1129.

Boyce, W. E., & DiPrima, R. C. (2008). *Elementary Differential Equations and Boundary Value Problems* (9th ed.). New York: Wiley.

Carlson, R. (2010). *Biology is Technology: The Promise, Peril, and New Business of Engineering Life.* Cambridge, MA: Harvard University Press.

Carroll, S. B., Grenier, J. K., & Weatherbee, S. D. (2005). *From DNA to Diversity: Molecular Genetics and the Evolution of Animal Design* (2nd ed.). Malden, MA: Blackwell Publishing.

Cheong, R., Hoffmann, A., & Levchenko, A. (2008). Understanding NF-κB signaling via mathematical modeling. *Molecular Systems Biology, 4,* 192.

Chubb, J. R., & Liverpool, T. B. (2010). Bursts and pulses: insights from single cell studies into transcriptional mechanisms. *Current Opinion in Genetics & Development, 20,* 478–484.

Cook, P. F., & Cleland, W. W. (2007). *Enzyme Kinetics and Mechanism.* London: Garland Science.

Cornish-Bowden, A. (1979). *Fundamentals of Enzyme Kinetics.* London: Butterworths.

Cornish-Bowden, A., & Wharton, C. W. (1988). *Enzyme Kinetics.* Oxford: IRL Press.

Cosentino, C., & Bates, D. (2011). *Feedback Control in Systems Biology.* London: CRC Press.

Curien, G., Ravanel, S., & Dumas, R. (2003). A kinetic model of the branch-point between the methionine and threonine biosynthesis pathways in *Arabidopsis thaliana. European Journal of Biochemistry, 270,* 4615–4627.

Curien, G., Bastien, O., Robert-Genthon, M., Cornish-Bowden, A., Cárdenas, M. L., & Dumas, R. (2009). Understanding the regulation of aspartate metabolism using a model based on measured kinetic parameters. *Molecular Systems Biology, 5,* 271.

Danchin, A. (2003). *The Delphic Boat: What Genomes Tell Us.* Cambridge, MA: Harvard University Press.

Danino, T., Mondragón-Palomino, O., Tsimring, L., & Hasty, J. (2010). A synchronized quorum of genetic clocks. *Nature, 463,* 326–330.

Davidson, E. (2006). *The Regulatory Genome: Gene Regulatory Networks in Development and Evolution.* Burlington, MA: Academic Press.

De Jong, H. (2002). Modeling and simulation of genetic regulatory systems: a literature review. *Journal of Computational Biology, 9,* 67–103.

Demin, O., & Goryanin, I. (2009). *Kinetic Modelling in Systems Biology.* London: CRC Press.

Edelstein-Keshet, L. (2005). *Mathematical Models in Biology.* Philadelphia: SIAM.

Eissing, T., Conzelmann, H., Gilles, E. D., Allgöwer, F., Bullinger, E., & Scheurich, P. (2004). Bistability analyses of a caspase activation model for receptor-induced apoptosis. *Journal of Biological Chemistry, 279,* 36892–36897.

Elowitz, M. B., & Leibler, S. (2000). A synthetic oscillatory network of transcriptional regulators. *Nature, 403,* 335–338.

Érdi, P., & Tóth, J. (1989). *Mathematical Models of Chemical Reactions: Theory and Applications of Deterministic and Stochastic Models.* Princeton, NJ: Princeton University Press.

Ermentrout, B. (2002). *Simulating, Analyzing, and Animating Dynamical Systems: A Guide to XPPAUT for Researchers and Students.* Philadelphia: SIAM.

Ermentrout, G. B., & Terman, D. H. (2010). *Mathematical Foundations of Neuroscience.* New York: Springer.

Fall, C. P., Marland, E. S., Wagner, J. M., & Tyson, J. J. (Eds.). (2002). *Computational Cell Biology.* New York: Springer.

Fell, D. (1997). *Understanding the Control of Metabolism.* London: Portland Press.

Ferrell, J. E., Jr. (1996). Tripping the switch fantastic: how a protein kinase cascade can convert graded inputs into switch-like outputs. *Trends in Biochemical Sciences, 21,* 460–466.

Garcia-Ojalvo, J., Elowitz, M. B., & Strogatz, S. H. (2004). Modeling a synthetic multicellular clock: repressilators coupled by quorum sensing. *Proceedings of the National Academy of Sciences of the United States of America, 101,* 10955–10960.

Gardner, T. S., Cantor, C. R., & Collins, J. J. (2000). Construction of a genetic toggle switch in *Escherichia coli. Nature, 403,* 339–342.

Gibson, M. A., & Bruck, J. (2000). Efficient exact stochastic simulation of chemical systems with many species and many channels. *Journal of Physical Chemistry A, 104*, 1876–1889.

Gilat, A. (2008). *MATLAB: An Introduction with Applications* (3rd ed.). New York: Wiley.

Gillespie, D. T. (2007). Stochastic simulation of chemical kinetics. *Annual Review of Physical Chemistry, 58*, 35–55.

Goldbeter, A. (1996). *Biochemical Oscillations and Cellular Rhythms: The Molecular Bases of Periodic and Chaotic Behaviour.* Cambridge, UK: Cambridge University Press.

Goldbeter, A., & Koshland, D. E., Jr. (1981). An amplified sensitivity arising from covalent modification in biological systems. *Proceedings of the National Academy of Sciences of the United States of America, 78*, 6840–6844.

Goodwin, B. C. (1965). Oscillatory behavior in enzymatic control processes. *Advances in Enzyme Regulation, 3*, 425–428.

Griffith, J. S. (1968). Mathematics of cellular control processes I. Negative feedback to one gene. *Journal of Theoretical Biology, 20*, 202–208.

Guevara, M. R. (2003). Dynamics of excitable cells. In A. Beuter, L. Glass, M. C. Mackey, & M. S. Titcombe (Eds.), *Nonlinear Dynamics in Physiology and Medicine* (pp. 87–121). New York: Springer.

Gunawardena, J. (2005). Multisite protein phosphorylation makes a good threshold but can be a poor switch. *Proceedings of the National Academy of Sciences of the United States of America, 102*, 14617–14622.

Haberman, R. (2003). *Applied Partial Differential Equations* (4th ed.). Upper Saddle River, NJ: Prentice Hall.

Hanselman, D. C., & Littlefield, B. L. (2004). *Mastering MATLAB 7.* Upper Saddle River, NJ: Prentice Hall.

Hartwell, L., Hood, L., Goldberg, M., Reynolds, A., & Silver, L. (2010). *Genetics: From Genes to Genomes* (4th ed.). New York: McGraw-Hill.

Hasty, J., Dolnik, M., Rottschäfer, V., & Collins, J. J. (2002). Synthetic gene network for entraining and amplifying cellular oscillations. *Physical Review Letters, 88*, 148101.

Heinrich, R., & Schuster, S. (1996). *The Regulation of Cellular Systems.* New York: Chapman & Hall.

Hersen, P., McClean, M. N., Mahadevan, L., & Ramanathan, S. (2008). Signal processing by the HOG MAP kinase pathway. *Proceedings of the National Academy of Sciences of the United States of America, 105*, 7165–7170.

Hill, A. V. (1969). The third Bayliss-Starling memorial lecture. *Journal of Physiology, 204*, 1–13.

Hoffmann, A., Levchenko, A., Scott, M. L., & Baltimore, D. (2002). The IκB-NF-κB signaling module: temporal control and selective gene activation. *Science, 298*, 1241–1245.

Iglesias, P. A. (2010). Spatial modeling. In P. A. Iglesias & B. P. Ingalls (Eds.), *Control Theory and Systems Biology* (pp. 45–68). Cambridge, MA: MIT Press.

Ingalls, B. (2008). Sensitivity analysis: from model parameters to system behaviour. *Essays in Biochemistry, 45*, 177–193.

Jacquez, J. A. (1985). *Compartmental Analysis in Biology and Medicine* (2nd ed.). Ann Arbor: The University of Michigan Press.

Jaeger, J., Surkova, S., Blagov, M., Janssens, H., Kosman, D., Kozlov, K. N., et al. (2004). Dynamic control of positional information in the early *Drosophila* embryo. *Nature, 430*, 368–371.

James, S., Nilsson, P., James, G., Kjelleberg, S., & Fagerström, T. (2000). Luminescence control in the marine bacterium *Vibrio fischeri*: an analysis of the dynamics of *lux* regulation. *Journal of Molecular Biology, 296*, 1127–1137.

Jones, D. S., Plank, M. J., & Sleeman, B. D. (2009). *Differential Equations and Mathematical Biology* (2nd ed.). London: CRC Press.

Kacser, H., Burns, J. A., & Fell, D. A. (1995). The control of flux: 21 years on. *Biochemical Society Transactions, 23*, 341–366.

Keener, J., & Sneyd, J. (1998). *Mathematical Physiology*. New York: Springer.

Kelly, K. (1995). *Out of Control: The New Biology of Machines, Social Systems, & the Economic World*. New York: Basic Books.

Khalil, A. S., & Collins, J. J. (2010). Synthetic biology: applications come of age. *Nature Reviews. Genetics*, *11*, 367–379.

Khammash, M. (2010). Modeling and analysis of stochastic biochemical networks. In P. A. Iglesias & B. P. Ingalls (Eds.), *Control Theory and Systems Biology* (pp. 29–44). Cambridge, MA: MIT Press.

Kholodenko, B. N. (2000). Negative feedback and ultrasensitivity can bring about oscillations in the mitogen-activated protein kinase cascades. *European Journal of Biochemistry*, *267*, 1583–1588.

Klamt, S., & Stelling, J. (2006). Stoichiometric and constraint-based modeling. In Z. Szallasi, J. Stelling, & V. Periwal (Eds.), *System Modelling in Cellular Biology* (pp. 73–96). Cambridge, MA: MIT Press.

Klipp, E., Liebermeister, W., Wierling, C., Kowald, A., Lehrach, H., & Herwig, R. (2009). *Systems Biology: A Textbook*. Weinheim, Germany: Wiley-VCH.

Koch, C., & Segev, I. (Eds.). (1998). *Methods in Neural Modeling: From Ions to Networks*. Cambridge, MA: MIT Press.

Konopka, R. J., & Benzer, S. (1971). Clock mutants of *Drosophila melanogaster*. *Proceedings of the National Academy of Sciences of the United States of America*, *68*, 2112–2116.

Korendyaseva, T. K., Kuvatov, D. N., Volkov, V. A., Martinov, M. V., Vitvitsky, V. M., Banerjee, R., et al. (2008). An allosteric mechanism for switching between parallel tracks in mammalian sulfur metabolism. *PLoS Computational Biology*, *4*, e1000076. doi:10.1371/journal.pcbi.1000076.

Kremling, A., Heerman, R., Centler, F., Jung, K., & Gilles, E. D. (2004). Analysis of two-component signal transduction by mathematical modeling using the *KdpD/KdpE* system of *Escherichia coli*. *BioSystems*, *78*, 23–37.

Krishna, S., Jensen, M. H., & Sneppen, K. (2006). Minimal model of spiky oscillations in NF-κB signaling. *Proceedings of the National Academy of Sciences of the United States of America*, *103*, 10840–10845.

Kruse, K., & Elf, J. (2006) Kinetics in spatially extended domains. In Z. Szallasi, J. Stelling, & V. Periwal (Eds.), *System Modelling in Cellular Biology* (pp. 177–198). Cambridge, MA: MIT Press.

Lazebnik, Y. (2002). Can a biologist fix a radio?—Or, what I learned while studying apoptosis. *Cancer Cell*, *2*, 179–182.

Leloup, J.-C., & Goldbeter, A. (2003). Toward a detailed computational model for the mammalian circadian clock. *Proceedings of the National Academy of Sciences of the United States of America*, *100*, 7051–7056.

Maeda, M., Lu, S., Shaulsky, G., Miyazaki, Y., Kuwayama, H., Tanaka, Y., et al. (2004). Periodic signaling controlled by an oscillatory circuit that includes protein kinases ERK2 and PKA. *Science*, *304*, 875–878.

Marks, F., Klingmüller, U., & Müller-Decker, K. (2009). *Cellular Signal Processing: An Introduction to the Molecular Mechanisms of Signal Transduction*. New York: Garland Science.

Martinov, M. V., Vitvitsky, V. M., Mosharov, E. V., Banerjee, R., & Ataullakhanov, F. I. (2000). A substrate switch: a new mode of regulation in the methionine metabolic pathway. *Journal of Theoretical Biology*, *204*, 521–532.

Martinov, M. V., Vitvitsky, V. M., Banerjee, R., & Ataullakhanov, F. I. (2010). The logic of the hepatic methionine metabolic cycle. *Biochimica et Biophysica Acta*, *1804*, 89–96.

Monod, J., Wyman, J., and Changeux, J.-P. (1965). On the nature of allosteric transitions: a plausible model. *Journal of Molecular Biology*. *12*, 88–118.

Morris, C., & Lecar, H. (1981). Voltage oscillations in the barnacle giant muscle fiber. *Biophysical Journal*, *35*, 193–231.

Murray, J. D. (2003). *Mathematical Biology II: Spatial Models and Biomedical Applications* (3rd ed.). New York: Springer.

Myers, C. J. (2010). *Engineering Genetic Circuits*. London: CRC Press.

Naderi, S., Meshram, M., Wei, C., McConkey, B., Ingalls, B., Budman, H., et al. (2011). Development of a mathematical model for evaluating the dynamics of normal and apoptotic Chinese hamster ovary cells. *Biotechnology Progress, 27*, 1197–1205.

Ndiaye, I., Chaves, M., & Gouzé, J. L. (2010). Oscillations induced by different timescales in signal transduction modules regulated by slowly evolving protein-protein interactions. *IET Systems Biology, 4*, 263–276.

Nelson, P. (2004). *Biological Physics: Energy, Information, Life.* New York: W. H. Freeman.

Nelson, D. L., & Cox, M. M. (2008). *Lehninger Principles of Biochemistry* (5th ed.). New York: W. H. Freeman.

Neuhauser, C. (2004). *Calculus for Biology and Medicine* (2nd ed.). Upper Saddle River, NJ: Prentice Hall.

Noble, D. (2004). Modeling the Heart. *Physiology, 19*, 191–197.

Othmer, H. G. (1997). Signal transduction and second messenger systems. In H. G. Othmer, F. R. Adler, M. A. Lewis, & J. C. Dallon (Eds.), *Case Studies in Mathematical Modelling: Ecology, Physiology, and Cell Biology* (pp. 99–126). Upper Saddle River, NJ: Prentice Hall.

Palsson, B. (2006). *Systems Biology: Properties of Reconstructed Networks.* Cambridge, UK: Cambridge University Press.

Perkins, T. J., Jaeger, J., Reinitz, J., & Glass, L. (2006). Reverse engineering the gap gene network of *Drosophila melanogaster. PLoS Computational Biology, 2*, e51. doi:10.1371/journal.pcbi.0020051.

Phillips, R., Kondov, J., & Theriot, J. (2009). *Physical Biology of the Cell.* New York: Garland Science.

Ptashne, M. (2004). *A Genetic Switch: Phage Lambda Revisited* (3rd ed.). Cold Spring Harbor, NY: Cold Spring Harbor Laboratory Press.

Purnick, P. E. M., & Weiss, R. (2009). The second wave of synthetic biology: from modules to systems. *Nature Reviews. Molecular Cell Biology, 10*, 410–422.

Raj, A., & van Oudenaarden, A. (2008). Nature, nurture, or chance: stochastic gene expression and its consequences. *Cell, 135*, 216–226.

Reed, M. C., Nijhout, H. F., Sparks, R., & Ulrich, C. M. (2004). A mathematical model of the methionine cycle. *Journal of Theoretical Biology, 226*, 33–43.

Reinitz, J., & Vaisnys, J. R. (1990). Theoretical and experimental analysis of the phage lambda genetic switch implies missing levels of co-operativity. *Journal of Theoretical Biology, 145*, 295–318.

Rinzel, J. (1990). Discussion: Electrical excitability of cells, theory and experiment: review of the Hodgkin-Huxley foundation and an update. *Bulletin of Mathematical Biology, 52*, 5–23.

Ross, S. M. (2007). *Introduction to Probability Models* (9th ed.). London: Academic Press.

Saltelli, A., Ratto, M., Andres, T., Campolongo, F., Cariboni, J., Gatelli, D., et al. (2008). *Global Sensitivity Analysis: The Primer.* New York: Wiley.

Santillán, M., Mackey, M. C., & Zeron, E. S. (2007). Origin of bistability in the *lac* operon. *Biophysical Journal, 92*, 3830–3842.

Sauro, H. M. (2009). Network dynamics. In J. McDermott, R. Samudrala, R. Bumgarner, K. Montgomery, & R. Ireton (Eds.), *Computational Systems Biology* (pp. 269–310). New York: Humana Press.

Sauro, H. M. (2011). *Enzyme Kinetics for Systems Biology.* Seattle: Ambrosius Publishing.

Savageau, M. A. (1976). *Biochemical Systems Analysis: A Study of Function and Design in Molecular Biology.* London: Addison-Wesley.

Segel, L. A., & Slemrod, M. (1989). The quasi-steady-state assumption: a case study in perturbation. *SIAM Review, 31*, 446–477.

Setty, Y., Mayo, A. E., Surette, M. G., & Alon, U. (2003). Detailed map of a cis-regulatory input function. *Proceedings of the National Academy of Sciences of the United States of America, 100*, 7702–7707.

Stathopoulos, A., & Levine, M. (2005). Genomic regulatory networks and animal development. *Developmental Cell, 9*, 449–462.

Strang, G. (2005). *Linear Algebra and its Applications* (4th ed.). Pacific Grove, CA: Brooks/Cole.

Stephanopoulos, G. N., Aristidou, A. A., & Nielsen, J. (1998). *Metabolic Engineering: Principles and Methodologies*. Burlington, MA: Academic Press.

Stricker, J., Cookson, S., Bennett, M. R., Mather, W. H., Tsimring, L. S., & Hasty, J. (2008). A fast, robust and tunable synthetic gene oscillator. *Nature, 456*, 516–519.

Strogatz, S. H. (2001). *Nonlinear Dynamics and Chaos: with Applications to Physics, Biology, Chemistry, and Engineering*. Boulder, CO: Westview Press.

Szallasi, Z., Stelling, J., & Periwal, V. (Eds.). (2006). *System Modelling in Cellular Biology*. Cambridge, MA: MIT Press.

Thomas, L. (1974). *The Lives of a Cell*. New York: Viking Press.

Thompson, D. (1917). *On Growth and Form*. Cambridge, UK: Cambridge University Press. Reprinted 1961.

Tözeren, A., & Byers, S. W. (2004). *New Biology for Engineers and Computer Scientists*. Upper Saddle River, NJ: Pearson Education.

Tyson, J. J., Hong, C. I., Thron, C. D., & Novak, B. (1999). A simple model of circadian rhythms based on dimerization and proteolysis of PER and TIM. *Biophysical Journal, 77*, 2411–2417.

Varma, A., Morbidelli, M., & Wu, H. (2005). *Parametric Sensitivity in Chemical Systems*. Cambridge, UK: Cambridge University Press.

Vilar, J. M. G., Kueh, H. Y., Barkai, N., & Leibler, S. (2002). Mechanisms of noise-resistance in genetic oscillators. *Proceedings of the National Academy of Sciences of the United States of America, 99*, 5988–5992.

Voit, E. O. (2000). *Computational Analysis of Biochemical Systems: A Practical Guide for Biochemists and Molecular Biologists*. Cambridge, UK: Cambridge University Press.

Voit, E. O. (2012). *A First Course in Systems Biology*. New York: Garland Science.

Wagner, G. P., Pavlicev, M., & Cheverud, J. M. (2007). The road to modularity. *Nature Reviews. Genetics, 8*, 921–931.

Waldherr, S., Eissing, T., Chaves, M., & Allgower, F. (2007). Bistability preserving model reduction in apoptosis. In *Proceedings of the 10th International Symposium on Computer Applications in Biotechnology*, Cancun, Mexico, 327–332. Laxenburg, Austria: International Federation on Automatic Control (IFAC).

Wiener, N. (1948). *Cybernetics: or Control and Communication in the Animal and the Machine*. New York: Wiley.

Weiss, R., Basu, S., Hooshangi, S., Kalmbach, A., Karig, D., Mehreja, R., et al. (2003). Genetic circuit building blocks for cellular computation, communications, and signal processing. *Natural Computing, 2*, 1–40.

Weiss, R., & Kaern, M. (2006). Synthetic gene regulatory systems. In Z. Szallasi, J. Stelling, & V. Periwal (Eds.), *System Modelling in Cellular Biology* (pp. 269–295). Cambridge, MA: MIT Press.

Weiss, R., & Knight, T. F. (2001). *Engineered Communications for Microbial Robotics. DNA Computing* (vol. 2054, pp. 1–16). Lecture Notes in Computer Science. Heidelberg: Springer.

White, D. (2000). *The Physiology and Biochemistry of Prokaryotes*. Oxford: Oxford University Press.

Wilkinson, D. J. (2006). *Stochastic Modelling for Systems Biology*. London: CRC Press.

Yi, T.-M., Kitano, H., & Simon, M. I. (2003). A quantitative characterization of the yeast heterotrimeric G protein cycle. *Proceedings of the National Academy of Sciences of the United States of America, 100*, 10764–10769.

You, L., Cox, R. S., III, Weiss, R., & Arnold, F. H. (2004). Programmed population control by cell-cell communication and regulated killing. *Nature, 428*, 868–871.

Index

Printed in the United States
by Baker & Taylor Publisher Services